The Freeman Library of Laboratory Separates

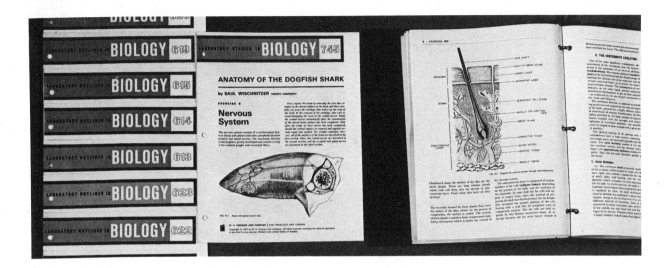

Freeman Laboratory Separates are self-bound, self-contained exercises. They are 8½ inches by 11 inches in format and are punched for a three-ring notebook. Separates may be ordered in any assortment or quantity at 25¢ each. Thus the instructor can construct an entire manual tailored to the design of his course, or he can select Separates to broaden the coverage of his present laboratory manual.

Order Separates by number. Students' sets are collated and packaged by the publisher before shipment to the bookstore.

Each exercise in this manual is available as a Separate (see below). On the facing page are listed other Freeman Laboratory Separates in Biology.

LABORATORY OUTLINES IN BIOLOGY, 846-882
25¢ each
by ABRAMOFF and THOMSON
*(Also available as a bound manual,
LABORATORY OUTLINES IN BIOLOGY-II)*

846. The Microscope
847. Organization of Cells
848. Cell Division
849. Plant Anatomy: Roots and Stems
850. Plant Anatomy: Leaves and Flowers
851. Vertebrate Anatomy: External Anatomy, Skeleton, and Muscles
852. Vertebrate Anatomy: Internal Organs
853. Movement of Materials Across Cell Membranes
854. Biologically Important Molecules: Proteins
855. Biologically Important Molecules: Carbohydrates and Lipids
856. Photosynthesis
857. Carbohydrate Metabolism
858. Cellular Respiration
859. Transport in Biological Systems
860. Biological Coordination: Plants
861. Biological Coordination: Animals
862. Mendelian Inheritance
863. Biochemical Genetics
864. Fertilization and Early Development of the Sea Urchin
865. Fertilization and Early Development of the Frog
866. Development of the Chick
867. Plant Growth and Development
868. Plant Development: Hormonal Regulation
869. Plant Development: Effect of Light
870. Procaryotes
871. Diversity in Plants: Fungi
872. Diversity in Plants: Algae
873. Diversity in Plants: Liverworts and Mosses
874. Diversity in Plants: Vascular Plants
875. Diversity in Animals: Protozoa
876. Diversity in Animals: Sponges and Coelenterates
877. Diversity in Animals: Flatworms, Roundworms, and Parasites of the Frog
878. Diversity in Animals: Annelids
879. Diversity in Animals: Arthropods
880. Diversity in Animals: Molluscs and Echinoderms
881. Diversity in Animals: Chordates
882. Appendices: Studies 846-881

W. H. FREEMAN AND COMPANY
660 Market Street, San Francisco, California 94104
58 Kings Road, Reading, England RG1 3AA

A SERIES OF BOOKS IN BIOLOGY

Editors: Donald Kennedy
Roderic B. Park

A 144-page instructor's handbook for use with this manual will be sent to instructors free of charge upon request.

Laboratory Outlines in Biology-II

PETER ABRAMOFF/ROBERT G. THOMSON

MARQUETTE UNIVERSITY

W. H. FREEMAN AND COMPANY
San Francisco

Copyright © 1962, 1963 by Peter Abramoff and
Robert G. Thomson

Copyright © 1964, 1966, 1972 by W. H. Freeman and
Company

No part of this book may be reproduced by any mechanical,
photographic, or electronic process, or in the form of a
phonographic recording, nor may it be stored in a
retrieval system, transmitted, or otherwise copied for
public or private use without written permission of the
publisher.

Printed in the United States of America

International Standard Book Number: 0-7167-0694-6

9-14-76
73-116556

5 6 7 8 9

Preface

Laboratory work in natural science, properly directed, constitutes an essential part of a liberal education. It furthers ingenuity and accuracy, enables the student to understand readily how the conclusions of the scientist proceed logically from experiment and observation, and allows the student to become intensely and personally involved in the knowledge he acquires. Laboratory work in biology conveys something of the meaning, scope, and excitement of biology as a significant perspective from which to view the world. It also provides an acquaintance with the world of living things, the relationship of one organism to others, and the structure and function of organisms.

It has been the basic assumption in preparing this manual that laboratory work is an integral part of any introductory biology program and should be considered as being complementary—not subservient—to the lectures. Furthermore, the opportunity to practice scientific inquiry is considered a significant part of any laboratory program. Thus it shall be an objective in the laboratory to lead the student to a deep and intelligent understanding of the main principles and concepts of biology as discoverable through the cooperative activity of the imagination, accurate and logical reasoning, and precise observation.

We do not believe that routine memorization of unrelated facts is a sound approach to biology, or to any other science. You will not be asked to set such a task for yourself here. However, you will be asked to come to grips with detailed information when that information is required to document, illustrate, or dramatize a basic biological principle. We have excluded descriptive material, except when such material is needed to illustrate the principles in question. The routine study of prepared slides, the preparation of highly detailed drawings, and dissections have not been emphasized.

More exercises are included in this manual than are generally used in a one- or two-semester introductory course in biology. Furthermore, individual exercises are not designed to meet the requirements of a specified number of laboratory hours. Most exercises are sufficiently comprehensive to allow subdivision and adaptation to any laboratory sequence, depending on the requirements of the course.

This laboratory manual has been designed to meet the needs of those institutions that have developed their beginning biology programs beyond the level covered in our original *Laboratory Outlines in Biology*. All of the exercises have been extensively revised and updated to include a more experimental approach, the use of more living materials, and the addition of more open-ended experiments. In doing so, we have attempted to keep the cost of the equipment and materials needed to conduct the exercises within the budgets of most colleges and universities. To facilitate the use of this manual, we have also prepared an extensively revised Instructor's Handbook that includes lists of all materials and supplies needed to conduct each of the experiments, detailed descriptions for the preparation of reagents and solutions and the handling of living materials, and additional reference and visual aid materials.

September, 1971

Peter Abramoff
Robert G. Thomson

Contents

EXERCISE **1. The Microscope** 1
    A. USE AND CARE OF THE COMPOUND MICROSCOPE 1
    B. USE AND CARE OF THE STEREOSCOPIC DISSECTING MICROSCOPE 6
    C. STUDY OF POND WATER 8

EXERCISE **2. Organization of Cells** 15
    A. PLANT CELLS 15
    B. ANIMAL CELLS 18
    C. CELL FINE STRUCTURE 19

EXERCISE **3. Cell Division** 27
    A. MITOSIS IN PLANT CELLS 28
    B. MITOSIS IN ANIMAL CELLS 31
    C. CHROMOSOME MORPHOLOGY 32

EXERCISE **4. Plant Anatomy: Roots and Stems** 37
    A. ROOTS 38
    B. STEMS 40

EXERCISE **5. Plant Anatomy: Leaves and Flowers** 47
    A. LEAVES 47
    B. FLOWERS 49

EXERCISE **6. Vertebrate Anatomy: External Anatomy, Skeleton, and Muscles** 53
    A. EXTERNAL ANATOMY OF THE FROG 53
    B. VERTEBRATE SKIN 54
    C. VERTEBRATE SKELETON 58
    D. VERTEBRATE MUSCULATURE 60

EXERCISE **7. Vertebrate Anatomy: Internal Organs** 65
    A. DIGESTIVE SYSTEM 65
    B. CIRCULATORY SYSTEM 69
    C. UROGENITAL SYSTEM 75

EXERCISE **8. Movement of Materials Across Cell Membranes** 81
    A. BROWNIAN MOVEMENT 82
    B. DIFFUSION 82
    C. MEASURING THE RATE OF OSMOSIS 86
    D. FACTORS AFFECTING MOVEMENT OF MATERIALS ACROSS CELL MEMBRANES 86

EXERCISE **9. Biologically Important Molecules: Proteins** 93
    A. QUALITATIVE TESTS TO DETECT PROTEINS 94
    B. QUANTITATIVE CHEMICAL DETERMINATION OF PROTEIN 95
    C. CHROMATOGRAPHIC SEPARATION OF AMINO ACIDS 96

EXERCISE **10. Biologically Important Molecules: Carbohydrates and Lipids** 99
    A. CARBOHYDRATES 99
    B. UNKNOWN CARBOHYDRATES 101
    C. LIPIDS 101

EXERCISE **11. Photosynthesis** 103
    A. ROLE OF LIGHT 104
    B. ROLE OF CARBON DIOXIDE 107
    C. ROLE OF CHLOROPLAST PIGMENTS 111

EXERCISE **12. Carbohydrate Metabolism** 121
- A. EFFECT OF TEMPERATURE ON THE ACTIVITY OF SALIVARY AMYLASE 122
- B. EFFECT OF pH ON THE ACTIVITY OF SALIVARY AMYLASE 124
- C. EFFECT OF SUBSTRATE CONCENTRATION ON THE ACTIVITY OF SALIVARY AMYLASE 125
- D. EFFECT OF ENZYME CONCENTRATION ON THE ACTIVITY OF SALIVARY AMYLASE 125
- E. ENZYMATIC SYNTHESIS OF STARCH 129

EXERCISE **13. Cellular Respiration** 133
- A. AEROBIC RESPIRATION 134
- B. BIOLOGICAL OXIDATION 135

EXERCISE **14. Transport in Biological Systems** 145
- A. TRANSPORT IN ANIMALS 145
- B. PERIPHERAL CIRCULATION 146
- C. HEART ACTION AND CONTROL 149
- D. TRANSPORT IN PLANTS 151

EXERCISE **15. Biological Coordination: Plants** 161
- A. COORDINATION IN PLANTS 161

EXERCISE **16. Biological Coordination: Animals** 173
- A. SPINAL CORD ANATOMY 174
- B. SPINAL NERVE ATTACHMENT TO SPINAL CORD 175

EXERCISE **17. Mendelian Inheritance** 181
- A. MEIOSIS 181
- B. MUTATION IN DROSOPHILA 186
- C. MUTATION IN CORN 187
- D. INHERITANCE OF HUMAN BLOOD GROUPS 190

EXERCISE **18. Biochemical Genetics** 195
- A. CHROMATOGRAPHIC SEPARATION OF DROSOPHILA EYE PIGMENTS 196
- B. INDUCTION OF MUTATION BY ULTRAVIOLET LIGHT 198
- C. MOLECULAR BASIS OF HEREDITY 201

EXERCISE **19. Fertilization and Early Development of the Sea Urchin** 207
- A. FERTILIZATION 207
- B. CLEAVAGE 209
- C. LATER STAGES IN DEVELOPMENT 209

EXERCISE **20. Fertilization and Early Development of the Frog** 213
- A. ARTIFICIAL INDUCTION OF OVULATION 213
- B. PREPARATION OF SPERM SUSPENSION 214
- C. FERTILIZATION 215
- D. UNFERTILIZED FROG EGGS 216
- E. CLEAVAGE STAGES 217
- F. LATER STAGES OF DEVELOPMENT 218
- G. FURTHER STUDIES ON FROG DEVELOPMENT 221

EXERCISE **21. Development of the Chick** 225
- A. UNFERTILIZED EGG 225
- B. FERTILIZED EGG 225
- C. GASTRULATION 226
- D. LATER DEVELOPMENTAL STAGES 227

EXERCISE **22. Plant Growth and Development** 233
- A. COMPARATIVE STUDY OF SEEDS 233
- B. GERMINATION 234
- C. MEASUREMENT OF PLANT GROWTH 239

EXERCISE **23. Plant Development: Hormonal Regulation** 245
- A. GIBBERELLIN 247
- B. PLANT GROWTH INHIBITORS 247
- C. DETECTION OF PLANT GROWTH HORMONES 250

EXERCISE **24. Plant Development: Effect of Light** 259
- A. PLANT GROWTH IN THE ABSENCE OF LIGHT 259
- B. EFFECT OF DAY LENGTH ON PLANT GROWTH 260

EXERCISE **25. Procaryotes** 267
- A. BACTERIA 267
- B. BLUE-GREEN ALGAE 272
- C. ALGAE IN WATER SUPPLIES 274

EXERCISE **26. Diversity in Plants: Fungi** 279
- A. SLIME MOLDS 279
- B. ALGALLIKE FUNGI (PHYCOMYCETES) 281
- C. SAC FUNGI (ASCOMYCETES) 284
- D. CLUB FUNGI (BASIDIOMYCETES) 286

EXERCISE **27. Diversity in Plants: Algae** 289
- A. UNICELLULAR GREEN ALGAE 289
- B. FILAMENTOUS GREEN ALGAE 291
- C. COLONIAL GREEN ALGAE 294
- D. RED, BROWN, AND GOLDEN BROWN ALGAE 295

Contents

EXERCISE **28. Diversity in Plants: Liverworts and Mosses** 297
- A. LIVERWORTS 297
- B. MOSSES 301
- C. ADVANCES IN COMPLEXITY SHOWN BY THE LIVERWORTS AND MOSSES 303

EXERCISE **29. Diversity in Plants: Vascular Plants** 305
- A. FERNS 305
- B. CONIFERS 307
- C. FLOWERING PLANTS 311
- D. ADVANCES IN COMPLEXITY SHOWN BY VASCULAR PLANTS 313

EXERCISE **30. Diversity in Animals: Protozoa** 317
- A. AMOEBA 317
- B. PARAMECIUM 319
- C. SYMBIOSIS 320
- D. OTHER PROTOZOA 321

EXERCISE **31. Diversity in Animals: Sponges and Coelenterates** 323
- A. SPONGES 323
- B. COELENTERATES 324

EXERCISE **32. Diversity in Animals: Flatworms, Roundworms, and Parasites of the Frog** 333
- A. FREE-LIVING FLATWORMS 333
- B. PARASITIC FLATWORMS 334
- C. ROUNDWORMS (NEMATODES) 339
- D. PARASITES OF THE FROG 341

EXERCISE **33. Diversity in Animals: Annelids** 345
- A. CLAMWORMS (CLASS POLYCHAETA) 345
- B. SEGMENTED WORMS (CLASS OLIGOCHAETA) 345
- C. LEECHES (CLASS HIRUDINEA) 350

EXERCISE **34. Diversity in Animals: Arthropods** 353
- A. PERIPATUS (CLASS ONYCOPHORA) 353
- B. CENTIPEDES (CLASS CHILOPODA) AND MILLIPEDES (CLASS DIPLOPODA) 353
- C. SPIDERS, SCORPIONS, AND HORSESHOE CRABS (CLASS ARACHNIDA) 354
- D. INSECTS (CLASS INSECTA) 354
- E. CRUSTACEANS 357

EXERCISE **35. Diversity in Animals: Molluscs and Echinoderms** 363
- A. MOLLUSCS 363
- B. ECHINODERMS 365

EXERCISE **36. Diversity in Animals: Chordates** 369
- A. TUNICATES (UROCHORDATA) 369
- B. AMPHIOXUS (CEPHALOCHORDATA) 369
- C. VERTEBRATES 371
- D. MORPHOLOGICAL ADAPTATIONS OF VERTEBRATES TO THE ENVIRONMENT 373

APPENDICES 383

APPENDIX **A. USE OF THE BAUSCH & LOMB SPECTRONIC 20 COLORIMETER** 385
- **B. COLORIMETRY** 389
- **C. CHROMATOGRAPHY** 393
- **D. USE OF LIVE ANIMALS IN THE LABORATORY** 397
- **E. RADIOISOTOPES** 399
- **F. ELEMENTARY STATISTICAL ANALYSIS** 401
- **G. SYRINGE-TYPE RESPIROMETER** 407

Laboratory Outlines in Biology-II

EXERCISE 1

The Microscope

The microscope is a major tool of biologists. Without the microscope the cell theory would not have been developed, and we would lack most of our present knowledge of living things too small to be seen by the unaided eye.

This exercise has been designed to familiarize you with the use and care of the microscope and to acquaint you with some of the variety of plant and animal life that you will be able to see with the microscope.

There are many kinds of microscopes in use today with a wide range of magnifications from as low as 2 or 3 to over 100,000. In your laboratory studies you will be using mainly two kinds of microscopes: the compound microscope with magnifications ranging from 40 to 450, depending on the particular make, and the stereoscopic dissecting microscope with magnifications ranging from 4 to 60.

A. USE AND CARE OF THE COMPOUND MICROSCOPE

A microscope is a precision instrument and should be treated accordingly. It may appear to be indestructible, but in reality the slightest jar may damage its working parts and optical system. Handle the microscope carefully at all times; repairs are costly and may require that the instrument be returned to the factory. Your instructor will demonstrate the proper care and use of the microscope, and you will then be assigned a microscope for which you will be responsible throughout the remainder of the semester.

1. Parts of the Microscope

After removing the microscope from its cabinet, place it on the table in front of you so you look at

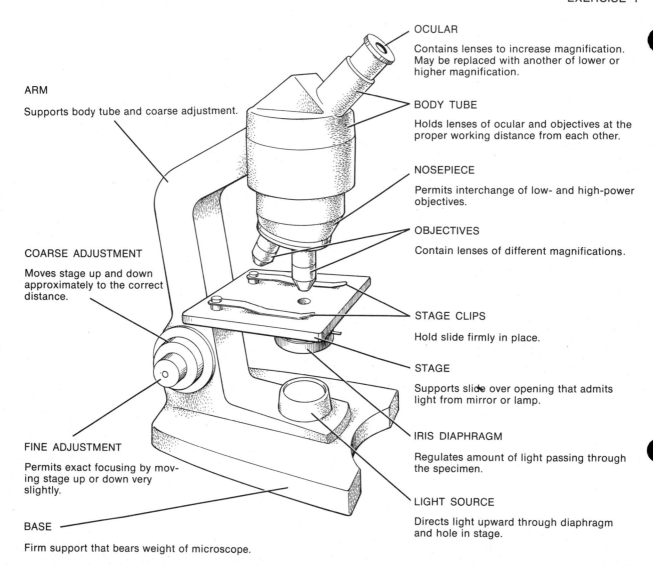

FIGURE 1-1
Parts of a compound microscope.

the side profile. As you examine the instrument, refer to Fig. 1-1 for help in locating the various parts described below.

The compound microscope is made up of a series of lenses that magnify objects lying beyond the range of ordinary vision. These lenses are located in the **ocular**, or eyepiece, and the **objectives**. A microscope may have one ocular—a **monocular** microscope—or it may have two oculars—a **binocular** microscope. The ocular and objective are kept a set distance apart by the **body tube**. Note that there are at least two objectives, attached to the body tube by means of a revolving **nosepiece**. Cautiously turn the nosepiece, taking care that the objectives do not hit the **stage**. How can you tell when the objective is correctly lined up with the ocular?

_____

_____

_____

Note that the objectives differ in length. The smaller one is the **low-power objective**, as contrasted to the larger, **high-power objective**.

Movement of either the body tube or the stage, up or down, enables you to focus the microscope. With the low-power objective in position turn the coarse adjustment knob one full turn in a clockwise

THE MICROSCOPE

direction. Does this raise or lower the stage or body tube?

Approximately how far did it move?

In a similar manner turn the fine adjustment knob. In contrast to the coarse adjustment, how far did it move this time?

Proper use of the microscope dictates that the coarse adjustment not be used when the high-power objective is in position.

In the center of the stage locate the opening through which light is reflected from a mirror or lamp. Notice that one surface of the mirror is flat and the other surface is concave. The surface acts as a lens to concentrate the light from the whole area of the mirror onto the microscope field. Daylight may be used for a light source; however, it is of such variable intensity that artificial light is generally preferred. An **iris diaphragm**, located below the stage, may be opened or closed, thereby regulating the amount of light entering the microscope.

The arm of the microscope is attached to the base upon which the microscope stands. When carrying the microscope, one hand should grasp the arm and the other hand should support the base in order to prevent any possibility of dropping the instrument.

2. Using the Microscope

Clean lenses are essential to working with the microscope. If the lenses are dirty, wet, or scratched, the image will be obscured, and observation of the object under study will be difficult. *Develop the habit of cleaning the ocular and objective lenses each time you use the microscope. Use only lens paper—never filter paper, paper toweling, or any other substitute material that may scratch the lenses.*

Rotate the nosepiece until the low-power objective clicks into position. Obtain a slide of the letter "e" (or a, k, or h) from your laboratory instructor. Place the slide on the stage so that the coverslip or paper "e" is located *on top* of the slide as it is inserted over the center of the circular opening in the stage. Otherwise the subject will not come into focus on high power and the objective may be rammed into the slide. Slowly lower the objective (or raise the stage) to a position about $\frac{1}{4}$ inch from the slide.

Observation of any specimen is difficult unless proper light is reflected through the lens systems in the objectives and ocular. While looking through the ocular, "aim" the mirror (not the microscope) toward a light source and reflect the light up through the diaphragm and opening in the stage. If this is done properly, a bright, evenly distributed circle of light—the field—will be visible. If the field is too bright, close the diaphragm to reduce the amount of light reflected from the mirror.

When proper illumination of the field has been achieved, slowly raise the body tube (or lower the stage) by means of the coarse adjustment knob until the letter comes into focus. It is helpful to move the slide slowly back and forth while doing this in order to help locate the material being studied. *In many microscopes you may become aware of a black line that appears to cut across the field. This is a pointer (usually made of hair or other fine material) and enables the student or the instructor to point out particular objects or regions of the material under observation.* Once the letter comes into focus, a small adjustment with the fine adjustment knob will bring it into clear, sharp focus.

When the letter is in sharp focus move the slide to the left. In which direction does the letter image appear to move?

Move the slide toward you. How does this affect the apparent direction of movement of the slide as you view it in the field?

Closely observe the letter under low power. Is it in the same position as when viewed with the naked eye, or is it inverted?

Slowly swing the high-power objective into position, making sure that as the objective moves into position it does not hit the stage or the slide. Many modern microscopes are parfocal—that is, once the image is brought into sharp focus under low power, it will remain in focus when the high-power objective is turned into position. If your instrument is not parfocal, a small adjustment of the fine-focus knob may be necessary. How does switching to high power affect the brightness of the field?

___

A minor adjustment in the position of the mirror will generally correct any loss of light that may be experienced. Open and close the diaphragm several times until you locate the position that provides the optimum light conditions for viewing under high power.

3. Precautions

In connection with the proper use of the compound microscope, there are several precautions that should be noted (Fig. 1-2).

a. When moved from one place to another, the microscope should always be held in an upright position.

b. Never touch the lens of the microscope with your hands. For cleaning the lenses in the ocular and objectives use only the special lens paper provided for this purpose.

c. If you should find that the microscope is out of order, do not attempt to make any repairs yourself. Notify your laboratory instructor.

d. Always locate the object to be studied with the low-power objective and then turn to the high-power objective if a more detailed study is desired.

e. Do not use the coarse adjustment when the high-power objective is in position. It is difficult to control the movement of this objective through small distances except by use of the fine adjustment knob.

f. Never focus downward while looking through the ocular. Rather, look at the objective from the side to prevent moving the body tube too far and possibly damaging the objective lens—to say nothing of the possibility of breaking the slide that is being studied.

4. Measuring Microscopic Objects

Frequently the professional biologist needs to know the dimensions of the object he is examining under the microscope. Described below are two methods you can use to obtain estimates of the size of microscopic objects.

Method A will enable you to obtain an approximate estimate of the size of an object by comparing it with the diameter of the field of view. To do this

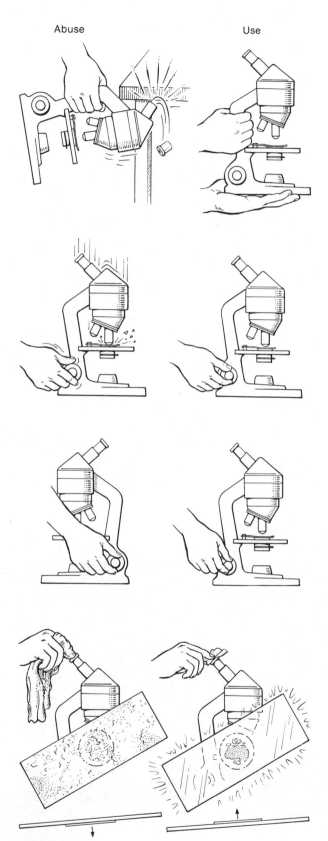

FIGURE 1-2
Common abuse and use of the compound microscope.

THE MICROSCOPE

it will first be necessary to measure the size of the field. For this measurement, as well as others you will be making in subsequent laboratory work, the metric system will be used.

a. Place a short, clear plastic ruler over the opening in the center of the stage so that the scale is visible through the microscope (Fig. 1-3A).

b. Line up one of the vertical lines so that it is just visible at the left side of the circular field of view (Fig. 1-3B).

c. The distance from the center of one line to the center of the next line is 1 **millimeter** (mm). Count the number of millimeters included from one side of the field to the opposite side. If the right side of the field does not coincide with one of the lines, you will have to estimate the fraction of a millimeter. What is the diameter (in millimeters) of the low-power field of view of your microscope?

For most microscopic measurements we need a much smaller unit than the millimeter. Scientists use the **micron** (μ), which is one-thousandth of a millimeter (0.001 mm). What is the diameter of of your low-power field in microns?

---

d. Turn the high-power objective into place. Note that the field of view is *less* than 1 mm. Instead of measuring this field directly it will be more accurate to obtain the diameter by the following method: divide the magnifying power of the high-power objective by the magnifying power of the low-power objective; then divide this quotient into the diameter of the low-power objective field previously obtained. What is the diameter of the high-power field in millimeters?

In microns?

---

e. Your instructor will supply you with various prepared slides of cells or other objects for you to measure. In each case find the length and width of the object in microns. For example, if the object's diameter is about one-third that of the low-power field, and you have calculated the diameter of the field as 1200 μ, then the diameter of the object is approximately 400 μ.

Method B: For more accurate measurements a microscope equipped with an **ocular micrometer**

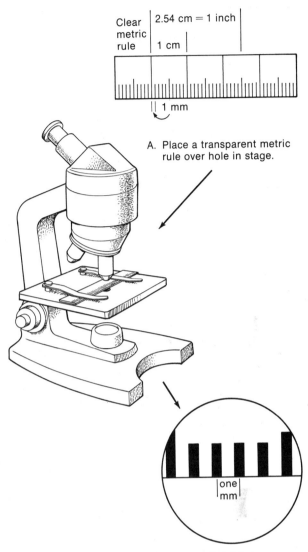

A. Place a transparent metric rule over hole in stage.

B. Line up one of the vertical lines so that it is visible at left edge of field. Count number of millimeter divisions from one edge of field to the other. In this example, the field diameter is approximately 6 mm.

FIGURE 1-3
Approximating the size of a microscope field (Method A).

can be used. This is a small glass disc on which uniformly spaced lines are etched. It is inserted into the eyepiece of the microscope (Fig. 1-4A). When an object is observed with a microscope having an ocular micrometer, the micrometer acts as a ruler that is superimposed on the object being measured. However, since microscopes vary in their actual magnification, it is first necessary to calibrate the micrometer for your specific microscope.

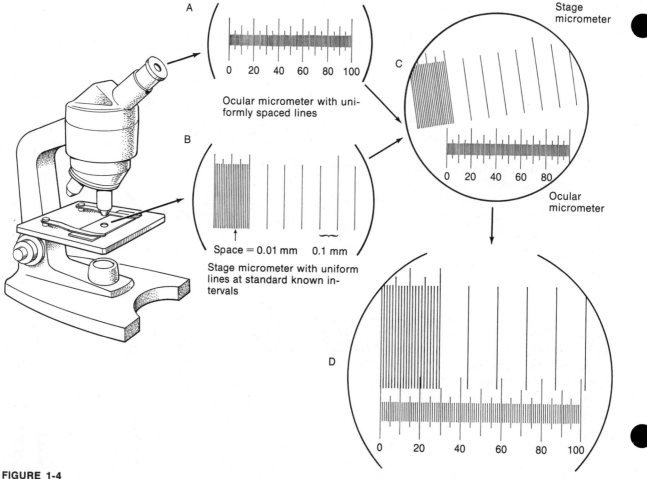

FIGURE 1-4
Ocular micrometer method for determining the size of microscopic objects (Method B).

a. Obtain a **stage micrometer** from your instructor; this is a special glass slide with uniform lines etched at standard known intervals. When you observe the stage micrometer with your microscope without the ocular micrometer in place, it will appear as shown in Fig. 1-4B. If you observe the stage micrometer with the ocular micrometer in place, it will appear as shown in Fig. 1-4C.

b. Turn the ocular in the tube until the lines of the ocular micrometer are parallel with those of the stage micrometer. Match the lines at the left edges of the two micrometers by moving the stage micrometer (Fig. 1-4D).

c. Calculate the actual distance in microns between the lines of the ocular micrometer by observing how many spaces of the stage micrometer are included within a given number of spaces on the ocular micrometer. Use the following formula: 10 spaces on the ocular micrometer = (X) spaces on the stage micrometer. Since the smallest spaces on the stage micrometer are known to be 0.01 mm apart, then

$$1 \text{ ocular micrometer space} = \frac{(X) \text{ times } 0.01 \text{ mm}}{10}$$

= _____ mm or

_____ microns (μ)

With this information you now can use this ocular micrometer in your microscope to measure various microscopic objects available in your laboratory.

B. USE AND CARE OF THE STEREOSCOPIC DISSECTING MICROSCOPE

The stereoscopic dissecting microscope (Fig. 1-5) has two distinct advantages over the compound microscope. It will (1) enable you to observe some

THE MICROSCOPE

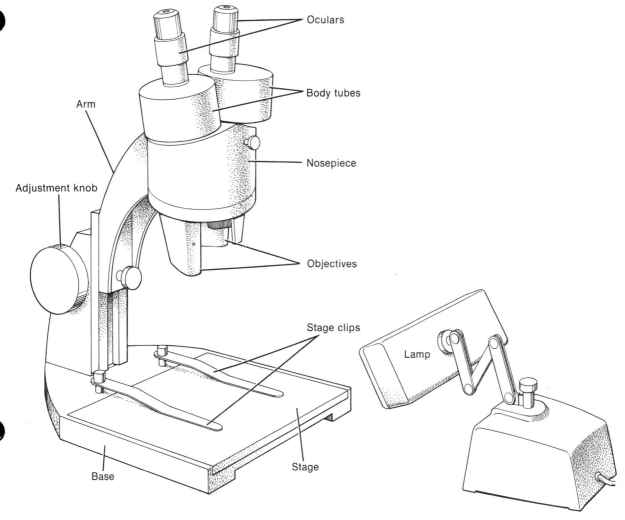

FIGURE 1-5
Parts of a stereoscopic microscope.

objects that are too large or too thick to see with higher magnifications, but too small for the unaided eye, and (2) give you an opportunity to observe objects in three dimensions. Magnifications obtained with the stereoscopic microscope commonly range between 4 and 60, depending on the lens combinations used.

Illuminate the stage of the microscope with a lamp or other similar light source. What is the low-power magnification of your microscope?

_____

The high-power?

_____

If there are any intermediate magnifications obtainable, what are they?

_____

Binocular microscopes are essentially two microscopes in one—a monocular microscope for each eye. The distance between the two oculars can be adjusted by moving the ocular tubes towards or away from each other. Looking through the microscope with the low-power objective in position, adjust the oculars to fit the distance between your eyes so that a single field of view is seen. Place a coin (or other object as directed by your instructor) on the center of the stage. Lower the objective as far as it will go. Looking through the oculars raise the objective until the object

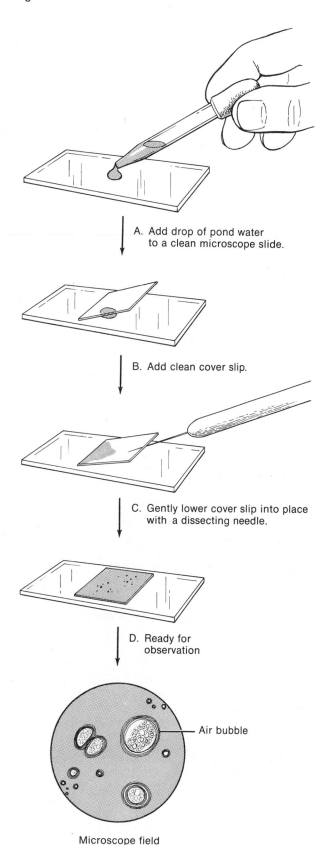

FIGURE 1-6
Preparation of a wet mount.

comes into focus. If the object seems out of focus it may be necessary to focus separately for each eye. If your microscope has separate focusing devices for each ocular, focus first with one ocular, and then the other. However, some microscopes have one ocular fixed in position. With this type of microscope you will first need to look through the ocular that cannot be individually focused. Then focus for this eye by turning the adjusting knob until the object is sharply outlined. Adjust the ocular for the other eye until the object is in focus for both eyes. If you are still having trouble, your instructor will help.

Move the object away from you. As you look through the microscope which way does the image move?

___

When the object is moved to the side, which way does the image move?

___

How does the direction of movement compare with that of the compound microscope?

___

Next observe the object at higher magnifications. As you increase the magnifications how does the area of the field of view change?

___

To get some practice with the stereoscopic microscope, observe various objects that are provided by your instructor. Try manipulating the object with dissecting needles while observing it.

C. STUDY OF POND WATER

In your laboratory work many of your observations with the microscope will be made on living organisms or on tissues or parts of organisms that you will want to keep alive. To allow them to dry out would greatly distort them, to say nothing of the effect death would have on a study of their movements. For observations of living material you will be making **wet mounts**. To prepare a wet mount, first obtain a clean microscope slide and cover slip (Fig. 1-6). The cover slip is very thin so that the objective lens of the microscope can be brought as close as possible to the specimens.

Using an eyedropper, add a drop of pond water to the center of the slide. The cover slip should be placed on the drop of water in the following way:

Lower one end of the cover slip so that it touches one side of the drop of water at about a 45° angle.

After the water has spread across the edge of the cover slip, carefully lower it by supporting the free end with a dissecting needle or the tip of your pencil. If this is done carefully it will prevent the accumulation of air bubbles under the cover slip. Air bubbles, which interfere with good observation, can be distinguished from other objects by their even, round shape and their heavy, dark outline.

Excess water at the edges of the cover slip can be soaked up by carefully placing a piece of paper toweling to the edge of the cover slip. However, if your preparation begins to dry out while under observation, add 1 drop of water at the edge of the cover slip.

Under low power and with reduced light, make a survey of the drop of pond water. Identify as many of the organisms as you can. Carefully study their differences in structure and their method of movement. Figure 1-7, 1-8, 1-9, and 1-10 should help you identify what you see.

Prepare additional wet mounts by taking samples from different parts of the jar of pond water. Do not be too hasty in discarding a slide as not containing any microorganisms; a systematic survey of the preparation is often necessary to locate the organisms. Why do the organisms often accumulate at the edge of the cover slip?

To identify the smaller organisms, it may be necessary to use the high-power objective. When your work is completed, clean and dry any slides and cover slips used. Wipe the lenses of the microscope with lens paper, clean the stage, and return the microscope to the cabinet.

REFERENCES

Anderson, M. D. 1965. *Through the Microscope*. Natural History Press, New York.

Benford, J. R. 1960. *The Theory of the Microscope*. Bausch & Lomb, New York.

Eddy, S., and A. C. Hodson. 1958. *Taxonomic Keys*. Burgess, Minneapolis.

John, T. L. 1949. *How to Know the Protozoa*. Wm. C. Brown, Dubuque, Iowa.

Johnson, G., and M. Bleisfeld. 1956. *Hunting with the Microscope*. Sentinel Books, New York.

Lenhoff, E. S. 1966. The Microscopes. In *Tools of Biology*. Ch. 1. Macmillan, New York.

Needham, G. H. 1958. *Practical Use of the Microscope*. Charles C Thomas, Springfield, Illinois.

Needham, J., and P. Needham. 1941. *Guide to the Study of Fresh-Water Biology*. Comstock, Ithaca, New York.

Palmer, M. C. 1959. *Algae in Water Supplies*. Public Health Service, Publication No. 657, Washington, D. C.

Taft, C. E. 1961. A Revised Key for the Field Identification of Some Genera of Algae. *Turtox News* **39**(4): 98–103.

Van Norman, R. W. 1963. Microscopy. *Experimental Biology*. Ch. 8. Prentice-Hall, Englewood Cliffs, New Jersey.

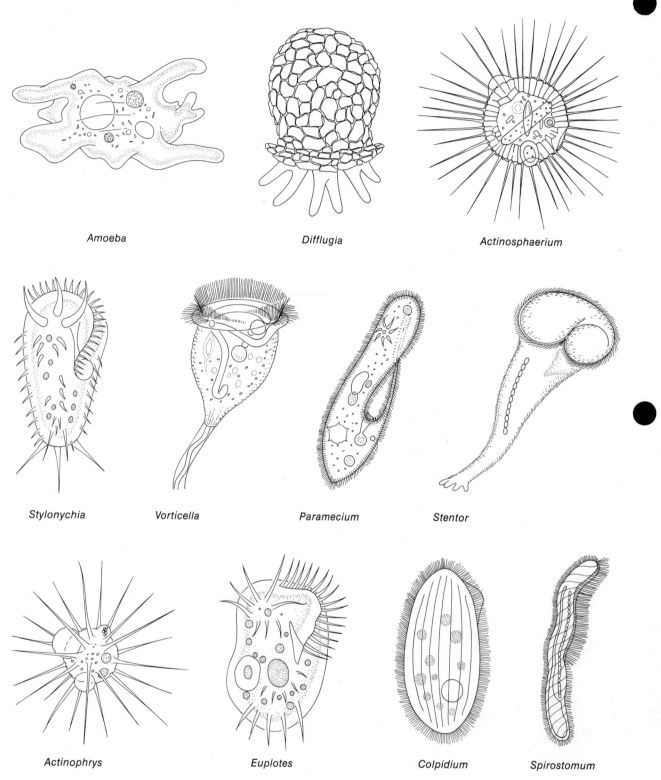

FIGURE 1-7
Protozoans commonly found in pond water.

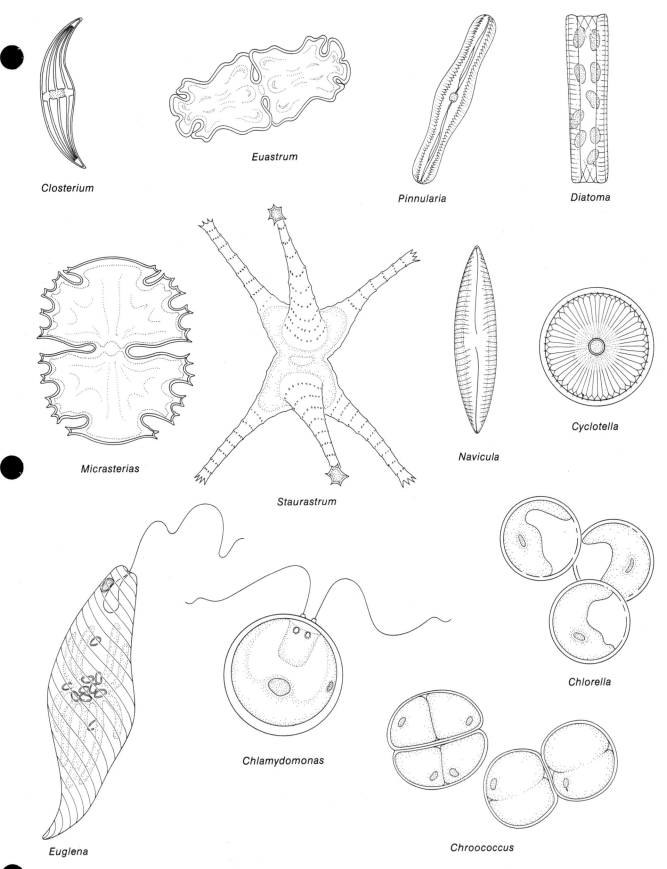

FIGURE 1-8
Unicellular algae commonly found in pond water.

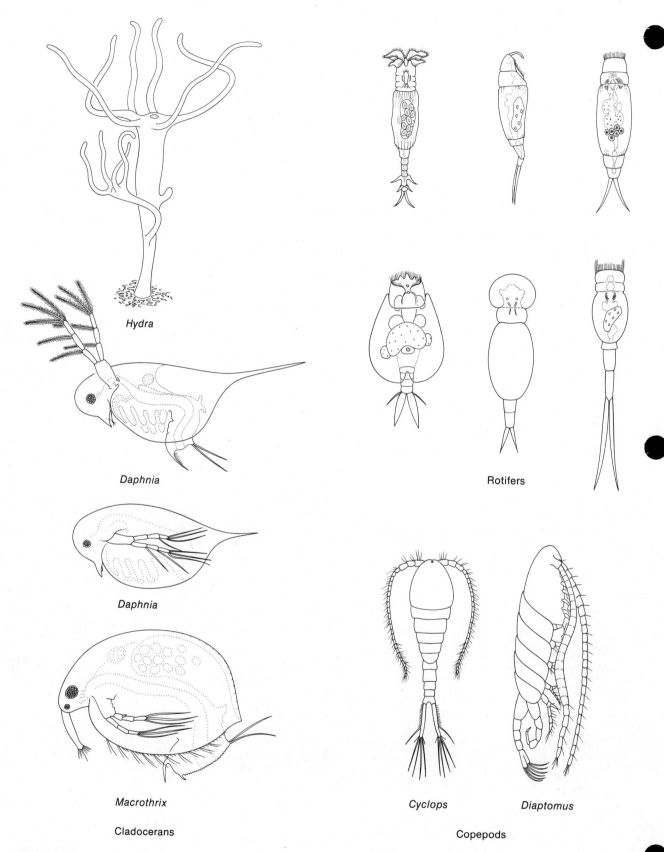

FIGURE 1-9
Invertebrates commonly found in pond water.

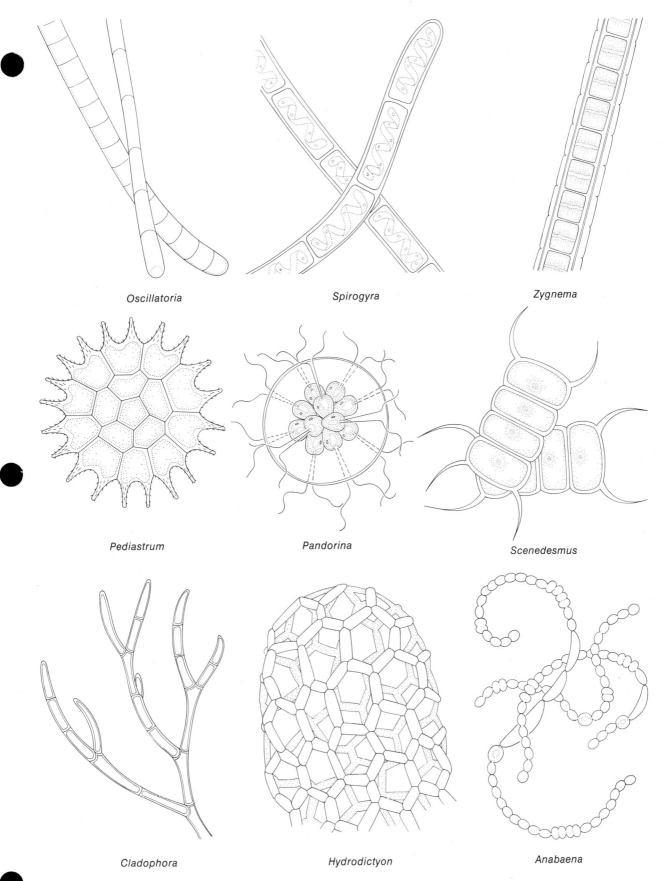

FIGURE 1-10
Multicellular algae commonly found in pond water.

EXERCISE 2

Organization of Cells

By the early part of the nineteenth century it was commonly accepted that all plants and animals were composed of cells. This cell theory has served as a basis for subsequent investigations of the cell.

Just as there is diversity of form in life, so there is in the forms and functions of cells that make up living organisms. Single cells, such as *Amoeba, Euglena,* or *Paramecium,* can be free-living organisms capable of carrying on an independent existence. Some cells live as part of a loosely organized colony of cells that move from place to place (such as *Volvox*). Others are immovably fixed as part of the tissues of higher plants and animals, depending upon the closely integrated activities of other cells for their existence.

Cells vary in size. For example, the length of many bacteria is roughly 1 micron (μ). On the other hand, the yolk of an ostrich egg, also a single cell, is the size of a small orange. Some cells, such as red blood cells, transport oxygen and carbon dioxide. Other cells have different specialities. Whatever its form or function, the cell is now recognized as the basic unit of living matter, containing all those properties and processes that we collectively call life.

The various parts of this exercise will help you to have a better understanding of the cell as a free-living, independent unit and as part of an integrated organism.

A. PLANT CELLS

1. Onion Cells

Although cells have diverse forms and functions, all cells are constructed according to a fundamental design and share certain common features. A basic knowledge of cell structure is indispensable to understanding the cell as an independent unit and its role in the life processes of higher plants and animals.

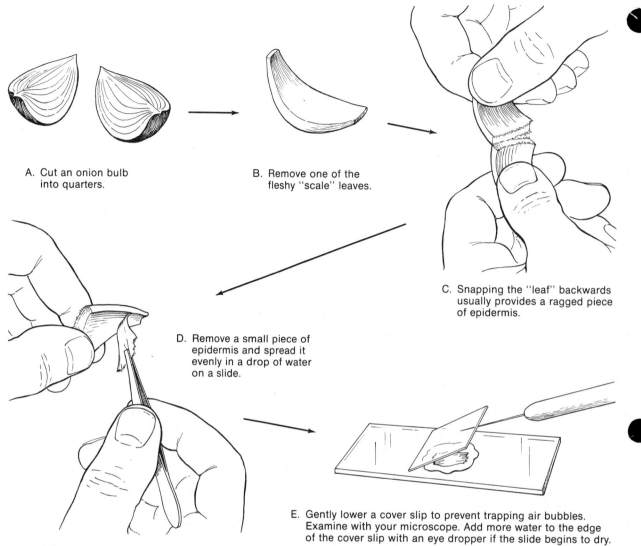

FIGURE 2-1
Procedure for studying living onion cells.

Following the procedure outlined in Fig. 2-1, prepare a wet mount of onion epidermal tissue. Examine this wet mount with the low-power objective (10×). What is the shape of the cells as seen in surface view?

_____

Next examine the cells under high power. The "lines" forming the network between the individual cells are nonliving cell walls composed chiefly of cellulose. The **cell wall** immediately surrounds the **cell membrane** (also called **plasma membrane**), which encloses the **cytoplasm**. The central part of many plant cells (difficult to observe in living cells) is taken up by a fluid-filled **vacuole** containing mostly water and salts.

Locate the **nucleus**, which appears as a dense body in the translucent cytoplasm. In some cells the nucleus seems to be lying in the central part of the cell and will look circular. In other cells, it appears compressed and pushed against the cell wall. Explain this apparent discrepancy in the shape and position of the nucleus.

_____

_____

_____

ORGANIZATION OF CELLS

The cytoplasm is separated from the central vacuole, nucleus, and cell wall by membranes, but the membranes are difficult to observe in this preparation.

2. Elodea Cells

In this study you will examine cells from the leaf of a water plant commonly called Elodea. These cells are green because they contain a pigment called **chlorophyll**. In **photosynthesis** this pigment absorbs light energy and converts it into chemical energy.

Remove a young leaf from the tip of the plant (Fig. 2-2). Place the leaf in a drop of water on a slide and add a cover slip. Examine the preparation with the low-power objective (10×) in position. Locate the nucleus, cytoplasm, and cell wall.

Examine a group of cells near the center of the leaf. Carefully switch to high power. Note that the green pigment is located in small structures in the cytoplasm. These structures are called **chloroplasts**. As you examine the chloroplasts, you should see them moving in the cell. Since chloroplasts cannot move by themselves, how would you account for the movement observed?

_____

_____

The plant cell is enclosed by a nonliving cell wall and a cell membrane that is difficult to observe because it is pushed tightly against the inside of the cell wall. You can make this membrane easier to see, however, by placing the cell in a saline (salt) solution that is more concentrated than the cytoplasm. The saline, being hypertonic to the cytoplasm, causes water to move out of the cell and results in the cell shrinking away from the wall, exposing the membrane.

Select another young Elodea leaf, mount it in a drop of water, and add a cover slip (Fig. 2-2C). Examine the preparation with the low-power objective. Along the edges of the leaf locate "spine" cells (Fig. 2-2D). Switch to high power and study the cell. Note that the cell membrane cannot be distinguished from the cell wall.

Add 1–2 drops of a concentrated salt solution to one edge of the cover slip. Then touch the liquid on the opposite side of the cover slip with a piece of lens paper (or paper toweling) so that the

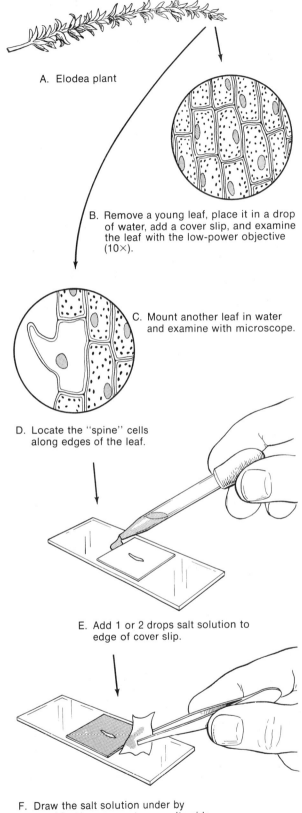

A. Elodea plant

B. Remove a young leaf, place it in a drop of water, add a cover slip, and examine the leaf with the low-power objective (10×).

C. Mount another leaf in water and examine with microscope.

D. Locate the "spine" cells along edges of the leaf.

E. Add 1 or 2 drops salt solution to edge of cover slip.

F. Draw the salt solution under by touching lens paper to opposite side of cover slip.

FIGURE 2-2
Preparation of Elodea cells for microscopic examination.

EXERCISE 2

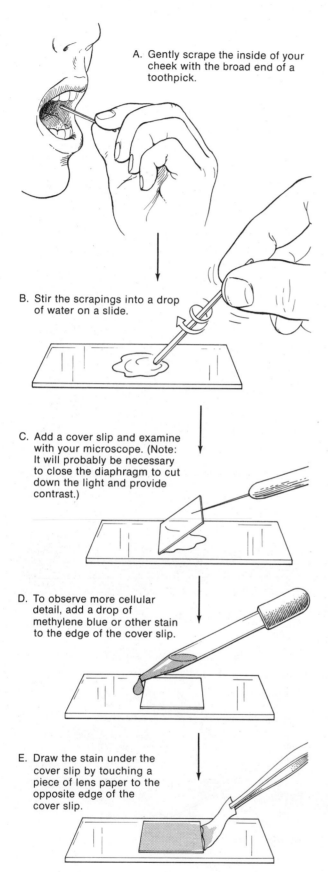

FIGURE 2-3
Procedure for studying human epidermal cells.

paper draws up the liquid. This will cause the saline to be drawn under the cover glass, replacing the liquid withdrawn by the paper (Fig. 2-2E and F). Repeat the above step two more times to be sure that the original water has been replaced by the saline. Examine the "spine" cell closely. Describe your observations.

_____

_____

_____

B. ANIMAL CELLS

The typical plant cell is characterized by having a nonliving cell wall and may contain chloroplasts. In this study you will examine human epidermal cells, which look different from plant cells but have many features in common.

Following the directions outlined in Fig. 2-3, examine cells obtained from the inner epidermal lining of the cheek. Locate the cells under high power and examine them carefully. What do these epidermal cells have in common with plant cells?

_____

_____

How are they different?

_____

_____

Some of the epidermal cells may have had their edges folded over. What does this indicate about the thickness of the cells?

_____

_____

Because it is difficult to observe structure in living cells, they are frequently stained with dyes to bring out cellular detail. Add a drop of methylene blue to the edge of the cover slip and draw it under as shown in Fig. 2-3D and E. What structure in the cell has been stained by this dye?

_____

ORGANIZATION OF CELLS

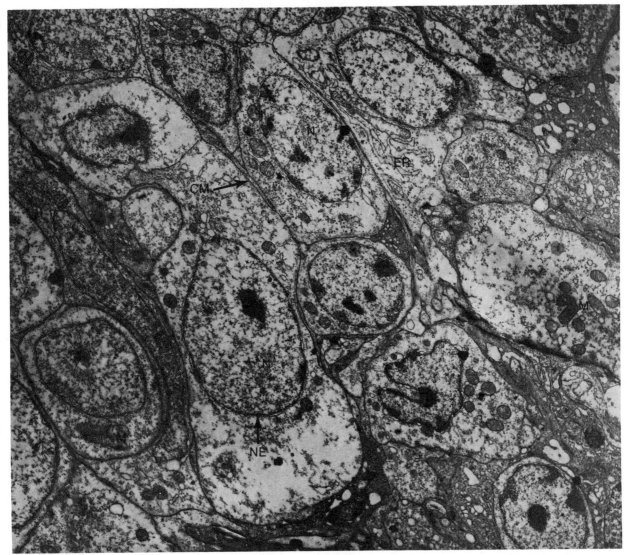

FIGURE 2-4
Fine structure of a typical animal cell. N, nucleus; NE, nuclear envelope; CM, cell membrane; M, mitochondrion; ER, endoplasmic reticulum.

If time permits, make new preparations of your cheek cells and treat them with other available stains.

C. CELL FINE STRUCTURE

With the development of the electron microscope, cells have been shown to contain a remarkable variety of membrane-bound, submicroscopic structures. Indeed, electron microscopic study of the cellular structure that is not visible with light microscopes has become one of the most active fields of research. In the following study you will become familiar with the "fine structure" (or "ultrastructure") of cells as seen in photographs taken through an electron microscope (called electron micrographs).

1. Basic Structure

Figs. 2-4 and 2-5 illustrate the basic fine structure of animal and plant cells, respectively. Examine

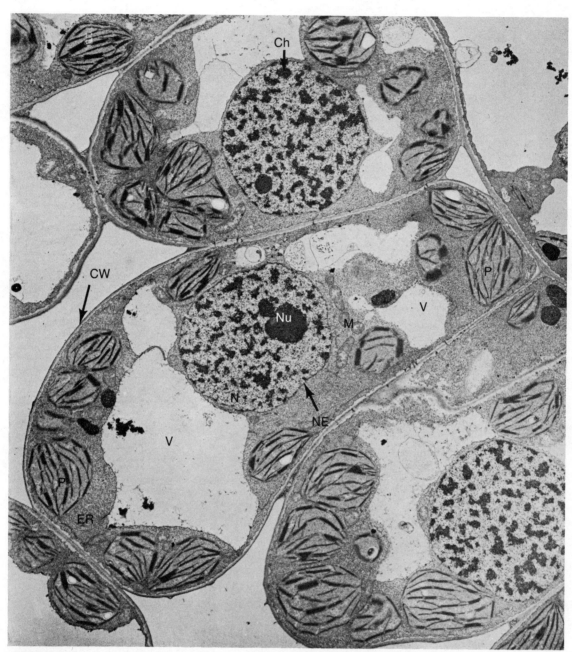

FIGURE 2-5
Structure of a higher plant cell. N, nucleus; Nu, nucleolus; NE, nuclear envelope; Ch, chromatin; CW, cell wall; V, vacuole; P, plastids; M, mitochondria; ER, endoplasmic reticulum. (Electron micrograph courtesy of Dr. Eugene L. Vigil, Marquette University.)

ORGANIZATION OF CELLS

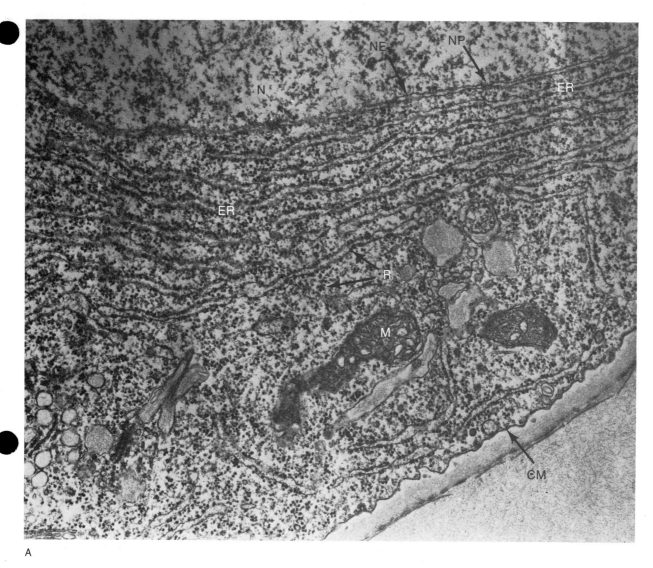

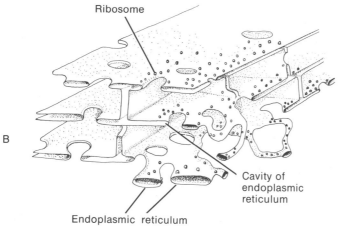

FIGURE 2-6
A. Electron micrograph showing cytoplasmic organization of cells. N, nucleus; NP, nuclear pore; NE, nuclear envelope; ER, endoplasmic reticulum; M, mitochondrion; R, ribosomes; CM, cell membrane. B. Diagrammatic representation of the structural organization of the endoplasmic reticulum. (Electron micrograph courtesy of Dr. Eugene L. Vigil, Marquette University.)

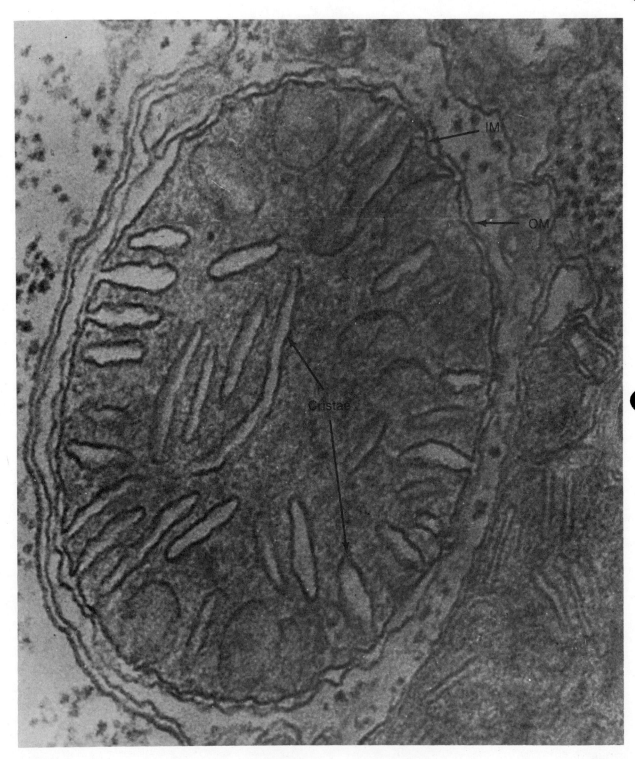

FIGURE 2-7A
Electron micrograph of a mitochondrion. OM, outer membrane; IM, inner membrane.

ORGANIZATION OF CELLS

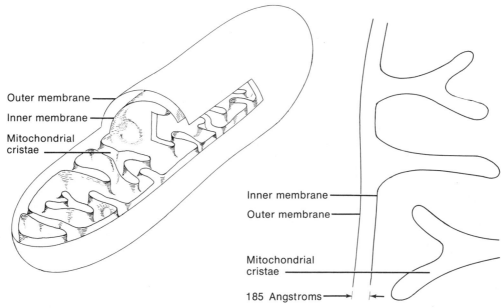

FIGURE 2-7B
Diagrammatic representation of the structural organization of mitochondria.

each photograph carefully. What submicroscopic features do plant and animal cells have in common?

_____

_____

_____

What are some obvious differences?

_____

_____

_____

_____

2. Cytoplasm

As seen in Fig. 2-6A, the cytoplasm is bounded on the outside by the cell membrane (plasma membrane) and internally by the nuclear envelope. Within the cytoplasm are found numerous membrane-bound organelles, each having its own peculiar ultrastructural organization.

  a. The **endoplasmic reticulum** (frequently called the ER) is common to many plant and animal cells (Fig. 2-6A). It consists of a complex, three-dimensional canalicular system extending throughout most of the cytoplasm (Fig. 2-6B). There may, however, be regional variations in its structure that form branching tubules or isolated vesicles.

Although little is known about the role of the endoplasmic reticulum, evidence suggests that in certain types of secretory cells the ER functions to transport the secretion products to regions of the cell where they are expelled.

  b. **Ribosomes** are the principal sites of protein synthesis in cells. These structures, composed of protein and ribonucleic acid (RNA), are frequently associated with the outer surfaces of the ER (Fig. 2-6A and B). Such an endoplasmic reticulum is referred to as rough or granular ER as compared with the smooth or agranular ER, which has no attached ribosomes.

  c. **Mitochondria** are subcellular structures that participate in cell respiration and contain those enzymes that take part in the Krebs cycle. Fine structure analysis of these organelles shows them to consist of two membranes: an outer membrane and an inner one that invaginates to the interior to form **cristae** (Fig. 2-7A and B).

  d. **Golgi bodies** (also called **dictyosomes**) are common organelles in cells, but are particularly abundant in secretory cells. The electron microscope has shown these structures to consist of stacks of platelike sacs (Fig. 2-8A and B). Note that at the margins the flattened plates expand into vesicles. Although the function of Golgi is not yet

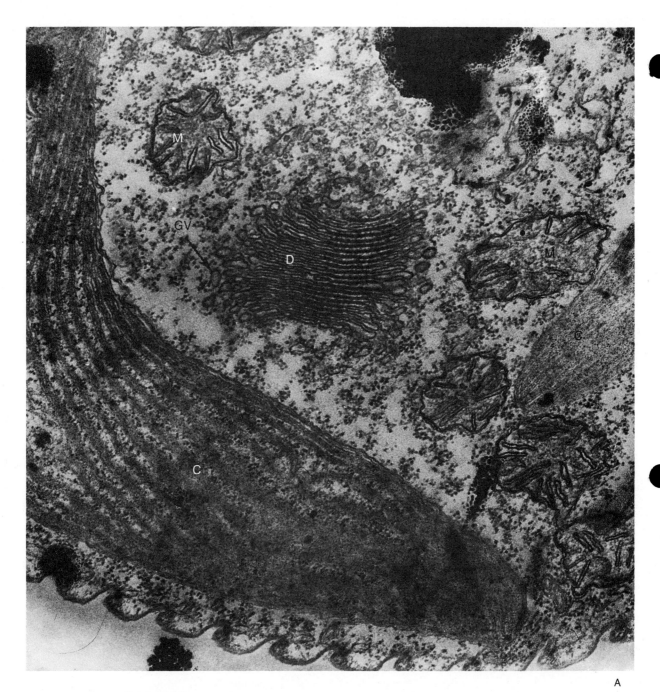
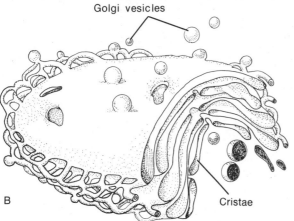

FIGURE 2-8
A. Electron micrograph of Golgi (dictyosomes) in the alga *Euglena*. D, dictyosomes; GV, Golgi vesicles; M, mitochondria, C, chloroplasts. B. Diagrammatic representation of the structural organization of Golgi.

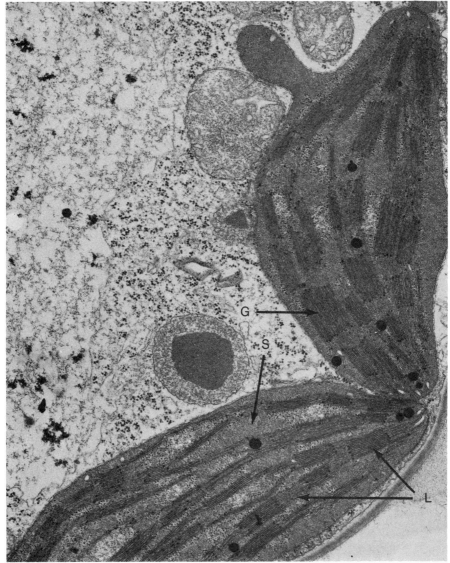

FIGURE 2-9
Electron micrograph of a chloroplast. S, stroma; G, granum; L, lamellae of stroma and granum. (Electron micrograph courtesy of Dr. Eugene L. Vigil, Marquette University.)

fully defined, it is thought that in secretory cells the products of secretion made in other parts are transported to the Golgi through the lumen of the ER. These products are then segregated in the vesicles at the margins of the Golgi apparatus, which are then pinched off and migrate elsewhere in the cytoplasm or the cell surface.

e. **Chloroplasts** are the sites of photosynthesis in green plant cells. As shown in Fig. 2-9 chloroplasts are surrounded by two membranes. Internally the chloroplast is divided into two main parts: a system of membranes called lamellae, and a matrix called the stroma. Locate the grana, which appear as concentrated layers of the lamellae. Note that some of the lamellae are connected to other grana.

REFERENCES

Brachet, J. 1961. The Living Cell. *Scientific American*, September (Offprint No. 90). *Scientific American* Offprints are available from W. H. Freeman and Company, 660 Market Street, San Francisco 94104, and 58 Kings Road, Reading RG1 3AA, England. Please order by number.

Butler, J. A. V. 1959. *Inside the Living Cell*. Basic Books, New York.

Fawcett, D. W. 1970. *An Atlas of Fine Structure: The Cell.* 2d ed. Saunders, Philadelphia.

Goldstein, L. 1966. *Cell Biology.* Wm. C. Brown, Dubuque, Iowa.

Hurry, S. W. 1966. *The Microstructure of Cells.* Houghton Mifflin, New York.

Jensen, W. A. 1965. *The Plant Cell.* Wadsworth, Belmont, California.

Jensen, W. A., and R. B. Park. 1967. *Cell Ultrastructure.* Wadsworth, Belmont, California.

Kennedy, D., Ed. 1965. *The Living Cell.* W. H. Freeman and Company, San Francisco.

Neutra, M., and C. P. Leblond. 1969. The Golgi Apparatus. *Scientific American,* February (Offprint No. 1134).

Nomura, M. 1969. Ribosomes. *Scientific American,* October (Offprint No. 1157).

Paul, J. 1965. *Cell Biology.* Stanford University Press, Stanford, California.

Pfeiffer, J., Ed. 1964. *The Cell.* Life Science Library, New York.

Rich, A. 1963. Polyribosomes. *Scientific American,* December (Offprint No. 171).

Robertson, J. D. 1962. The Membrane of the Living Cell. *Scientific American,* April (Offprint No. 151).

Sharon, N. 1969. The Bacterial Cell Wall. *Scientific American,* May (Offprint No. 1142).

Stern, H., and D. L. Nanney. 1965. *The Biology of Cells.* Wiley, New York.

Swanson, C. P. 1970. *The Cell.* 3d ed. Prentice-Hall, Englewood Cliffs, New Jersey.

Wessells, N. K., and W. J. Rutter. 1969. Phases in Cell Differentiation. *Scientific American,* March (Offprint No. 1136).

# EXERCISE 3

Cell Division

The tremendous diversity of form and function that cells assume is even more remarkable when you consider that multicellular organisms begin life as a single fertilized cell, the **zygote**. This cell, through repeated divisions, gives rise to all the cells that make up the organism. The complex series of events that encompasses the life-span of an actively dividing cell is termed **mitosis**, or the **mitotic cell cycle** (Fig. 3-1). The actual replication of chromosomes including the synthesis of deoxyribonucleic acid (DNA), the main structural component of chromosomes, occurs during **interphase**. The role of DNA as the molecular basis of heredity will be examined in another exercise. The doubling of DNA during the synthesis or **S** period provides a full complement of DNA for the daughter cells that will result from the next active mitotic division. There are two nonmitotic and non-DNA-synthetic periods (gaps) during interphase: the one preceding DNA synthesis is G_1; the one following synthesis and just preceding mitotic division is G_2. During G_1 (prior to DNA synthesis) a cell may follow a pathway leading to differentiation, rather than continue in active division. This possibility is indicated in Fig. 3-1 by the arrow leading out of the cell cycle. The time required for one mitotic cycle varies from tissue to tissue and from organism to organism. The actual mitotic process usually occupies only 10% of the total time of the cell cycle. The rest of the time is spent in interphase. It is important to distinguish between several parts of mitosis. Thus, chromosomal duplication and separation, i.e., nuclear division, is more properly termed **karyokinesis**. Division of the cell body, i.e., cytoplasm and its organelles, is more properly termed **cytokinesis**.

There are certain structural differences in plant and animal cells; the two vary to some extent also in the process of mitosis. However, in all organisms this process is essentially the same. It is the objective of this exercise to examine the essential steps in mitosis and to characterize the similarities and differences in this process in plant and animal cells.

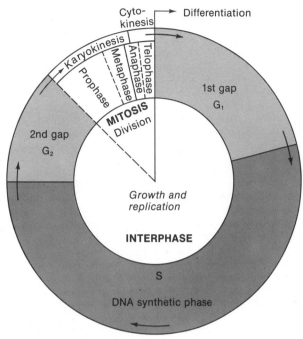

FIGURE 3-1
Stages of the mitotic cycle.

A. MITOSIS IN PLANT CELLS

The onion root tip is one of the most widely used materials for the study of mitosis, since it is available in quantity and preparations of the dividing cells are easily made. The chromosomes are relatively large and few in number and hence easier to study than the cells of many other organisms. There are regions of rapid cell division in root tips; therefore, the chances are good that within such tissues one can identify every stage in mitosis. There are several reasonably distinct stages in cell division, although the process is continuous and there is some gradation among the various steps. These steps, in sequence, are prophase, metaphase, anaphase, and telophase (Fig. 3-2).

Obtain a slide of onion root tips. Hold the slide above a sheet of white paper, and note a series of dark streaks on the slide. Each streak is a very thin longitudinal section through an onion root tip.

Place the slide on the stage of your microscope and locate one of the sections under low power. It is often possible, under low power, to determine whether a given section shows good mitotic stages. Since each section is very thin, not all will be equally good for study. After preliminary examination under low power, change to high power, being very careful not to break the slide. Keep in mind the sequence in which the different stages occur, but do not try to find them in sequence. Thus, if you happen to find an anaphase first, study it before proceeding to another stage. Chances are that most of the cells will be in interphase. The next largest number will be in prophase, and only a few will be seen in metaphase, anaphase, and telophase. The reason is that these cells remain in interphase and prophase longer than in the other stages.

1. Prophase

During **prophase** the chromosomes become distinguishable in the nucleus. The nuclear membrane breaks down, and the chromosomes become distributed haphazardly through the cytoplasm. At this stage in the onion root tip, the chromosomes often appear as a coiled mass. These elongated chromosomes later become condensed into shorter chromosomes and the nuclear membrane disappears. Even at this early stage each chromosome has probably doubled, although this will be difficult to see on the slides.

2. Metaphase

During **metaphase** the chromosomes become arranged near the center of the cell. This stage is apparently a preparation for the equal division of chromosomes between the daughter cells, a process that begins in the next phase. During or somewhat before metaphase, small, threadlike structures (**spindle fibers**) form in the dividing cell. Some of them appear to be attached to the chromosomes and seem to provide the machinery for the movement of the chromosomes, although the way in which this is accomplished is not yet understood. These fibers usually appear most clearly in anaphase.

The composition of the spindle fibers is not known. It has been suggested that they form by the aggregation of protein molecules. Under the electron microscope these fibers appear as fine, straight, hollow tubules (Fig. 3-3). Although they lengthen and then shorten during mitosis, they do not appear to get thicker or thinner. This suggests that they do not stretch or contract but that new material is added to the fiber or removed from it as the spindle changes shape.

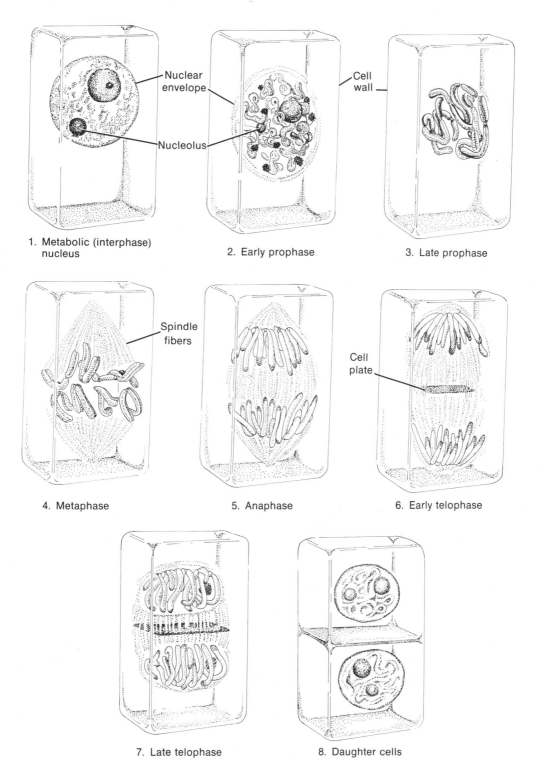

FIGURE 3-2
Diagrammatic representation of the mitotic stages in the onion root tip.

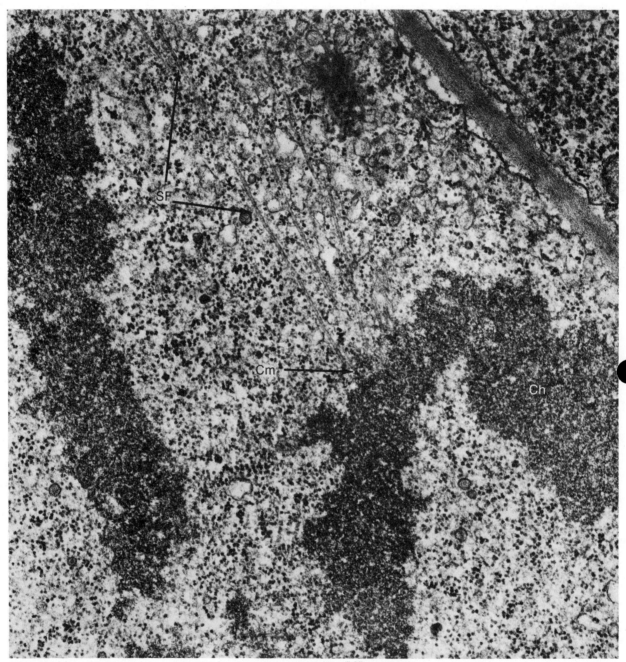

FIGURE 3-3
Spindle fibers (SF) attached to a chromosome (Ch) at the centromere (Cm). Note the tubular structure of the spindle fibers. (Electron micrograph courtesy of Dr. Eugene L. Vigil, Marquette University.)

CELL DIVISION

3. Anaphase

At the beginning of **anaphase** the two members of each of the previously doubled chromosomes separate and move toward opposite ends of the cell. This stage can be recognized in the onion by the two groups of V-shaped chromosomes on opposite sides of the cell. The sharp end of the V is pointed toward the cell wall. The onion has sixteen chromosomes; hence it is seldom possible to see all of them at one time. Reduce the light coming through the objective of your microscope and see if you can find any spindle fibers near the center of the cell. They will appear as very fine lines between the two groups of chromosomes, but they are often not visible in a study of this kind.

4. Telophase

Cell division is completed during **telophase**, and reorganization of the cell contents of the two daughter cells begins. It is often difficult to distinguish between late anaphase and early telophase in the cells of the onion root tip. During telophase, however, a **cell plate** starts to form as a fine line across the center of the cell. When complete, the cell plate will divide the original cell into two daughter cells. In some cells it will be indistinct. As telophase progresses, the nuclei begin to reorganize and the chromosomes become indistinct.

In both plant and animal cells, the daughter cells resulting from mitotic division have the same number and kinds of chromosomes as the original cell from which they came. Thus, in the onion each daughter cell has sixteen chromosomes, just as the original cell had. This is not so evident in the slides you are studying, because specially prepared slides and considerable experience are necessary before this many chromosomes can be accurately counted.

One difference in mitosis between most plant and animal cells should be pointed out at this time. When plant cells divide, a cell plate forms, as we saw in the onion root tip. Animal cells do not have a cell wall but are surrounded only by the plasma membrane. When most animal cells divide, there is a constriction near the center of the original cell. This constriction, which goes completely around the cell, deepens until the cell cleaves into two daughter cells.

B. MITOSIS IN ANIMAL CELLS

Mitosis is readily observed in animal cells by examining a prepared slide of a whitefish blastula — the early developmental stage formed by successive cell divisions following fertilization of the egg by the sperm (Fig. 3-4). The behavior of the chromosomes in animal cells is essentially the same as that observed in plant cells. In whitefish blastula cells, however, the chromosomes are smaller and more numerous than in onion root-tip cells. The most obvious difference in mitosis between plant and animal cells is in the formation of the spindle. In animal cells, the formation of the spindle seems to be related to the **centrioles** and their movement. At or near the beginning of prophase, the centrioles begin to move apart as if they repelled one another. A group of fibers that radiate from each centriole like spokes from the hub of a wheel then appears. This configuration is known as the **aster** and the radiating fibers are called **astral rays**. Some of the fibers reach from one centriole to another, stretching as the pair separates. The centrioles continue to migrate until they lie at opposite sides of the nucleus. Their positions mark the **poles** towards which the chromosomes will move. When the nuclear envelope breaks down at prophase, the region between the centrioles becomes clearly visible as a relatively transparent region called the **spindle**. As the centrioles reach the poles, some of the spindle fibers extend to reach from pole to pole. Other spindle fibers attach to the **centromeres** (kinetochore) of the chromosomes (Fig. 3-3) and appear to guide them to the equator. When all the centromeres are lined up along the equator, the centromeres divide and the chromosomes separate and move toward the opposite poles. They appear to be pulled along by the spindle fibers; the centromeres, to which the fibers are attached, move first, while the two "arms" of the chromosomes drag behind. Locate the various events described above on your slides.

Now find a cell that has almost completely divided into two cells. The two groups of chromosomes have reached the opposite ends of the spindle and have become aggregated near the centrioles. The centrioles soon become indistinct, and the chromosomes become less distinctly stained. The division of cytoplasm in animal cells is accomplished through a **cleavage furrow** rather than through the formation of a cell plate, as in plant cells. This cleavage furrow is a constriction

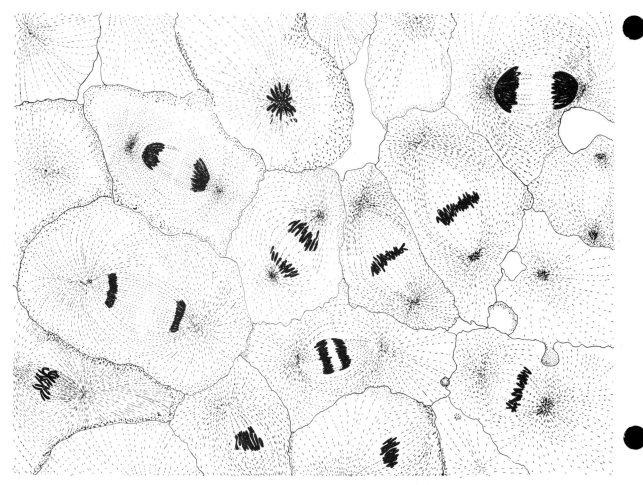

FIGURE 3-4
Mitotic stages in the whitefish blastula.

that cuts through the cytoplasm and spindle in the vicinity of the equator and eventually cuts the cell in two.

C. CHROMOSOME MORPHOLOGY

1. Salivary-Gland Chromosomes

Examine a culture of the fruit fly, *Drosophila melanogaster*, and locate wormlike larvae. Select one of the larger, slower-moving ones, preferably one that is crawling up the sides of the culture container. Gently remove it with a forceps and place it on a glass slide. Examine it with your dissecting microscope. Note that it has a blunt rear end and a pointed head end that contains black mouth parts. To obtain the salivary glands, the head end must be dissected away from the rest of the body (Fig. 3-5).

a. Add a drop of saline to the slide just to the left (or right) of the larva. Then, with a finely pointed needle, pierce the head as close to the anterior end as possible. The larva wriggles so you will probably have to make several attempts before you are successful.

b. After the head has been secured, grasp the rear end with a pair of finely pointed forceps and slowly stretch the larva until the mouth parts are torn off and pulled into the drop of saline. This can all be done while observing the operation with the dissecting microscope.

c. The salivary glands, when pulled out, will probably be accompanied by the digestive tract and fat bodies. When you are sure that you have the salivary glands, remove the digestive tube

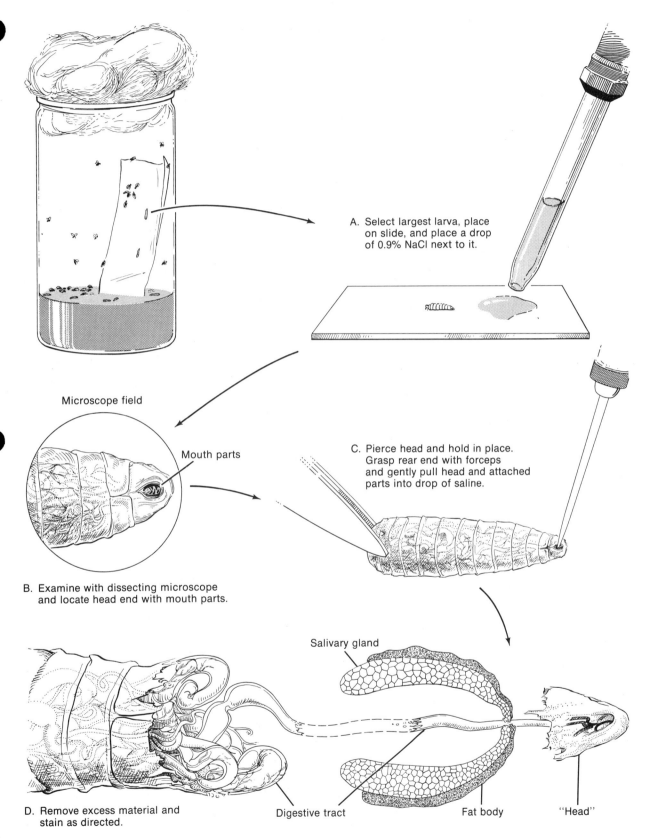

FIGURE 3-5
Procedure for removing *Drosophila* salivary glands.

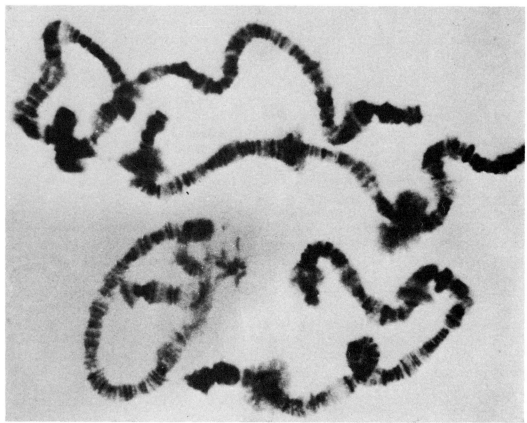

FIGURE 3-6
Salivary gland chromosomes of *Sciara coprophila* showing distinct longitudinal differentiation. (Photomicrograph courtesy of Dr. Ellen M. Rasch, Marquette University.)

and any other undesirable parts with a sharp razor blade or scalpel. Discard all parts, so that only the salivary glands are left on the slide.

d. The salivary glands are now ready to be stained. Remove the excess saline solution from the slide by soaking it up with a small piece of paper toweling or filter paper. Try not to touch it to the glands as they might adhere to the paper. Cover the glands with a drop of aceto-orcein stain.

e. After staining for 5 minutes, carefully place a plastic coverslip over the glands. Then place a small piece of paper toweling over the cover slip and press down hard with your thumb. Examine the preparation with your compound microscope. If properly squashed, the cells of the gland will be separated from each other. You should be able to see the nucleus in most of the cells.

f. If your preparation has been adequately stained you should be able to see banding along the length of the chromosomes. These may be better observed if the chromosomes are released from the nuclei. To do this, firmly tap the cover slip with a pencil eraser several times. Examination under the high power of the microscope should now show the chromosomes separated and the bands sharply stained. Compare your preparation with that of *Sciara* chromosomes shown in Fig. 3-6.

2. Onion Root-Tip Chromosomes

Chromosomes from living onion root-tip cells may be obtained by the following procedure (Fig. 3-7).

a. Obtain a root from the base of an onion provided by your instructor. Select roots 3–4 cm in length.

b. Cut a 2- to 3-mm piece from the tip of the root and place it on a clean glass slide. Cover it with several drops of 1 N hydrochloric acid and heat gently over a flame for 1 minute. Do not boil.

c. Remove the HCl by blotting it with filter paper or paper toweling. Then cover the root with several drops of 0.5% aqueous toluidine blue. Heat gently, without boiling, for 1 minute.

CELL DIVISION

d. Blot away excess stain and then add 1 drop of fresh toluidine blue. Apply a cover slip. Then place a small piece of paper toweling over the cover slip and press down with your thumb.

e. Examine your preparation microscopically and locate cells showing various stages of mitotic activity.

REFERENCES

Anderson, N. G. 1956. Cell Division. *Quart. Rev. Biol.* **31**:169–199, 243–269.

Bass, A. D. 1959. Chemical Influences on Cell Division and Development. *Ann. Rev. Physiol.* **21**:49–68.

Bottura, C., and I. Ferrari. 1960. A Simplified Method for the Study of Chromosomes in Man. *Nature* (London) **186**:904–905.

Conger, A. D., and L. M. Fairchild. 1953. A Quick-Freeze Method for Making Smear Slides Permanent. *Stain Technol.* **28**:281–283.

Furness, F. N., E. W. White, and P. R. Gross, Eds. 1960. Second Conference on the Mechanisms of Cell Division. *Ann. N.Y. Acad. Sci.* **90**:345–613 (Art. 2).

Giese, A. C. 1968. *Cell Physiology.* 3d ed. Saunders, Philadelphia.

Gillette, N. J. 1965. A Demonstration of Mitosis. *Turtox News* **43**(3):90–91.

Jacobson, W., and M. Webb. 1952. Nucleoproteins and Cell Division. *Endeavour* **11**:200–207.

Levine, L., Ed. 1963. *The Cell in Mitosis.* Academic Press, New York.

Mazia, D. 1961. How Cells Divide. *Scientific American,* September (Offprint No. 93). *Scientific American* Offprints are available from W. H. Freeman and Company, 660 Market Street, San Francisco 94104, and 58 Kings Road, Reading RG1 3AA, England. Please order by number.

Mazia, D. 1961. Mitosis and the Physiology of Cell Division. In *The Cell.* J. Brachet and A. E. Mirsky, Eds. Vol. 3, pp. 77–412. Academic Press, New York.

McLeish, J., and B. Snoad. 1958. *Looking at Chromosomes,* Macmillan, New York.

Patau, K. 1960. The Identification of Individual Chromosomes, Especially in Man. *Am. J. Human Genet.* **12**:250–276.

Swann, M. M., and J. M. Mitchison. 1958. The Mechanism of Cleavage in Animal Cells. *Biol. Rev.,* **33**:103–135.

Taylor, J. H. 1958. The Duplication of Chromosomes. *Scientific American,* June (Offprint No. 60).

Wilson, G. B. 1966. *Cell Division and the Mitotic Cycle.* Reinhold, New York.

Wolf, F. E., and R. Fournier, Jr. 1966. A Simplified Method for Making Chromosome Squashes. *Turtox News* **44**(9):235.

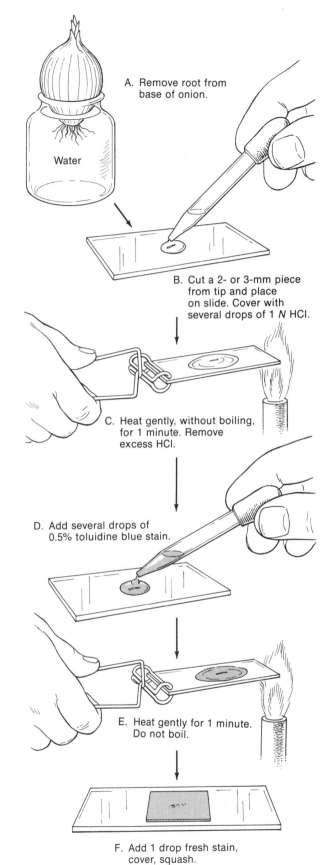

FIGURE 3-7
Procedure for studying root-tip chromosomes.

EXERCISE 4

Plant Anatomy: Roots and Stems

In order to comprehend the complex structure of higher plants and the interrelationships between structure and function, it is necessary to study the anatomy of the organism under consideration.

Higher plants, like animals, are made up of groups of cells organized into tissues that become specialized for different tasks. However, the tissues of plants are different from those of animals in many respects. A striking difference is seen in the way they grow. Animals, in general, reach their full growth after a period of time and then stop growing. Plants, however, continue to produce new organs (e.g., roots and leaves) throughout their growth period. Certain regions at the tips of the roots and stems remain embryonic and, through cell division, produce new cells that differentiate into the tissues of the stems, leaves, and roots. These regions of embryonic tissue are called **meristems** in contrast to the fully developed permanent tissues.

From a functional point of view, a plant consists of the following types of cells.

Meristematic cells have the ability to divide and give rise to new cells of the plant. In general, meristematic cells have cubelike configurations, thin walls, and numerous small vacuoles. The nucleus is quite large in relation to the rest of the cell.

Parenchymatous cells are large, usually thin-walled, many-sided cells. They are relatively undifferentiated and form the bulk of softer plant parts—most parts of the fruit, soft parts of stems and leaves, and so on. They may contain chloroplasts or may be adapted for the storage of water and reserve foods such as starch.

Supporting and conducting cells are all relatively elongated, but otherwise are greatly different from one another. One group, whose walls have large quantities of lignin, are dead at maturity. Another type of supporting cell, **collenchyma**, is thickened only at the corners of the cell. **Sieve tube** elements are the main pipeline cells of the phloem and function in transporting organic materials such as sugars. Vessels, composed of perforate, nonliving cells lined end-to-end, are the predominant water carrying cells of the xylem.

Protective cells have flattened surfaces with a waxy, waterproof coating, which prevents excessive water loss from the plant's inner tissues. There are many different kinds of specialized cells that form the outer protective "skin" of a plant. Most important among these protective cells are those forming the **epidermis**.

A. ROOTS

Characteristically, the root is described as that part of the plant growing beneath the surface of the soil. The principal function of roots is the absorption of water and soluble minerals and the transport of these substances to the above-ground parts—the stem, leaves, and flowers. In addition, roots anchor the plant in the soil and serve as storage organs for the reserve food of the plant, as in sweet potatoes, carrots, and turnips. Roots may function in food manufacture. For example, aerial roots of orchids contain chlorophyll, thereby supplementing the leaves in the photosynthetic process.

Roots may be broadly classified by form into two groups: **tap roots** and **fibrous roots**. In tap roots the main root becomes many times larger than the branch roots (those that arise from the main root) and penetrates some distance into the soil. In some cases the tap root may be greater in diameter than the stem. Examine the tap root of a carrot plant for the presence of branch roots. In fibrous root systems the primary root and the branch roots are approximately the same length and diameter. Examine the fibrous root system of a bean plant. Why do you think the fibrous root system would be more efficient in terms of absorption of water and minerals than the tap root?

_____

_____

1. Root Apex

Just as the above-ground parts of the plant become extensive, so do roots grow and ramify throughout the soil. Obtain a germinating radish or grass seed from the instructor. Mount it in a drop of water and examine the young root with a dissecting microscope. Locate a cone-shaped mass of loosely arranged cells covering the root tip. This **root cap** covers the meristem and protects it from damage as the root passes through the soil. Locate the root **apical meristem**, which appears as a relatively dense, opaque region at the tip of the root. Somewhat behind the root tip, note the presence of root hairs, special absorbing cells on the root. These generally persist for a short time and then die. New root-hair cells formed near the root tip replace those that are lost. In what way do root hairs increase the efficiency of absorption of water and minerals by the root?

_____

_____

Examine a prepared slide of a longitudinal section through an onion root tip and locate the structures described above (Fig. 4-1A and B).

2. Primary Tissues of the Root

Using a compound microscope examine a prepared slide of a cross section of the buttercup (*Ranunculus*) root, cut through a region in which the cells have differentiated (Fig. 4-1D). Three general regions are readily seen: the epidermis, the cortex, and the vascular cylinder. Starting from the outside, the primary tissues (those tissues having their origin from cells produced in the apical meristem) consist of an outer **epidermis**, which is a single layer of cells covering the outside of the root. In some instances the epidermis dies and the outer cells of the cortex function as a special epidermal covering called the **hypodermis**. The **cortex** in most roots consists largely of parenchymatous cells. Note the presence of starch grains in many of the cortical cells, and the large intercellular spaces formed where the cells abut one another.

Limiting the cortex on the inside, and considered to be part of it, is the **endodermis**. It is represented by a single layer of cells separating the parenchyma of the cortex from the vascular cylinder. The endodermis, present in all roots, is believed to function in regulating the flow of water into the vascular cylinder from the cortex.

The **vascular cylinder** consists of several tissues lying internal to the cortex. The **pericycle**, a unicellular layer of cells adjacent to the endodermis,

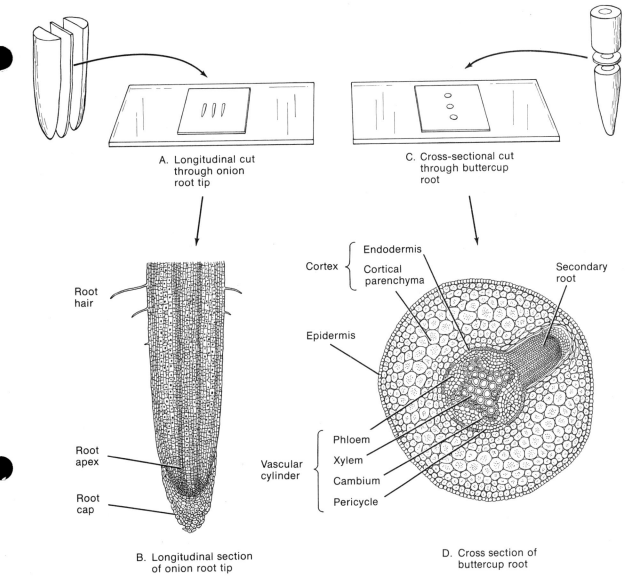

FIGURE 4-1
Study of root anatomy.

has the ability to become meristematic and to initiate the growth of lateral roots. The **xylem** is represented by three, four, or five radiating arms or plates. Alternating with the xylem arms are groups of **phloem** cells. This alternate arrangement is peculiar to roots and anatomically distinguishes roots from stems. Locate the region between the phloem and the xylem where the **cambium** will develop. The cambium is a lateral meristem responsible for growth in the diameter of a root. The tissues present in this cross section are all primary tissues, that is, they originate in the apical meristem of the root. When the cambium divides it gives rise to cells that will differentiate as xylem and phloem, which are then called secondary xylem and phloem since they originated in meristems other than an apical meristem.

Obtain a carrot root. Using a sharp razor, cut a thin, transverse section from the root and place it on a slide. Add 1–2 drops of iodine and a cover slip. Examine microscopically. Iodine reacts with starch to form a deep blue-black color. Where is the starch located in the root?

_____

Prepare a second cross section of carrot and stain with phloroglucinol-HCl. (CAUTION: This solution contains concentrated hydrochloric acid.) This chemical reacts with a substance called **lignin** in the cells that make up the vascular tissue. How does the organization of vascular tissue of the carrot root differ from that of the *Ranunculus* root examined above?

_____

_____

3. Branch Roots

As previously stated, the pericycle may give rise to branch or lateral roots. In lateral root formation a new meristem complete with root cap is formed. Eventually, the young branch root grows, ruptures through the cortex to the outside, and continues growing. Examine demonstration slides of developing branch roots (Fig. 4-1D). If available, remove young roots from water lettuce plants (*Pistia*) and examine developing branch roots.

B. STEMS

The stem is that part of the plant which bears leaves and is commonly aerial and upright. Stems are like roots in general structure; they have an epidermis, cortex, and vascular cylinder, although there are variations within the overall plan. Stems differ from roots in the fundamental vascular structure and in possessing lateral appendages, the leaves.

Several basic functions may be ascribed to the stem. It is the framework supporting the leaves, flowers, and fruits. It provides for the transport of water and solutes absorbed by the roots and serves to transport and distribute the foods manufactured by the leaves to the places where it is used or stored. Some stems are green and carry on a limited amount of photosynthesis. Many stems function as food storage areas—sometimes to the extent of being economically useful (e.g., sweet potatoes and sugar cane).

1. Primary Growth

a. Shoot Apex

All primary stem tissues originate in a terminal meristem, the **shoot apex**. The shoot apex, like the root apex, is a region of active cell division. The activity of this meristem brings about the increase in the length of stems. The shoot apex also gives rise to **leaf primordia** and branch primordia (buds) that may develop into lateral branches.

Examine a *Coleus* plant and note where the branches arise along the stem. How are the leaves arranged on these branches—that is, are they opposite or alternate in position?

_____

What is the arrangement of the leaves on the main stem?

_____

What structures are found in the axils of the leaves?

_____

_____

What will these structures develop into?

_____

_____

Remove as many of the leaves as possible from the tip of one of the branches. What indication is there of leaves smaller than those already removed?

_____

How do you account for the difference in size between the mature leaves and the young leaves?

_____

Because of the small size of the shoot tip, the appearance of the very young leaf and branch primordia can only be seen with the aid of a microscope. Examine a prepared slide of a *Coleus* stem tip that has been cut lengthwise (Fig. 4-2). For purposes of orientation, first examine the slide with a dissecting microscope. How does the arrangement of the leaf primordia on the stained section compare with the arrangement of the leaves on the plant?

_____

Examine the slide under the low power of a compound microscope and locate the apical meristem. What criteria would you use to find this region?

_____

Locate the youngest leaf primordia at the sides of the apex. Next, examine the older leaf primordia located successively lower on the stem. In the angle that the leaf primordia make with the stem locate small, dense, rounded projections. These are **branch primordia** and contain meristems that may produce a new stem and leaves. In the herbaceous (nonwoody) stem the younger leaves enclose the meristem and prevent it from drying out. In woody plants, the meristem (at definite times of the year) stops producing leaves and begins producing structures called **bud scales** that protect the delicate growing point over the winter. What stimulus might initiate the change from one structure to another?

Obtain a bud from a woody maple stem. With a finely pointed forceps remove the bud scales while examining the bud under a binocular microscope. What other structures are present in the bud besides bud scales?

b. Primary Tissues of a Stem

Obtain a portion of a geranium stem from the instructor. With a razor blade make a fresh cut in a transverse plane to obtain the thinnest section possible. Prepare a fresh mount of this material and examine under the low power of the microscope. Locate the outer epidermal layer and the numerous hairs associated with it. Located beneath the epidermis is a layer of tissue composed of parenchymatous cells. This is the cortex. Note the presence of numerous chloroplasts in the cells. Describe a function for this tissue in the living plant?

The cortex is separated from the centrally located pith by a band of vascular tissue. Note that the vascular cylinder is not of uniform thickness, but is composed of more or less separate vascular bundles (strands of vascular tissue). The vascular cylinder may be more readily observed if the tissue is stained with phloroglucinol-HCl. Mount the

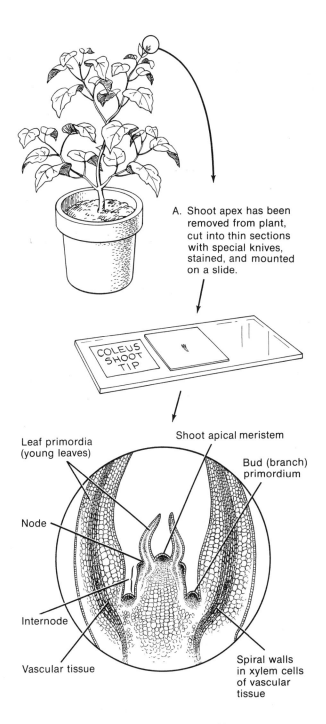

FIGURE 4-2
Study of the *Coleus* shoot tip.

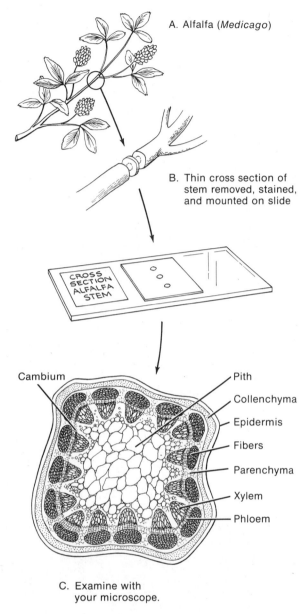

FIGURE 4-3
Study of alfalfa stem anatomy.

which appear in cross section as separate vascular bundles. A mature vascular bundle consists of three main tissues—xylem, phloem, and cambium. The xylem is located on the inner side toward the pith. Adjacent to the xylem, appearing as small rectangular cells having the long axis at right angles to the radius of the stem, is the cambium. Division of cambial cells results in the formation of secondary tissues, which increase the diameter of the stem. How does the vascular arrangement in the stem contrast with that of the root?

Capping each vascular bundle is a group of thick-walled fibers. Morphologically, these fibers are part of the primary phloem tissue and are properly called **phloem fibers**. The parenchymatous cells lying between the vascular bundles are **pith rays**.

The cortex lies immediately outside the phloem fibers and consists of two distinguishable tissues—cortical parenchyma and collenchyma (mechanical tissue). The cortical parenchyma consists of loosely packed parenchymatous cells, many of which contain chloroplasts. In many stems the collenchyma may be found exterior to the cortical parenchyma. In alfalfa there is considerable development of the collenchyma at the corners.

The outermost layer of cells constitutes the epidermis. How many cell layers thick is the epidermis?

The outer walls of epidermal cells usually have become thickened, and during their development a waxy substance (**cutin**) is secreted. Close the diaphragm on your microscope so that you can observe this layer of cutin. It will appear as a faint, pink, noncellular layer covering the epidermis. What effect does this waxy layer have on the loss of water from the stem?

---

section in phloroglucinol solution. Wait about 1 minute and reexamine the section. How does the arrangement of the vascular tissue differ from that of the root?

For a detailed examination of the primary tissues obtain a prepared slide of a cross section of the alfalfa (*Medicago*) stem (Fig. 4-3). The cells in the central part of the stem differentiate into **pith cells**. Encircling the pith are vascular strands,

2. Secondary Growth

The apical meristem produces cells that elongate and add to the growth in length of the root or stem. The elongation of the cells produced at the stem tip is restricted to a region behind the apical meristem; little or no further increase in length occurs beyond this region of elongation.

The cells and tissues formed as a result of the activity of the apical meristem are called **primary tissues**. Some plants consist only of primary tissues. Many plants, however, have **secondary tissues**, which are formed as a result of the activity of lateral meristems called the **cambium** and **cork cambium**. The cambium divides and produces secondary xylem (or wood) and secondary phloem. The cork cambium produces cells that make up the outer bark of the stem.

a. Cambium

Before examining secondary growth, it will be helpful to recall the position of the cambium in a stem having little or no secondary thickening.

Reexamine the prepared slide of a cross section of the alfalfa stem (Fig. 4-3). Locate the **vascular bundles**. The cambium will be found as a band of narrow cells between the xylem and the phloem. Examine closely the pith rays that extend between the vascular bundles. What indications are there that the cambium extends between the vascular bundles in this region?

_____

Examine a slide of an older alfalfa stem showing secondary thickening. Note that the vascular bundles are not as distinct as in the younger stem and that the cambium that was located between the vascular bundles has formed secondary xylem and secondary phloem.

Obtain a prepared slide having cross sections of a one-, two-, and three-year-old basswood stem (Fig. 4-4A). Examine the one-year-old stem. Although these cross sections are complex in their organization, they will be easier to study if the cambium is located before attempting to identify any other tissues. The cambium will be found between the xylem and phloem and appears as a band of narrow, thin-walled cells having their long axes oriented at right angles to the radius of the stem. By a series of tangential divisions the cambium has produced a considerable quantity of secondary xylem and phloem. Examine the secondary xylem (wood), which is composed of vessels, tracheids, fibers, and xylem parenchyma cells. The vessels (in a cross-sectional view) may be distinguished from tracheids and fibers by their larger diameter.

External to the cambium, the secondary phloem appears as pyramidal masses of thick- and thin-walled cells. The thick-walled cells are phloem fibers. Situated among these fibers will be found sieve tubes and other phloem cells. Locate the thin strands of tissue that radiate outward through the secondary xylem and phloem. These are rays and terminate in wedge-shaped masses of parenchymatous cells, which alternate with the secondary phloem. Suggest a function for these rays.

_____

Compare the one-, two-, and three-year-old basswood stems and note the changes that have occurred as a result of cambial activity. Note that the secondary xylem appears to be made up of concentric rings. These are more commonly called **annual rings** because each layer represents a seasonal increment of growth in those plants that have an annual alternation of growing and dormant periods. A close examination of the annual rings in basswood will reveal that each ring consists of an inner layer composed of large cells formed in the spring when growth resumed and an outer layer of smaller cells formed in the summer. What is a possible explanation for the variations in the width of different annual rings?

_____

_____

Examine cut sections or prepared slides of other woody stems and locate the annual rings. Examine these with a hand lens or dissecting microscope and identify the "spring" and "summer" wood. What predominant cell type would you expect to find in the spring wood and why?

_____

_____

Determine the age of at least one specimen and confirm your answer with the instructor.

b. Cork Cambium

In those plants that have only primary growth, desiccation (drying) of tissues is prevented by

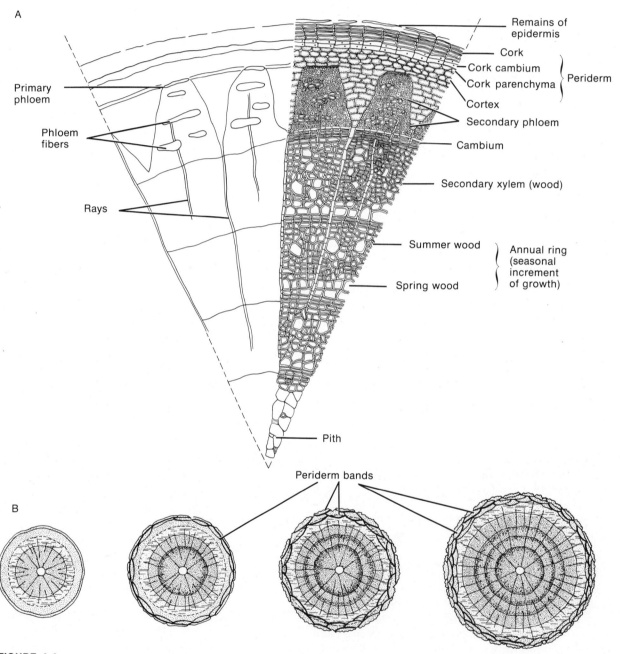

FIGURE 4-4
A. Diagrammatic cross section of a basswood stem. B. Periderm development.

cutinization (waxy deposition) of the epidermis. In plants that have secondary growth, the epidermis is soon lost as a result of the increase in stem circumference. Consequently, new protective tissue is formed that prevents water loss. This protective tissue is known as the **periderm** and consists of three layers: the cork cambium, cork, and cork parenchyma (Fig. 4-4A).

Locate the cork cambium on the slide of the one-year-old basswood stem. It is located just below the epidermis and appears as a narrow band of thin-walled cells, oriented in a manner similar to the cambium. Through a series of tangential divisions, the cork cambium gives rise to cork cells externally and to cork parenchyma internally.

Locate the cork layer. It appears as a band of cells, which are stained deep red and have relatively thick walls impregnated with suberin. **Suberin**, like cutin, is highly impervious to water and certain gases. Describe any changes in the amount of cork produced in the two- and three-year-old basswood stems.

___

___

Internal to the cork cambium, locate the cork parenchyma, which appears in cross section as a band of cells having their radial and outer walls somewhat compressed. Separating the cork parenchyma from the phloem is a band of large, thin-walled parenchyma cells, many of which contain crystals.

When first formed, the periderm is a cylinder of cells located just beneath the epidermis. As the stem diameter increases through secondary growth, the original periderm ruptures and new periderm bands are formed progressively deeper toward the center of the stem (Fig. 4-4B). As a consequence, the tissues outside are unable to obtain food and water and subsequently die, forming an outer crust made up of overlapping layers of cork. All the tissues lying outside of the cambium, including this "crust," make up what is commonly called the bark. Examine the bark of several specimens of woody plants. Why is it a serious matter when rabbits "girdle" the bark around the circumference of a tree?

___

___

REFERENCES

Arzee, T., N. Lipschitz, and Y. Waisel. 1968. The Origin and Development of the Phellogen in *Robinia pseudoacacia* L. *New Phytologist* **67**:87–93.

Carlquist, S. J. 1961. *Comparative Plant Anatomy.* Holt, Rinehart & Winston, New York.

Cutter, E. G. 1969. *Plant Anatomy: Experiment and Interpretation.* Addison-Wesley, Reading, Massachusetts.

Esau, K. 1960. *Anatomy of Seed Plants.* Wiley, New York.

Esau, K. 1965. *Vascular Differentiation in Plants.* Holt, Rinehart & Winston, New York.

Fisk, E., and W. F. Millington. 1959. *Atlas of Plant Morphology.* Portfolio I. Photomicrographs of Root, Stem and Leaf. Burgess, Minneapolis.

Foster, A. S. 1940. *Practical Plant Anatomy.* 2d ed. Van Nostrand, New York.

Foster, A. S., and E. M. Gifford, Jr. 1959. *Comparative Morphology of Vascular Plants.* W. H. Freeman and Company, San Francisco.

Salisbury, F. B., and R. V. Park. 1964. *Vascular Plants: Form and Function.* Wadsworth, Belmont, California.

EXERCISE 5

Plant Anatomy: Leaves and Flowers

We have already discussed two plant organs: the root, which functions chiefly as an absorbing organ, and the stem, which serves to conduct water and soluble materials. A third organ, the leaf, is specialized to carry on photosynthesis, a physicochemical process that converts light energy to chemical energy. All forms of life, with the exception of some types of bacteria, depend upon the process of photosynthesis at some point in their food chain.

Flowers are reproductive branches consisting of the stem axis and lateral appendages called sepals, petals, stamens, and carpels. The significance of the flower lies in its role in sexual reproduction.

A. LEAVES

Leaves consist basically of three parts: an expanded or flattened portion—the **blade**; a thin stemlike portion—the **petiole**; and small, paired, lobelike structures at the base of the petiole—the **stipules**. The petiole and stipules may not be present in some plants.

Leaves may be classified according to the type of **venation** (arrangement of vascular tissue) or to the arrangement of the leaves on the stem. Thus leaves are parallel veined when the larger vascular strands traverse the leaf without apparent branching, and net veined when the main branches of the vascular system form a network. According to the position on the stem, leaves may be alternate or opposite, if they are borne singly or in pairs at each node. Examine the plants on demonstration and become familiar with the parts of the leaf and the simple classification given above.

1. Epidermis

Obtain a leaf from a bean or geranium plant. Remove the lower epidermis by ripping the leaf as shown in Fig. 5-1A. Mount the thin, colorless piece of tissue in a drop of water on a slide and then add a cover slip (Fig. 5-1B). Observe under low and high power. Note that many of the epidermal cells have an irregular shape. Scattered throughout

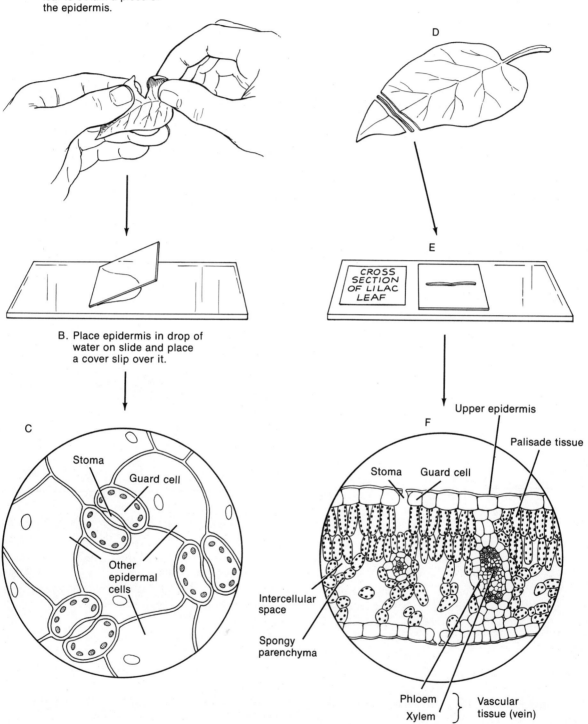

FIGURE 5-1
Study of leaf anatomy.

PLANT ANATOMY: LEAVES AND FLOWERS

the epidermis will be found openings called **stomata** (singular, **stoma**) (Fig. 5-1C). Each stoma is surrounded by two bean-shaped **guard cells**. Suggest a function for the stomata.

_____

_____

_____

_____

Which of the epidermal cells contain chloroplasts?

_____

What relationship is there between the function of the chlorophyll-containing cells and the presence of chlorophyll?

_____

_____

_____

_____

_____

2. Internal Anatomy

Obtain a prepared slide of a cross section of a lilac (*Syringa*) leaf for further examination of the internal tissues (Fig. 5-1F). Below the upper epidermis is a region of **palisade tissue**, consisting of one or more layers of elongated cells with the long axis of the cell perpendicular to the surface of the leaf. Palisade cells contain numerous chloroplasts. What function do you ascribe to this tissue?

_____

Beneath the palisade tissue locate a region of rounded, parenchymatous cells, which constitute the **spongy tissue**. The palisade tissue and spongy parenchyma are collectively called leaf **mesophyll**. Note the presence of numerous intercellular spaces in the mesophyll. Locate the stomata. What is the relationship between the stomata and the intercellular spaces of the mesophyll tissue?

_____

_____

The veins of a leaf are vascular bundles (strands) that are continuous with the vascular tissues of the petiole and stem. A vein contains xylem and phloem and has the same cellular elements as the stem. What is the function of the mechanical (supporting) tissue associated with the vascular bundles?

_____

_____

Examine demonstration slides of a corn leaf, a parallel-veined leaf. Describe any differences in the arrangement of tissues between this leaf and the net-veined lilac leaf.

_____

_____

_____

_____

B. FLOWERS

Thus far, the plant has been studied as an individual composed of roots, stems, and leaves. These structures develop and function so as to insure the successful existence of the plant. Ultimately, however, the plant dies. It is through a constant succession of new individuals that the species survives. This is accomplished by a process called **reproduction** in which the plant produces a group of offspring.

There are two main types of reproduction—**asexual** (or vegetative) and **sexual**. Flowering plants can reproduce asexually by numerous means. For example in strawberry plants, the tip of a branch may arch down, come into contact with the soil, and develop leafy branches that root and form new plants. The most common method of reproduction, however, is by sexual means. To understand sexual reproduction in

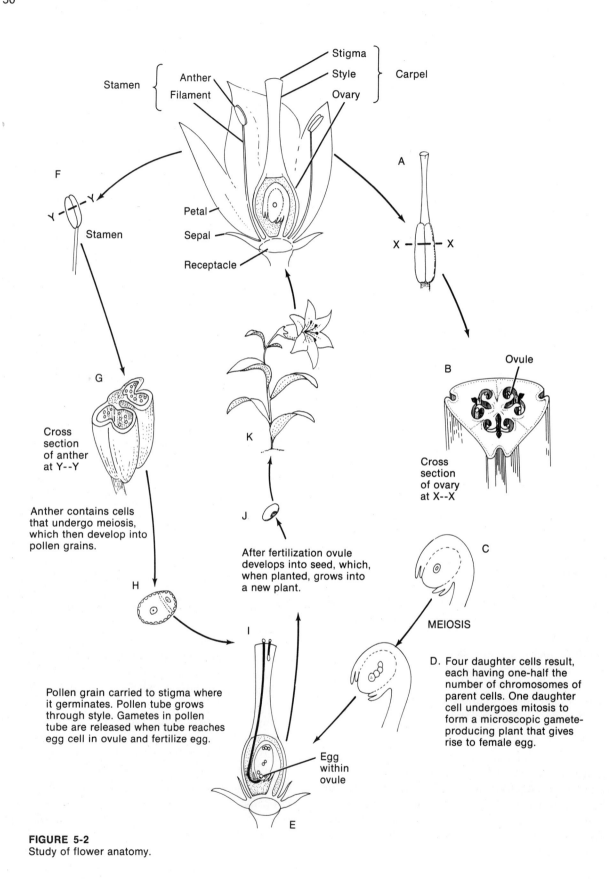

FIGURE 5-2
Study of flower anatomy.

plants, it is necessary to study the anatomy of the flower.

Your instructor will provide you with two or three different flowers that you may use for this part of the exercise.

With the help of Fig. 5-2 locate the following parts of these flowers:

- **Sepals** (which are modified leaves) are the outermost structures of the flower. They are typically green, although they may be another color. In some cases they may be absent.
- **Petals** lie to the inside of the sepals and are often brightly colored. Both the sepals and petals are attached to the enlarged end of the branch—the **receptacle**.

Carefully remove the sepals and petals. In the center of the flower locate a stalklike structure (Fig. 5-2A and B). This is the female part of the flower—the **carpel** (pistil). It is composed of a swollen base, the **ovary**, and an elongated style that is terminated by a **stigma**. There may be more than one carpel in a flower.

The ovary contains one or more ovules. These may be seen if the ovary is cut lengthwise and mounted in a drop of water on a slide and examined with a dissecting microscope. One of the cells in the young developing ovule undergoes a special type of cell division, called **meiosis**, where the daughter cells that are formed have *one-half* (haploid) the number of chromosomes possessed by the parent cell. One of these haploid daughter cells develops into a microscopic haploid plant that will produce the female **gamete**, the egg cell (Fig. 5-2C, D and E). Why is it necessary that meiosis occur during the formation of the gametes?

Locate the **stamens** that surround the carpel (Fig. 5-2). These are the male parts of the flower and consist of a terminal capsule—the **anther**—attached to a slender **filament**.

The anthers contain cells that also undergo meiosis, producing cells that eventually develop into microscopic gamete-producing plants commonly called **pollen grains** (Fig. 5-2G and H). Crush one of the anthers in a drop of water on a slide and add a cover slip. Examine it with your microscope and locate the pollen grains. During pollination, the pollen grains are transferred to the stigma—the sticky surface of the carpel. The pollen grain germinates and the **pollen tube** grows through the style to the ovary and enters the ovule (Fig. 5-2I). It is known that many organisms, or parts of organisms, grow toward or away from various stimuli such as light, chemicals, gravity, and so forth. Suggest a mechanism that could direct the growth of the pollen tube toward the ovules.

During the growth of the pollen tube, two male gametes (sperm) are produced. These are released when the tube enters the ovule. One sperm fertilizes the egg cell and the other unites with the nucleus of another cell in the ovule. This second fertilization results in the formation of a special tissue—the **endosperm**—which functions as a nutrient tissue to the developing embryo.

Fertilization of the egg initiates extensive changes in the carpel. The ovary enlarges and develops into the fruit; in some cases, other floral parts also become part of the fruit. The ovules develop into seeds, which now contain the embryo of a new plant (Fig. 5-2J and K). The development of the embryo and regulation of plant growth will be considered in other exercises.

REFERENCES

Carlquist, S. J. 1961 *Comparative Plant Anatomy.* Holt, Rinehart & Winston, New York.

Eames, A. J., and L. H. MacDaniels. 1947. *An Introduction to Plant Anatomy.* 2d ed. McGraw-Hill, New York.

Esau, K. 1960. *Anatomy of Seed Plants.* Wiley, New York.

Fisk, E., and W. F. Millington. *Atlas of Plant Morphology.* 2 portfolios. Portfolio I. Photomicrographs of Root, Stem and Leaf, 1959. Portfolio II. Photomicrographs of Flower, Fruit, and Seed, 1962. Burgess, Minneapolis.

Salisbury, F. B., and R. V. Park. 1964. *Vascular Plants: Form and Function.* Wadsworth, Belmont, California.

Wardlaw, C. W. 1968. *Morphogenesis in Plants.* Methuen, London.

EXERCISE 6

Vertebrate Anatomy: External Anatomy, Skeleton, and Muscles

Frogs are probably the most commonly studied animals in general biology and zoology courses because they are easy to obtain and dissect, and their structure clearly shows the major features of the organ systems characteristic of the backboned animals, or vertebrates. Furthermore, the frog is one of the most widely used animals in vertebrate physiology and in other experimental work. It is therefore an animal with which students should be familiar.

Vertebrates are of particular interest to us because we too are members of the subphylum Vertebrata of the phylum Chordata. Other chordate subphyla include a few soft-bodied marine animals such as sea squirts and amphioxus. As members of the class Amphibia, frogs have characteristics that are between those of the lower vertebrates (fishes) and those of the advanced classes of terrestrial vertebrates (reptiles, birds, and mammals).

Contemporary amphibians are grouped into three orders: frogs and toads, salamanders, and legless, worm-shaped caecilians of the tropics. All reproduce in the water, or in a very moist place on land. Most go through the aquatic larvae stage, commonly called the tadpole. This period of development is followed by a metamorphosis to a terrestrial adult. Adult amphibians have only a rudimentary ability to conserve body water; hence they must live in damp habitats on the land and must often return to the water. As a result of their phylogenetic position between fishes and reptiles and their double mode of life in water and on land, amphibians have a mixture of aquatic and terrestrial attributes.

The most widespread of our North American frogs is the leopard frog, also known as the grass or meadow frog. Its scientific name is *Rana pipiens*. Another frog, which is often studied because of its large size, is the bullfrog, *Rana catesbeiana*.

A. EXTERNAL ANATOMY OF THE FROG

The body of a frog can be divided into a **head**, which extends posteriorly to the shoulder region, and a **trunk** (Fig. 6-1). Fishes, which were ancestral

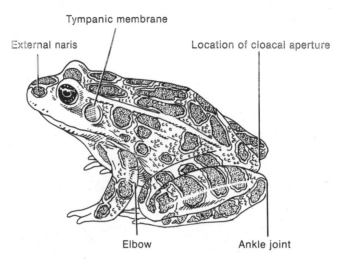

FIGURE 6-1
Lateral view of a leopard frog, *Rana pipiens*. (From *Dissection of the Frog* by Warren Walker, Jr. W. H. Freeman and Company. Copyright © 1967.)

to amphibians, have a powerful tail used in swimming. Aquatic tadpoles retain such a tail, but it is lost in the adult frog, which moves instead by means of its powerful hind legs. The much smaller front legs are used primarily to keep the front of the body raised from the ground. Notice that a distinct neck is absent. This is a retention of a characteristic of fishes, for which an independent motion of the head and trunk would be disadvantageous during swimming.

The combined orifice of the digestive and urogenital tracts, the **cloacal aperture**, is at the posterior end of the trunk just dorsal to the junction of the hind legs. The anus of human beings and other mammals, since it is the opening of just the digestive tract, is only partly comparable to the cloacal aperture.

A large **mouth**, a pair of nostrils, or **external nares** (singular **naris**), and the **eyes** will be recognized on the head. The upper **eyelid** is a simple fold of skin. The lower lid is a transparent membrane that can be drawn across the surface of the eyeball. The disc-shaped area posterior to each eye is the ear drum, or **tympanic membrane**. Although it is the same size in male and female leopard frogs, it is considerably larger in male bullfrogs than in females. This is one example of **sexual dimorphism** that shows a structural difference between the sexes. Look carefully at the top of the head between the eyes, and you may see a small, light **brow spot**, which is about the diameter of a pin. This is a remnant of a light-sensitive eye that characterized primitive groups of fishes and amphibians.

The appendages of a frog have the same parts as our own. In the front leg, or **pectoral appendage**, notice the upper arm, the elbow, the forearm, the wrist joint, and the hand. There are only four fingers. Although the finger closest to the body is comparable to our second digit, it is often called the thumb because it is stouter than the others. During the breeding season, the thumb in the male is particularly stout and darkly pigmented—another example of sexual dimorphism. In the hind limb, or **pelvic appendage**, observe the thigh, the knee, the shank, the ankle joint, and the very long foot. Two elongated ankle bones lie within the proximal part of the foot, and the distal part bears five toes with a conspicuous web between them. The first toe is the smallest and the one closest to the body, and is comparable to our great toe. Technically it is called the **hallux**, and the small spur at its base is the **prehallux**. The prehallux is much larger in toads, which use their hind feet in burrowing.

B. VERTEBRATE SKIN

All organisms are covered by some sort of external covering. In one-celled organisms this may be only a thin cell membrane; in higher organisms the coverings are complex structures consisting of several layers of cells. The skin of vertebrates is a highly functional organ that provides physical protection, excludes disease organisms, acts as a means of water absorption, and in some forms serves an essential role in respiration. In frogs both the adult and tadpole stages absorb oxygen and eliminate carbon dioxide through the skin (**cutaneous respiration**). During the winter, when the frog lies buried in the mud, the skin becomes practically the only respiratory organ. It functions as a respiratory organ both in water and in air. Shortly after hatching, the tadpole develops gills, which also function in the absorption of oxygen. This **branchial respiration** lasts until the tadpole begins to change into a frog. Adult frogs absorb oxygen and give off carbon dioxide through the lining of the mouth (**buccal respiration**) and through the lungs (**pulmonary respiration**). Even in full-grown frogs, however, three or four times as much oxygen can be absorbed through the skin as through the lungs. In warm-blooded mammals, the skin helps to regulate the temperature of the body. The skin of some vertebrates has been modified into hair (mammals), feathers (birds), scales (fish), and teeth (sharks).

1. Amphibian Skin and Pigmentation

Remove a small piece of skin from a lightly pigmented portion of a frog, such as the throat, and make a wet mount. The small, branching, dark spots that you see with a microscope are the pigment cells, or **chromatophores**. These particular cells contain granules of a dark pigment known as **melanin**. The granules can migrate out to the processes of the chromatophores, making the frog darker, or they can concentrate near the center of the cells, giving the frog a lighter color. To see this difference in hues, examine a frog that has been kept for several hours on a light background and another on a dark background. Some pigment cells contain a yellow pigment, and others contain refractive crystals of guanine, which disperse light to give a blue effect. The combination of yellow and blue causes the green colors.

The color pattern of the frog is said to be **cryptic**, because it conceals the frog in its natural habitat. The greenish color of the back blends with the natural background, and the dark spots and blotches form a disruptive pattern, which tends to obscure the shape of the animal. The dark back and the light belly—dark where the body is highlighted by natural illumination and light where there is normally a shadow—reduces the appearance of solidity or mass when the frog is viewed from the side. Changes in color result from both external conditions and internal states; low temperature produces darkening, whereas under higher temperatures, drying, or increased light, the color pales.

Examine a stained slide of a cross section of frog skin (Fig. 6-2). The stains used in preparing the slide were **hematoxylin**, which stains nuclei blue or purple, and **eosin**, which stains most cytoplasmic structures pink or shades of red. The skin consists of two distinct regions. The outer layer, or **epidermis**, is predominantly stained purple and contains large numbers of cell nuclei. The inner layer, or **dermis**, is predominantly pink and contains only a few nuclei. Rounded structures in the dermis are sections through glands. The dark layer between the epidermis and the dermis contains the chromatophores. Note that the cells of the uppermost layer in the epidermis are thin and lie parallel to the surface (pavement layer). The deeper epidermal cells are more nearly cuboidal. The cells in the basal layer may be even longer (columnar in shape), lying perpendicular to the skin surface. The cell membranes are often not clearly visible, but the shapes of the nuclei reveal the shapes of the cells. If the nucleus is long in one direction, the cell is ordinarily long in the same direction. The epidermis, with its gradation from pavement to columnar cells in successive layers, is an example of **stratified squamous epithelium**. The basal layer is often called the **germinative layer**, because it continually produces new cells that push out toward the surface, becoming steadily flatter and cornified. Monthly, during the summer, a new layer forms beneath and the old covering is molted or sloughed off.

The dermis consists mainly of fibrous connective tissue composed of nonliving, intercellular substances in the form of densely interwoven fibers produced by scattered cells called **fibroblasts**. The other spongy portion of the dermis contains two types of glands: **mucous** (**slime**) **glands** and **poison glands**. These glands produce useful fluids or secretions that pass out onto the epidermis through fine ducts. Both types are round, hollow glands, bordered by a layer of cuboidal secretory cells. The poison gland can be identified by its larger size and the presence of granular material in the

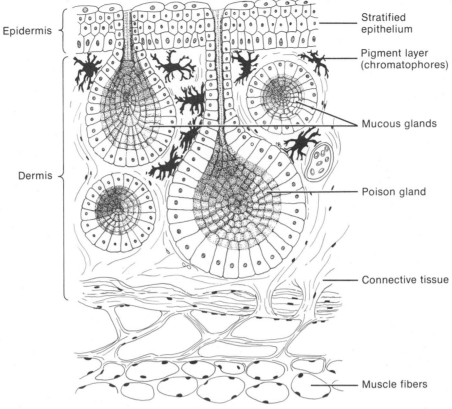

FIGURE 6-2
Diagram of a vertical section through frog skin.

cavity of the gland. If a frog is roughly handled, the poison glands pour out a thick whitish secretion with a burning taste that partially protects the animal from enemies. The smaller mucous glands secrete a colorless, watery fluid that keeps the skin moist, glistening, and sticky. Locate a mucous gland that shows its duct opening to the surface. The surface opening of these ducts can be widened or narrowed by the contraction of a specialized cell (**stoma cell**), thereby regulating the amount of slime excreted. Examine a slide of frog squamous epithelium. Find a stoma cell and note the shape of its opening.

The dermis contains numerous small blood vessels that transport food to the skin and take back oxygen absorbed through surface layers from the air. Nerve fibers and muscle fibers are also present in the dermis but will not be readily seen in your slides.

2. Mammalian Skin

Examine a slide of mammalian skin. The epidermis consists of stratified squamous epithelium, usually composed of many more layers of cells than in the frog and showing more flattened, horny layers at the surface (Fig. 6-3).

The dermis is composed mainly of densely interwoven fibrous connective tissue. Note the abundance of small blood vessels, whose contractions and dilations play an important role in the control of body temperature in mammals.

Pigment cells, when present, are usually confined to the basal layer of the epithelium. At various levels in the dermis are the hair follicles, nerves, and sweat and **sebaceous (oil) glands**. Locate a **hair follicle** and notice that each hair arises in a tubular invagination of the horny layer of the epidermis. The root of the hair develops into the hair shaft, the free end of which protrudes beyond the surface of the skin. One or more sebaceous glands is associated with each hair follicle. These are branched glands that lie in the dermis and open into the follicle. They secrete an oily substance onto the hair or upon the surface of the epidermis. What would be the function of this secretion?

_____

_____

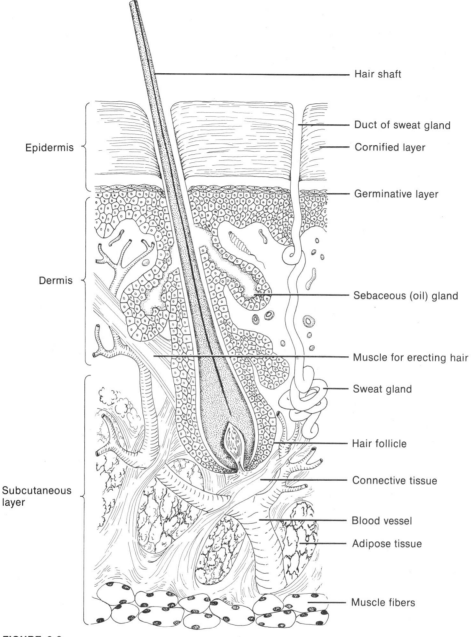

FIGURE 6-3
Diagram of a vertical section through mammalian skin.

Distributed along the surface of the skin are the openings of the sweat glands. These are long, tubular glands whose ends coil deeply into the dermis or subcutaneous layer. The secretion formed by these glands flows onto the surface of the skin, where, by the process of evaporation, the surface is cooled. The activity of these glands is another body-temperature regulating mechanism that is under the control of the nervous system.

The subcutaneous layer is composed of various numbers of fat cells (**adipose tissue**), depending on the portion of the body and the nutrition of the organism. In your slide the fat cells will appear empty, since the method of preparing this slide has dissolved the fat droplet that occupied the greater portion of the cell, leaving only a thin film of cytoplasm with its compressed nucleus. The fat cells are held together by fine fibrous connective tissue. At intervals between the fat areas heavy strands of fibrous connective tissue traverse the subcutaneous layer and bind the layers. The subcutaneous layer is especially important in heat conservation by acting as an insulating layer.

C. VERTEBRATE SKELETON

One of the more significant evolutionary advancements of the vertebrates over the invertebrates is the possession of an internal skeleton (**endoskeleton**). The external skeleton (**exoskeleton**) of the invertebrates has the disadvantage of limiting the ultimate size of an organism and in several cases is so heavy as to restrict the movements of some animals. The endoskeleton of vertebrates, on the other hand, permits relatively unrestricted development in size, as evidenced by such vertebrates as the whale and elephant.

The vertebrate skeleton, in addition to providing protection and support for the internal organs of the body, provides a highly efficient supporting structure for the attachment of muscles. Furthermore, the flexibility provided by the large number of separate bones, coupled with the strength and relative lightness of vertebrate skeletons, has provided even the largest animals with much mobility.

The skeletal systems of all animals with an endoskeleton have a basic pattern consisting of two major parts, the axial and appendicular skeletons. The **axial skeleton** consists of the skull, the vertebral column, and the sternum (breastbone); the **appendicular skeleton** consists of the pelvic (hip) and pectoral (shoulder) girdles and the limbs.

1. Axial Skeleton

a. The **skull** is composed of the **cranium**, which houses the brain; the olfactory, optic, and auditory capsules for the organs of smell, sight, and hearing; and the **visceral skeleton**, which comprises the jaws and supports for the gills. In sharks and rays the skull is cartilaginous, but in higher forms most of the cartilage is replaced by bone. In land vertebrates, the visceral skeleton has undergone a great transformation due to the development of an entirely different method of breathing. Parts of the skeleton are converted to other structures, such as the bones of the middle ear and lower jaw and the cartilages of the larynx. Primitive forms tend to have a larger number of skull bones than higher forms. Some fish may have 180 skull bones; amphibians and reptiles, 50 to 95; and mammals, 35 or fewer. Man has 29.

Examine a frog skull and refer to Fig. 6-4. The skull is attached to the anterior end of the vertebral column. Note the large jaws and the very small brain case or cranium, which is roofed mainly by the **frontoparietals**, the **nasal** bones (which cover the nasal capsules), the **prootics** (which house the inner ears), and the **exoccipital** bones (each of which has a rounded **occipital condyle**). The two condyles fit depressions in the first vertebra, permitting slight movements of the head on the spinal column. Between the condyles is a large opening, the **foramen magnum**, through which the spinal cord passes. Between the prootics and the maxilla are the **squamosal** and **pterygoid** bones, which form the lateral borders of the cranium. The upper jaw consists of the small **premaxillary** bones in front, and the larger **maxillary** bones extending posteriorly to join with the pterygoid. On the margin of the upper jaw are the **maxillary teeth**, and on the roof of the mouth are the **vomerine teeth**.

b. The **vertebral column** is composed of vertebrae, which vary greatly with different animals and with different regions of the vertebral column in the same animal. The vertebral column in fish is differentiated into trunk vertebrae and caudal vertebrae; the column in many of the other vertebrates, into neck (or **cervical** vertebrae), chest (**thoracic**), back (**lumbar**), pelvic (**sacral**), and tail (**caudal**). In birds, and also in man, the caudal vertebrae are reduced in number and size, and the sacral vertebrae are fused. The number of vertebrae varies among animals. The python has the largest number—435. In man there are 7 cervical, 12 thoracic, and 5 lumbar vertebrae, plus 5 that have fused to form the **sacrum** and **coccyx**. How many vertebrae does the frog have?

The first cervical vertebra—the **atlas**—is modified for articulation with the skull. The vertebral column terminates in a long bone, the **urostyle**.

Note that the vertebrae are much alike. The lateral projections are called **transverse processes**. In higher vertebrates, ribs are attached to these processes. Ribs show many variations among the vertebrates. The basic plan seems to have been a pair of ribs for each vertebra from head to tail, but the tendency has been to reduce the number from the lower to the higher forms. Ribs, however,

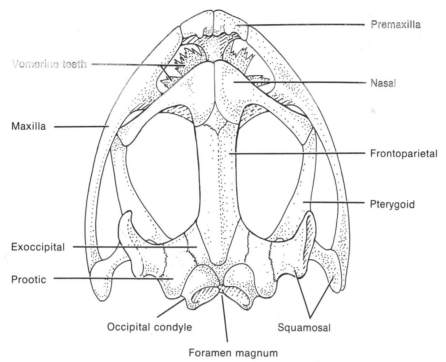

FIGURE 6-4
Dorsal view of a frog skull.

are not universal among vertebrates; many, including the frog, do not have them at all. Man has 12 pairs of ribs.

2. Appendicular Skeleton

Most vertebrate animals have some form of paired appendages, which are supported by **pectoral** (shoulder) and **pelvic** (hip) **girdles**. Among the various vertebrates there are many modifications in the girdles, limbs, and digits, which enable the animals to meet the requirements of their special modes of life. Examine the skeletons in the laboratory to determine how the appendages are adapted for each animal's particular mode of existence.

_____

_____

_____

Whatever the modification, in forms above the fish, the girdles and appendages are all built according to the same plan. Both the cartilaginous and bony fishes have pectoral and pelvic fins, which are supported by the pectoral and pelvic girdles, respectively. Close to the body, in the frog's legs are single strong bones—the **humerus** in the anterior limb (Fig. 6-5) and the femur in the posterior (Fig. 6-6). Distal (away from the body) to these are two roughly parallel bones—the **radius** and **ulna** in the forelimb and the **tibia** and **fibula** in the hind limb. In the frog the bones of the forearm have fused to form the tibio-fibula. The humerus is attached to the pectoral girdle at the **glenoid fossa** by means of ligaments. The bone passing dorsally, the **scapula**, has a broader extension folded toward the midline called the **suprascapula**. Ventrally, the posterior larger bone is the **coracoid**. The anterior smaller one is the **clavicle**. At the point of junction of the two clavicles is the **sternum**, which continues forward as the **omosternum** and **episternum** and posteriorly as the **mesosternum** and **xiphisternum**.

The femur attaches to the pelvic girdle (Fig. 6-6) in a socket, the **acetabulum**, which is formed by the fusion of three bones. The feet and hands are also built according to a common pattern, with a number of wrist (**carpal**) or ankle (**tarsal**) bones, followed by a group of elongated foot (**metatarsal**) and hand (**metacarpal**) bones, and then the bones of the toes and fingers (**phalanges**). The two large

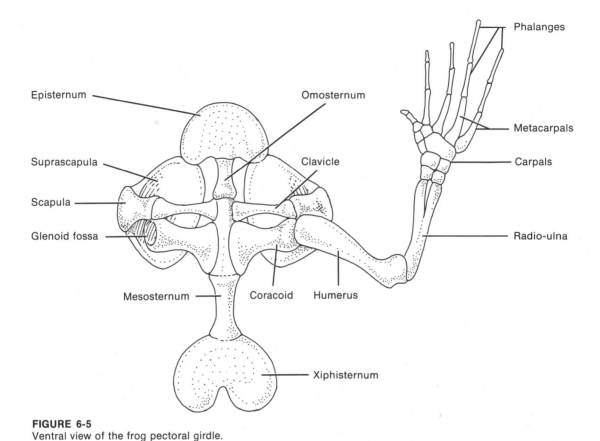

FIGURE 6-5
Ventral view of the frog pectoral girdle.

tarsal bones joining the **tibio-fibula** of the frog are called the **tibiale** (smallest and inside the foot) and **fibulare** (largest and on the outside of the foot).

D. VERTEBRATE MUSCULATURE

1. Joints

Before the muscles of the frog are studied, it is advisable to learn something about the joints between bones, for it is these joints that allow the muscles to perform their various functions of bending, twisting, and turning the various parts of the body. Become familiar with the following classification of joints before proceeding with the muscle dissection.

 a. **Ball and socket** joints allow movement in any direction, including rotation as in the shoulder and hip.

 b. **Condyloid** joints allow movement in any direction except rotation, as between metacarpals and phalanges.

 c. **Hinge** joints allow bones to bend in only one direction, as in the knee, and between flat surfaces of bones, as between the vertebrae.

 d. **Plane** joints allow sliding movements between flat surfaces of bones, as between the vertebrae.

 e. **Radial** joints permit rotation of one bone or another, as between the proximal ends (ends nearest the humerus) of the radius and ulna.

2. Muscles

Muscles form the bulk of a vertebrate's body. By holding the bones in proper relationship to each other, they have an important function in supporting the body. They are responsible for the movement of the body as a whole, the passage of food down the digestive tract, the ventilation of the lungs, the circulation of the blood, and the movements of most of the materials in the body.

Each end of most muscles is attached to a skeletal element, or, possibly, to another muscle. An attachment may be **fleshy**, that is, the muscle fibers

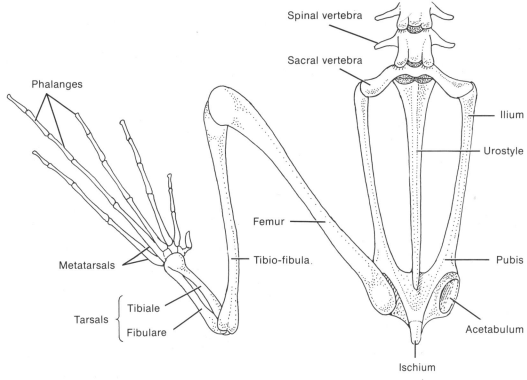

FIGURE 6-6
Dorsal view of the frog pelvic girdle.

attach directly to the bone; or it may be a cordlike, connective-tissue **tendon**; or it may be a broad, connective-tissue sheet known as an **aponeurosis**. When a muscle contracts, one end is fixed in place, so that the contraction causes the other end to move and pull a bone or structure toward the stabilized end. The stabilized end is the **origin** of the muscle; the other end is its **insertion**. For limb muscles, the origin is proximal to the insertion. The part of a muscle between its attachments is its **belly**.

Muscles perform work only by contracting. A particular muscle moves a limb, for example, in only one direction; the contraction of an antagonistic muscle is required to move the limb back to its original position. Terms that describe the major antagonistic actions are defined here: (1) **Flexion** is a movement that causes a bending, or decrease in the angle, between two parts. The term is most appropriately applied to movements at the elbow, wrist, knee, ankle, and digits. However, bending of the head or trunk toward the ventral side is also called flexion, and lateral bending of the trunk is known as lateral flexion. **Extension** is the antagonistic action. (2) **Protraction** is a forward movement of the entire limb at the shoulder or hip; **retraction** is the opposite movement. (3) **Adduction** is the movement of a part toward a point of reference; the movement away from the reference point is **abduction**. For limb movements, the point of reference is the midventral line of the body; hence movement of the limb ventrally is adduction.

Our knowledge of the actions of frog muscles has been inferred largely from observing their attachments and pulling upon them in cadavers. Their actions in a living animal are doubtless far more complex.

The names used in this exercise for the individual muscles are those used in reference works. Many of these are terms for human muscles that have been applied to comparable muscles in the frog, but the homology is not always as exact as this practice implies.

In this study the student will dissect the muscles of the hind leg of the frog in order to understand in some detail the relation of muscles to body movement.

Begin your dissection by cutting through the skin around the frog's abdomen. Grasp the skin

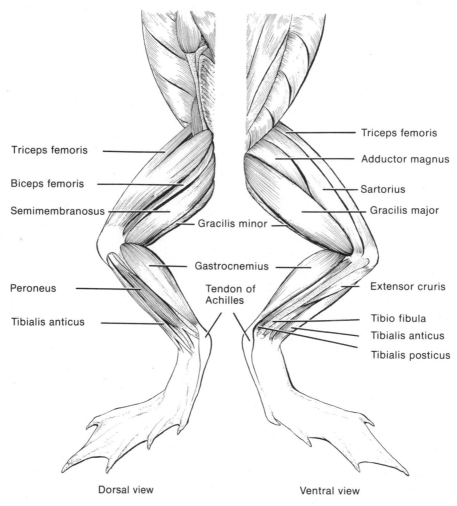

FIGURE 6-7
Muscles of the frog hind leg.

TABLE 6-1
Function of the muscles of the frog hind leg.

Triceps femoris	Extensor and abductor of leg
Biceps femoris	Flexor of leg and adductor of thigh
Semimembranosus	Flexor of leg and adductor of thigh
Gracilis minor	Adductor of leg and extensor of foot
Gracilis major	Adductor of leg and extensor of foot
Adductor magnus	Adductor and abductor of thigh
Sartorius	Flexor of shank and adductor of thigh
Gastrocnemius	Extensor of foot and flexor of leg
Peroneus	Extensor of leg and flexor of foot
Tibialis anticus	Flexor of foot and extensor of leg
Tibialis posticus	Flexor and extensor of foot
Extensor cruris	Extensor of foot

enclosed by a shiny white sheath of connective tissue, the **fascia**. Using a probe, carefully separate the muscles of the hind leg of the frog. Identify each of the muscles outlined in Fig. 6-7.

From the descriptions of the various muscles listed in Table 6-1, which pairs of muscles would you consider to be antagonistic in their actions?

_____

_____

_____

_____

_____

firmly and pull it posteriorly over the legs, turning it inside out, until the abdomen and legs are completely free of skin. Note that the muscles are

REFERENCES

Cochran, D. C. 1961. *Living Amphibians of the World.* Doubleday, Garden City, New Jersey.

Conant, R. 1958. *A Field Guide to Reptiles and Amphibians.* Houghton Mifflin, Boston.

Gilbert, S. G. 1965. *Pictoral Anatomy of the Frog.* University of Washington Press, Seattle.

Holmes, S. J. 1934. *The Biology of the Frog.* 4th ed. Macmillan, New York.

Moore, J. A., Ed. 1964. *Physiology of the Amphibia.* Academic Press, New York.

Noble, G. K. 1954. *The Biology of the Amphibia.* Dover, New York.

Romer, A. A. 1962. *The Vertebrate Body.* 3d ed. Saunders, Philadelphia.

Walker, W. F., Jr. 1967. *Dissection of the Frog.* W. H. Freeman and Company, San Francisco.

Wright, A. H., and A. A. Wright. 1949. *Handbook of Frogs and Toads of the United States and Canada.* 3d ed. Comstock, Ithaca, New York.

Young, J. A. 1962. *The Life of Vertebrates.* 2d ed. Oxford University Press, New York.

EXERCISE 7

Vertebrate Anatomy: Internal Organs

In this exercise you will study three of the major internal organ systems of vertebrates. The systems selected for study are the digestive, circulatory, and reproductive systems of the frog.

A. DIGESTIVE SYSTEM

1. Buccopharyngeal Cavity

Open the **mouth** by cutting directly back through the angle of the jaw on each side. Extend the cut to a point slightly posterior to the tympanic membrane (Fig. 7-1). Pull the floor of the mouth ventrally, thereby exposing the buccal cavity and the pharynx (Fig. 7-2). The **buccal**, or **oral**, **cavity** is the cavity of the mouth and it is bounded by the jaws; the **pharynx** is a transitional region between the buccal cavity and the gullet, or **esophagus**, which leads to the stomach. The **auditory**, or **eustachian**, **tubes** indicate the approximate anterior border of the pharynx.

A large **tongue** lies on the floor of the buccal cavity. It is attached anteriorly in such a way that its forked posterior end can be quickly flicked out of the mouth toward an insect or other small animals upon which it feeds. As it is thrust out, the tongue scrapes off a sticky secretion produced by glands in the roof of the mouth, so the prey sticks to the tongue and is pulled back into the mouth.

Two patches of **vomerine teeth** in the roof of the buccal cavity help to hold the prey, which is swallowed whole. The lower jaw has no teeth, and those on the upper jaw are very small.

Paired nasal passages open into the roof of the mouth by **internal nares**, one of which is lateral to each patch of vomerine teeth. Air enters the lungs by way of the **glottis**, a short longitudinal slit in the floor of the pharynx.

If your specimen is a male, look for the openings of the paired **vocal sacs**, one on each side of the

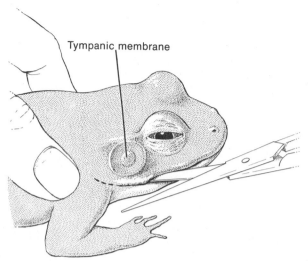

FIGURE 7-1
Procedure for dissection of the frog mouth.

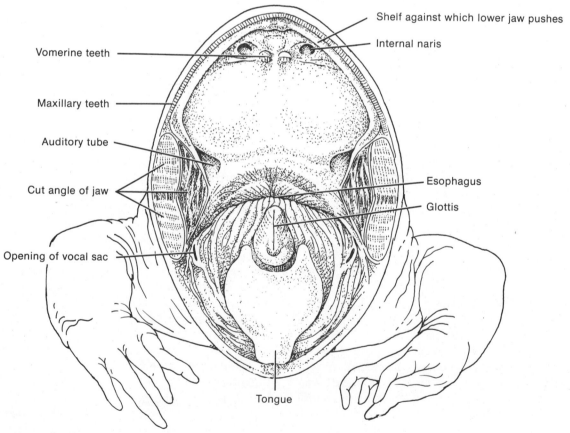

FIGURE 7-2
Interior view of the buccopharyngeal cavity of a male frog. The angles of the jaw have been cut so that the mouth can be opened widely. (Adapted from *Dissection of the Frog* by Warren Walker, Jr. W. H. Freeman and Company. Copyright © 1967.)

floor of the pharynx and slightly medial to the angle of the jaw.

2. Body Cavity

Make a longitudinal incision (X–Y) that is slightly to the left of the midventral line and extends from the pectoral to the pelvic girdle (Fig. 7-3). Cut through the muscle layers of the abdominal wall and then lift up the wall to expose the **ventral abdominal vein**. Carefully pull this blood vessel away from the midventral line and cut it so it can be pushed aside. Now make transverse incisions (A–B and A'–B') through the skin and muscle just anterior to each hind leg and posterior to each front leg. Continue the cuts to the back, and turn the flaps of the body wall, which may then be pinned to a dissecting pan. Lift the pectoral girdle and cut through the bone to expose the cardiac cavity.

The body cavity, or **coelom**, in which the visceral organs are located, is completely lined by a shiny layer of epithelium, called **parietal peritoneum**. The epithelium covering the walls of the viscera is called **visceral peritoneum**. Membranes of this epithelium, the **mesenteries**, extend from the dorsal body wall to the organs as well as between many of the organs. The coelom is a space that facilitates the expansion, contraction, and other functional movements of the internal organs. Mesenteries limit these movements to some extent and provide pathways for blood vessels and nerves.

The coelom of the frog is divided into two chambers: a large **pleuroperitoneal cavity** containing the lungs and most of the other viscera, and a small **pericardial cavity** surrounding the heart. The coelomic epithelium that lines the pericardial cavity is called the **pericardium**.

3. Visceral Organs

The large trilobed organ occupying much of the anterior part of the body cavity is the liver (Fig. 7-4). Lift it up and find the **gall bladder**, a small dark sac attached to the dorsal surface of the liver. The **stomach** lies dorsal to the liver and to your right as you view the body cavity. The **lungs** also lie in the anterior part of the body cavity. Lift up the stomach and locate a small, dense, round body, the **spleen**, on the left side of the mesentery supporting the intestine. The spleen is intimately related to the circulatory system, because, in adult

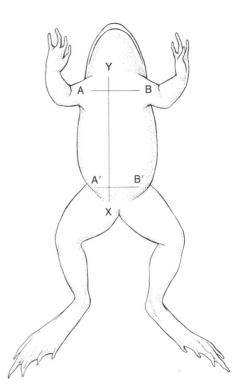

FIGURE 7-3
Procedure for dissection of the abdominal body wall.

frogs, it is the most important site for the production of both red and white blood cells. In the spleen, some blood cells are stored, and senescent (aging) red cells are destroyed. It is also a major site for the production of antibodies (protein molecules produced in response to foreign organisms as a defense mechanism). The coiling small intestine leads from the posterior end of the stomach to the large intestine, or **colon**. Lying in the loop between the small intestine and stomach is a whitish tissue, the **pancreas**. The colon is a short segment from which some water and ions are absorbed, and where fecal matter is temporarily stored. It has a large anterior diameter, but tapers posteriorly as it passes through the pelvis to enter the **cloaca**. The cloaca is the terminal part of the digestive tract, which also receives the urinary and genital ducts. The **testes** of the male are a pair of small, oval bodies located on each side of the mesentery that supports the intestine. The fingerlike lobes attached to the testes or kidney are the **fat bodies**. The ovaries are located in the same position in the female, except that they vary greatly in size depending upon the female's reproductive state. Just before ovulation, the ovaries are filled with ripe eggs and occupy all

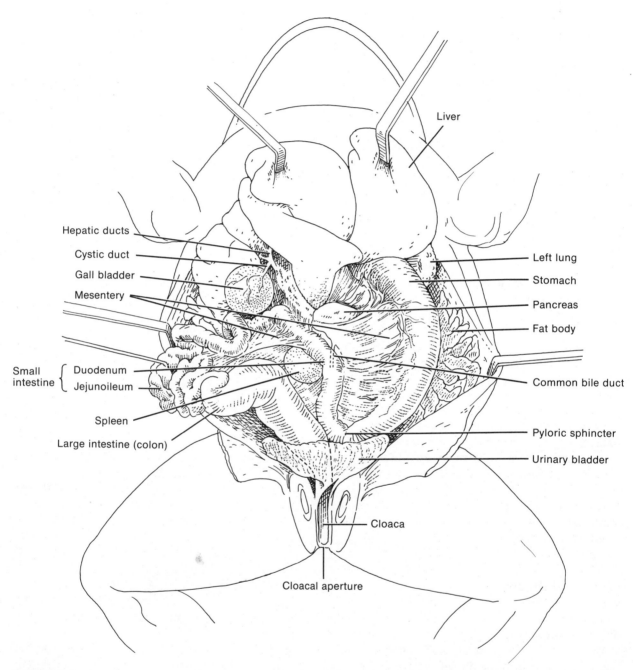

FIGURE 7-4
Ventral view of the digestive tract and associated organs. The liver has been pulled forward, and the pelvic region cut open. (Adapted from *Dissection of the Frog* by Warren Walker, Jr. W. H. Freeman and Company. Copyright © 1967.)

available space in the pleuroperitoneal cavity. Dorsal to each **gonad** (testis or ovary) is an elongated **kidney**. In females, a long, coiled, white tube —the oviduct—will be found lateral to each kidney. Males may have a small vestigial oviduct. The **urinary bladder** is a bilobed sac lying ventral to the large intestine.

4. Digestive Tract

Connecting the pharynx to the stomach is a short **esophagus**. The **stomach** is a large, saclike organ in which food is stored and digested. Cut it open and observe that its lining forms longitudinal folds that help to grind the food. At the posterior end of the stomach is the **pyloric sphincter**, whose contraction keeps food in the stomach until it is partly digested by the gastric juices and broken up mechanically by the churning action of the stomach folds.

Digestion is completed and food is absorbed in the small intestine. The first part of the small intestine, the **duodenum**, curves up toward the liver; the rest of it is comparable to the mammalian **jejunum** and **ileum**. Make a longitudinal slit in the small intestine and note the irregular folds, which greatly increase the internal surface area for the absorption of digested food materials.

Most of the secretions that act in the small intestine come from the **pancreas**, which produces enzymes that act on most food substances. The **bile salts** produced in the liver are secreted into the duodenum where they serve to emulsify fats, thereby facilitating the action of the enzyme lipase. The pancreas releases its secretions only when food is in the duodenum, but bile, which is secreted continuously, backs up into the gall bladder where it is stored until food enters the intestine.

Besides their digestive roles, the liver and the pancreas have other functions. The **bile pigments** are excretory products derived from the breakdown in the liver of hemoglobin from red blood cells that are destroyed in the spleen and liver. The liver also helps to convert many absorbed foods carried to it by the circulatory system. Excess sugars, fats, and amino acids are stored, largely in the form of glycogen, and the excretory product urea is synthesized from ammonia derived from amino acids. Although most of the pancreas produces digestive enzymes, it also contains microscopic clumps of endocrine tissue, **the islets of Langerhans**, which produce the hormone **insulin**.

Insulin enters the circulatory system and is carried to all parts of the body. It is essential for normal carbohydrate metabolism.

B. CIRCULATORY SYSTEM

The circulatory system includes the heart, the blood vessels, and the blood and lymph, which the vessels conduct throughout the body. It is a transport system that distributes food, oxygen, and other materials from sites of intake to all of the cells and carries carbon dioxide, urea, and other waste products of metabolism to sites of removal. It also carries hormones from the glands in which they are produced to the various target organs. In addition to its transport functions, the blood helps to regulate the amount of water and salts in the tissues, defends the body against the invasion of disease organisms, and generally helps the organism to maintain a constant internal environment.

1. Frog Heart

The frog heart lies in the pericardial cavity. Carefully dissect away the wall of the pericardial cavity to see the chambers of the heart and the vessels that enter and leave it. The posterior, cone-shaped part of the heart, which is more muscular than the rest, is the **ventricle** (Fig. 7-5). A tubular chamber of the heart, the **conus arteriosus**, receives blood from the ventricle and carries it anteriorly to a pair of large arteries (each is a **truncus arteriosus**), which distribute it to the lungs and body. The **right** and **left atrium** form the anterior part of the heart. Because their walls are thinner than those of the ventricle, the blood within them makes the atria appear darker in color. The conus arteriosus passes between them. Lift up the heart, and note the large, somewhat triangular chamber of the heart leading to the right atrium, the **sinus venosus**. It receives the drainage from all of the body except that from the lungs, and it transfers this blood to the right atrium. A large **posterior vena cava** emerges from the liver and enters it posteriorly. A pair of smaller **anterior venae cavae** drain the arms and head and enter the sinus venosus anteriorly. The lungs are drained by **pulmonary veins**, which enter the left atrium directly.

Study the beating of the heart on a demonstration dissection of a pithed frog. Notice that the sinus venosus is the "pacemaker." It contracts

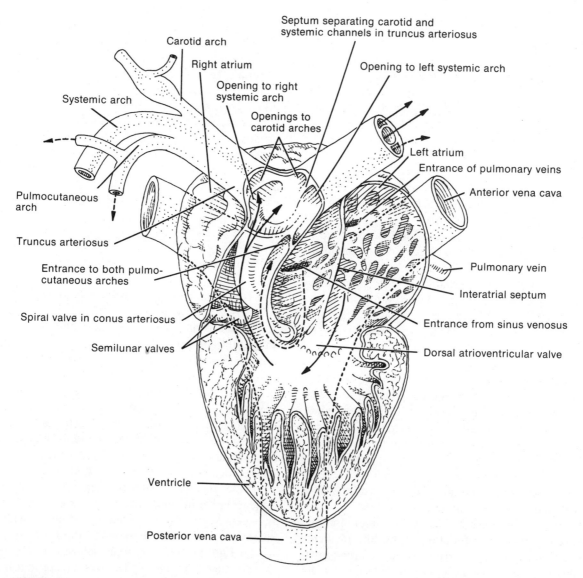

FIGURE 7-5
Ventral view of the frog heart. The position of the sinus venosus on the dorsal surface of the heart is shown by broken lines. (Adapted from *Dissection of the Frog* by Warren Walker, Jr. W. H. Freeman and Company. Copyright © 1967.)

VERTEBRATE ANATOMY: INTERNAL ORGANS

first, sending blood into the right atrium. Both atria contract next, sending their blood into the single ventricle. Contraction of the ventricle and conus greatly increases the blood pressure and drives the blood to the lungs and body. Valves located between the various chambers prevent a backflow of blood. They, together with other internal features of the heart, are shown in Fig. 7-5, but unless you are dissecting an exceptionally large frog it will be difficult to find them in your specimen. Because of this, you will study the sheep heart, which is approximately the size of the human heart.

2. Sheep Heart

The sheep heart lies in a tough, membranous sac, the **pericardium**, the inner surface of which is a membrane that faces the outer surface, or **epicardium**, of the heart. The pericardial space, between these facing membranes, contains a serous pericardial fluid. In removing the pericardium, you will observe that it is attached only around the base of the heart, where the large vessels emerge. The cavities of the heart are lined with **endocardium**, which is the same type of tissue as the endothelium lining all of the blood vessels. Between the epicardium and endocardium is the bulk of heart tissue, the **myocardium** or heart musculature.

The right and left sides of the heart are identified in two ways: (1) the pointed end, or **apex**, is entirely a part of the left ventrical, and the right and left divisions are indicated superficially by a diagonal furrow followed by coronary vessels; (2) when feeling the ventricular musculature, the left side appears firm and muscular, whereas the right ventricle feels soft and flabby. The auricles appear as earlike projections at the base of the heart. The cavity within each auricle is referred to as an **atrium**, a reception chamber for incoming blood.

a. Make a long incision through the **right auricle** in line with the superior vena cava. Lift the edges of the flaps and observe the wide mouths of the great veins (**superior** and **inferior vena cava**) as they enter the atrium. With scissors, carry the incision downward toward the ventricle and determine the relationship between the two cavities. Note the irregular bands of muscle lining the interior of the auricular wall, the **pectinate muscles**. Find the **coronary sinus**, which receives venous blood directly from the heart musculature and enters the auricle as a wide cavity. Note that the partition separating the auricles is membranous rather than muscular. Locate the thinnest portion of the wall between the two auricles. This area is the **fossa ovalis**, an oval depression marking the position in fetal circulation where blood was carried directly to the left side of the heart and thus bypassed the pulmonary circuit.

b. Carry the incision from the right auricle in a straight line through the lateral wall of the **right ventricle**. Note the three rounded flaps of membranous tissue suspended into the ventricle and held in place by tendinous cords. These flaps constitute the **tricuspid valve**. Study in detail its position, structural character, and attachments. If necessary, wash out both cavities. Note that pointed columns of ventricular muscle (**papillary muscles**) are continuous with the wall of the ventricle and with the strong fibrous cords (**chordae tendineae**) that extend to the edges of the cusps, or flaps, of the valve. Observe the heavy muscular ridges within the ventricle, the **columnae carneae**. Find the exit of blood from the right ventricle, considering the tricuspid valve as closed between auricle and ventricle. Carry an incision upward through the wall of this exit (**pulmonary artery**) and note that the mouth of the artery is surrounded by three membranous pockets (the pulmonary **semilunar valves**).

c. Open the **left auricle** in the same manner as the right side. Before cutting on down through the ventricle, push your finger from the auricle into the ventricles and distend the auriculoventricular opening. Determine the number of openings draining into the auricle from the lungs. These openings are the mouths of **pulmonary veins**.

d. On the **left ventricle**, make an incision to expose the cavity. Study the details of the **mitral**, or **bicuspid (two-parted), valve** between this auricle and ventricle. Explore the thickness of the septum between the two ventricles. With your finger, find the outlet of the ventricle into the aorta. Open the aorta to expose its semilunar valve. Find the openings of the two coronary arteries just above the valve—that is, within the pockets. Trace these to the walls of the heart. Observe the tough ligamentous connection, which is usually covered with a conspicuous pad of fat between the pulmonary artery and aorta. This is a remnant of a vessel that connected the pulmonary artery and aorta in the embryo. If the vessel remains persistently open, a mixing of oxygenated and unoxygenated blood occurs.

In studying the dissected heart, emphasis should be placed on the structural arrangement of the parts

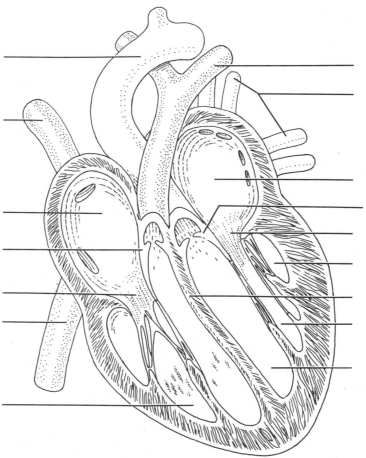

FIGURE 7-6
Schematic representation of the mammalian heart.

and their functional continuity. The mammalian heart is essentially a double pump equipped with specially designed valves. It serves as the propelling force for two circuits of blood that do not mix in the heart. The circuit from the heart to the lungs and back is the **pulmonary circuit**, and the circuit from the heart to all the body tissues and back is the **systemic circuit**. These circuits act concurrently and interdependently, since no more blood can be sent through the pulmonary circuit than is delivered to it by the systemic circuit. In Fig. 7-6 indicate the directions of blood flow.

3. Arterial System

Arteries are vessels that carry blood away from the heart to the capillaries in the tissues. Only the major arteries of the frog are described in this part of the exercise. The **truncus arteriosus** curves dorsally and laterally, soon dividing into three branches. These are, from anterior to posterior, the **carotid arch**, the **systemic arch**, and the **pulmocutaneous arch**.

The carotid arch divides into an **external carotid artery**, which supplies capillaries in the floor of the pharynx and mouth, and an **internal carotid artery**, which continues dorsally to supply much of the top of the head, including the major sense organs and brain (Fig. 7-7). The small, oval swelling on the junction of the two carotids is the **carotid body**. Its functions have been a matter of considerable dispute. It may help to regulate blood pressure and blood flow. Other investigators have suggested that it contains chemical receptors sensitive to the partial pressure of oxygen in the blood.

A small, oval nodule of tissue lies in the angle formed by the truncus arteriosus and external carotid. This is the **thyroid gland**. Its hormone, **thyroxine**, is essential for metamorphosis from

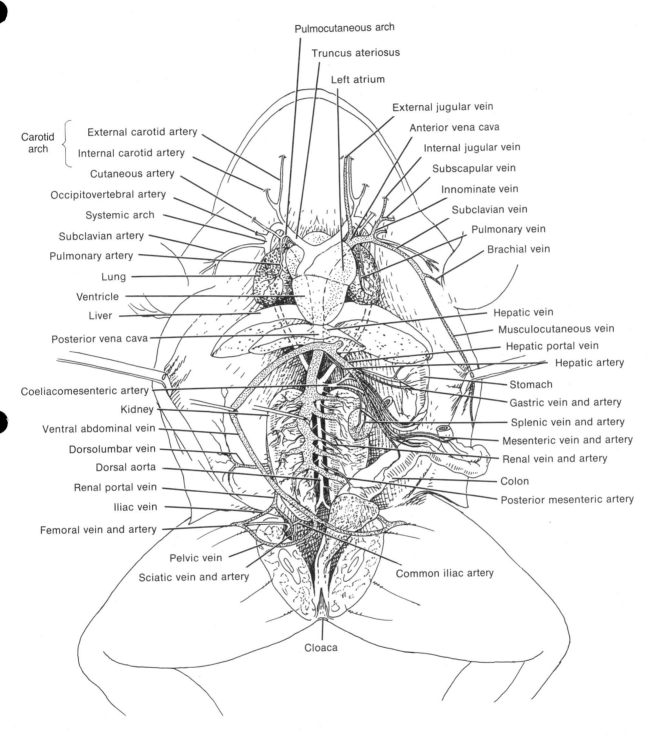

FIGURE 7-7
Semidiagrammatic ventral view of the circulatory system of the frog. Many of the viscera have been removed. Anterior veins are shown only on the right side of the drawing. The position of the sinus venosus dorsal to the atria and ventricle is shown by dotted lines. (From *Dissection of the Frog* by Warren Walker, Jr. W. H. Freeman and Company. Copyright © 1967.)

tadpole to frog, because it increases the rate of metabolism at this time. Its function in adult frogs in unknown.

Next, trace the pulmocutaneous arch, which is the major vessel carrying depleted blood to respiratory surfaces where gas exchange takes place. It divides into two branches: the larger branch, the **pulmonary artery**, continues posteriorly to the lung; the smaller branch, the **cutaneous artery**, continues dorsally and laterally to supply certain trunk muscles and much of the skin.

The systemic arch supplies the rest of the body. It curves dorsally and posteriorly and, upon reaching the back, gives off two branches. The anterior branch is the **occipitovertebral artery**, which supplies muscles on the posterodorsal surface of the head, and much of the head itself. The posterior branch is the **subclavian artery**. It extends ventrally and laterally, giving off small branches to muscles in the shoulder region and body wall. Its main continuation into the arm is the **brachial artery**. After giving off these branches, the systemic arch continues posteriorly and medially along the back, finally uniting with the opposite systemic arch to form the **dorsal aorta**.

The first branch of the aorta, which arises from the site at which the two systemic arches unit, is the **coeliacomesenteric artery**. This vessel divides into two arteries: a **coeliac artery**, which in turn divides into branches supplying the stomach, pancreas, liver, and gall bladder; and an **anterior mesenteric artery**, which supplies most of the intestine and the spleen. The dorsal aorta next gives off four to six small, paired **urogenital arteries**, which supply the kidneys, gonads, fat bodies, and oviducts. From the same region of the aorta two or three small **lumbar arteries** extend dorsally to the muscles of the back. A small, median **posterior mesenteric artery** leaves near the posterior end of the aorta to supply part of the large intestine. Shortly after this, the aorta branches and gives rise to a pair of **common iliac arteries**, which give off small branches to the urinary bladder, the cloaca, the skin on the thigh, and a small **femoral artery** to the lateral side of the thigh. The main part of the common iliac continues into the leg as the **sciatic artery**.

4. Venous and Lymphatic Systems

Water, food, oxygen, and other materials leave the blood from the **vascular capillaries** and enter the tissue fluid, which bathes the cells of the body. Much of the excess water and most of the various waste products of metabolism reenter the capillaries to be carried toward the heart by **veins**. But a large amount of water, together with some plasma proteins that leak out of the vascular capillaries, enters a separate system of **lymphatic capillaries**. Exceptionally low pressures in the lymphatic capillaries, combined with a greater permeability of their walls, permit the entry of such relatively large molecules as the plasma proteins and, in the intestinal region, fat molecules, which are absorbed by this route. Lymph is moved to larger lymphatic vessels, and eventually to certain of the larger veins, by the contraction of surrounding body muscles, aided (in amphibians) by pulsating parts of the lymphatic vessels known as **lymph hearts**.

Most of the lymphatic system is inconspicuous, but some of the lymphatic vessels lead to large lymphatic sacs and certain of these are easily seen. Frogs have a particularly large **subvertebral lymph sac**, into which the kidneys protrude, and several **subcutaneous lymph sacs**. The presence of the latter lymph sacs is the reason the skin is so loosely attached to the muscles. When the frog is on land, much of the body water it loses to the environment is lost from the subcutaneous lymph sacs; conversely, when the frog is in a pond, much of the water it takes up from the environment first enters these lymph sacs.

The pulsation of the pair of **posterior lymph hearts** can be seen in a pithed and skinned frog. One sac lies on each side of the body between the posterior end of the urostyle and the pelvic girdle. These hearts move lymph from certain of the subcutaneous sacs into veins of the hind leg.

The head, the arms, and much of the skin are drained by a pair of **anterior venae cavae**, which enter the sinus venosus (Fig. 7-7). It is formed, just posterior to the branching of the truncus arteriosus, by the joining of three tributaries. The largest and posterior tributary is the **subclavian vein**, which extends towards the base of the arm. Here it is formed by the union of a **brachial vein**, which emerges from the arm, with a large **musculocutaneous vein**, which carries blood from the musculature of the body wall and from the skin. Blood from the skin is, of course, rich in oxygen and low in carbon dioxide.

The **external jugular vein** is the anterior tributary of the anterior vena cava. It extends anteriorly and receives a tributary that comes from the tongue and the floor of the mouth and another one from the area of the lower jaw.

The last of the tributaries of the anterior vena cava is the **innominate vein**. It lies between the other two and extends dorsally. Before entering the anterior vena cava, it receives a **subscapular vein**, which drains the shoulder region, and an **internal jugular vein**, which drains the brain and the major sense organs of the head.

The rest of the venous blood from the body enters the sinus venosus through the **posterior vena cava**. As you trace it posteriorly, observe that it first disappears in the liver, but soon re-emerges on the dorsal surface of the liver and continues posteriorly between the kidneys. The posterior vena cava receives several paired **renal veins** from the kidneys and a number of small **gonadial veins** from the reproductive organs (Fig. 7-8). As it passes through the liver, it receives the drainage of that organ through many **hepatic veins**. Some of them can be seen by carefully dissecting away a bit of the anterior part of the liver.

Portal veins differ from other veins in that, after draining one region, they lead to capillaries, or to **sinusoids** (microscopic passages) in another organ, rather than continuing on to the heart. The digestive tract and the spleen are drained by the hepatic portal system because veins from them lead to sinusoids in the liver. The entry into the liver cells of materials in this blood that are absorbed from the digestive tract makes possible the various metabolic roles of the liver.

The main **hepatic portal vein** enters the liver near the junction of the liver and the pancreas. The ventral abdominal vein joins it at this point. Trace the hepatic portal vein posteriorly through the pancreas. Its main tributaries are **gastric veins** from the stomach and **mesenteric veins** from the intestine and spleen.

Trace the **ventral abdominal vein** posteriorly. It receives tributaries from the body wall and the urinary bladder, and then is formed by the joining of two **pelvic veins** that extend dorsally and laterally and join **femoral veins** that come from the legs. The common trunk thus formed is the **iliac vein**. A **sciatic vein**, which accompanies the corresponding artery in the leg, joins the iliac, and this common vessel, the **renal portal vein**, continues forward along the dorsolateral edge of the kidney. As the name implies, blood in it enters capillaries in the kidney, and these, in turn, are drained by the posterior vena cava. A **dorsolumbar vein**, which drains a large part of the back, also enters the renal portal.

This complex arrangement of vessels allows some of the blood from the hind leg to flow forward through the ventral abdominal vein and the liver, and some to flow through the renal portal system and the kidney. The functional significance of a renal portal system, which is present in most of the lower vertebrates, is not completely understood. It does, however, ensure a greater flow of blood through the kidneys than is provided by the rather small renal arteries.

C. UROGENITAL SYSTEM

The urogenital system consists of the kidneys and the reproductive organs. The kidneys are excretory organs that eliminate nitrogenous waste products of protein metabolism, but they also help to regulate the water and salt contents of the body fluids. The reproductive organs obviously have a very different function, but it is convenient to consider certain aspects of these two systems together, since the reproductive cells of the male are discharged through urinary ducts.

1. Excretory System

Remove the stomach and small intestine to get a better view of the urogenital organs. Do not remove the anterior part of the liver of the lungs in a female lest you destroy the anterior portion of the oviduct. Each **kidney** is an elongated, oval organ located against the back. The kidneys are composed of several thousand kidney tubules (about 2,000 in *Rana pipiens*, and about 7,000 in *Rana catesbeiana*), which produce the urine.

The urine from each kidney is drained by a **Wolffian duct**, which can be seen along its lateral edge (Fig. 7-8). The duct is larger in the male, because it also carries sperm. Trace one posteriorly, and find the site at which it joins the dorsal surface of the cloaca. Because the bilobed **urinary bladder** attaches to the ventral surface of the cloaca, urine must flow across the cloaca to enter the bladder.

Notice the irregularly shaped band of light tissue lying on the ventral surface of each kidney. This is the **suprarenal**, or **adrenal**, **gland**. Some of its cells produce **adrenalin**, a hormone that increases blood sugar levels, increases the rate of heartbeat, and in general helps the body to adjust to conditions of stress. Other cells produce a variety of **hormones**, which affect carbohydrate metabolism, mineral metabolism, and the water and salt balances of the body. For example, a certain amount of water uptake and an active inward transport of sodium ions normally occur through the skin, but frogs that have the adrenal glands removed show an increase of water uptake through the skin and a loss of sodium ions.

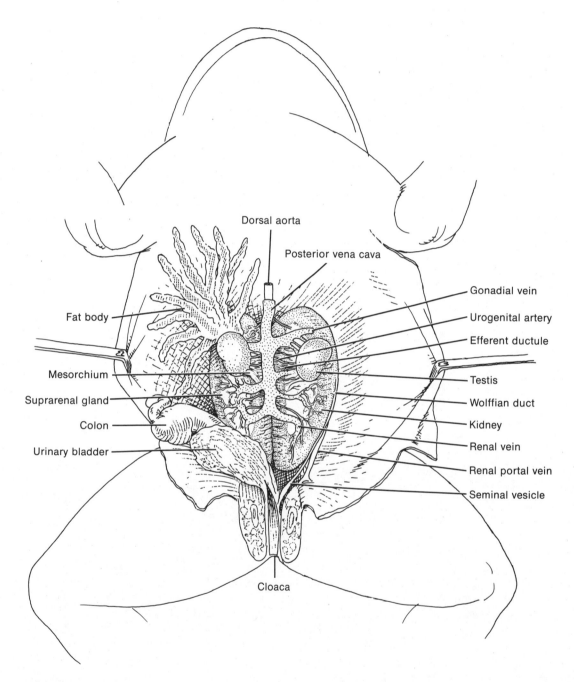

FIGURE 7-8
Ventral view of the urogenital system of a male frog. One fat body has been removed. (Adapted from *Dissection of the Frog* by Warren Walker, Jr. W. H. Freeman and Company. Copyright © 1967.)

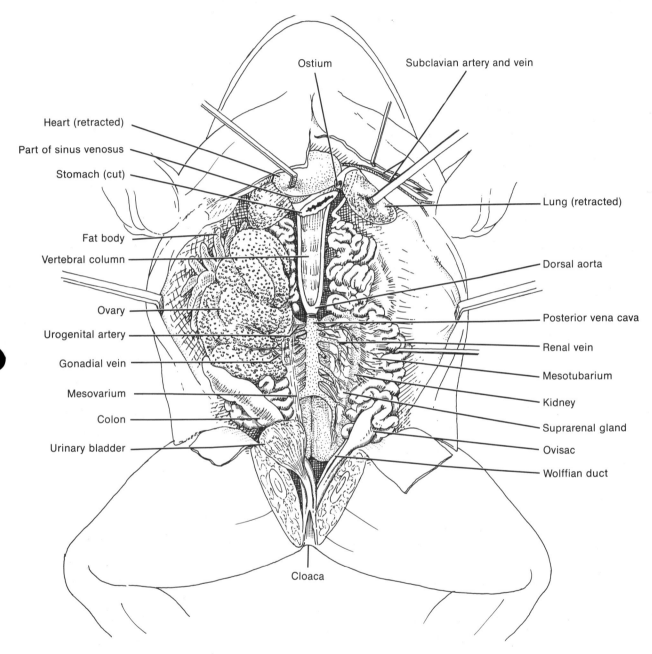

FIGURE 7-9
Ventral view of the female urogenital system. One ovary and one fat body have been removed. Because this drawing is based on a specimen taken in midsummer, the ovaries and fat bodies have not reached their full size. (Adapted from *Dissection of the Frog* by Warren Walker, Jr. W. H. Freeman and Company. Copyright © 1967.)

2. Reproductive System

After studying the reproductive system of your specimen, examine one of the opposite sex by exchanging specimens with another student. Each student should dissect his specimen with particular care so that another student may see all parts clearly.

a. Male Reproductive Organs

Each **testis** is a small oval organ attached by a mesentery, the **mesorchium**, to the ventral surface of a kidney (Fig. 7-8).

A **fat body**, which consists of long, fingerlike lobes of tissue, lies at the anterior end of the testis. The fat bodies are largest during the fall of the year. Some of the food stored in them is used during hibernation, and the rest is rapidly depleted during the spring spawning season, when the frogs become very active.

Sperm are produced in the testis during the summer and are stored there during the fall and winter. In the spring they leave the testis by way of minute **efferent ductules**, which extend through the mesorchium to the kidney. Within the kidney the ductules branch, and each one connects with the proximal ends of eight to twelve of the anterior kidney tubules. These kidney tubules convey the sperm to the Wolffian duct and then to the **seminal vesicle**. Sperm are stored here temporarily until the male frog mounts the back of the female—an embrace called **amplexus**. Sperm are discharged through the male's **cloaca** as the eggs are laid.

During development, the embryos of frogs and other vertebrates pass through a sexually indifferent stage, during which the primordia of both male and female reproductive ducts are present and the embryo has the morphological possibility of developing into either sex. Normally one set of passages continues to develop and those of the other set atrophy, but in some adult male frogs a rudimentary oviduct may be found lateral to each kidney.

b. Female Reproductive Organs

Each **ovary** is a multilobed organ attached by a mesentery, the **mesovarium**, to the ventral surface of a kidney (Fig. 7-9). The ovaries vary greatly in size according to the season of the year. They are smallest after ovulation in the spring. During the summer, oogonia develop and accumulate yolk. The ovaries attain their maximum size during the fall and winter, when they contain thousands of ripe eggs (about 2,000 in the leopard frog, and up to 20,000 in the bullfrog) and fill all the available space within the pleuroperitoneal cavity. If you have a female specimen in this condition, remove one ovary in order to see other organs more clearly. A fat body, consisting of long, fingerlike lobes of tissue, is attached to the anterior border of the mesorchium.

A long, convoluted **oviduct** lies against the back of each side of the body. Trace one forward. At the anterior end of the pleuroperitoneal cavity, it curves ventrally beside the pericardial cavity and ends in a small opening, the **ostium**, which is located at the anterior end of the liver.

In the spring of the year, and under the stimulus of pituitary hormones, the eggs are **ovulated**; that is, they are discharged from the ovary and enter the pleuroperitoneal cavity. A large part of the coelomic epithelium of the female is ciliated at this time, and ciliary currents carry the eggs toward the ostium and into the oviduct. As the eggs move down the oviduct, they accumulate layers of jelly, consisting largely of albumen and mucoprotein, and undergo meiosis I. The eggs may be stored for a day or two in a thin-walled expansion of the posterior end of the oviduct known as the **ovisac**. The ovisac enters the dorsal surface of the **cloaca** beside the entrance of the Wolffian duct. The stimulus of amplexus is required for the eggs to be discharged from the ovisac into the cloaca. As the eggs emerge through the cloacal aperture, sperm penetrate them, and the jelly layers imbibe water and swell up. The jelly layers help to protect the developing embryo from mechanical injury, fungi, and bacterial infections, and they help to conserve the heat produced by the metabolism of the embryo.

REFERENCES

Cochran, D. C. 1961. *Living Amphibians of the World.* Doubleday, Garden City, New Jersey.

Conant, R. 1958. *A Field Guide to Reptiles and Amphibians.* Houghton Mifflin, Boston.

Gilbert, S. G. 1965. *Pictorial Anatomy of the Frog.* University of Washington Press, Seattle.

Griffin, D. R. 1962. *Animal Structure and Function.* Holt, Rinehart & Winston, New York.

Holmes, S. J. 1934. *The Biology of the Frog.* 4th ed. Macmillan, New York.

Moore, J. A., Ed. 1964. *Physiology of the Amphibia.* Academic Press, New York.

Noble, G. K. 1954. *The Biology of the Amphibia*. Dover, New York.

Romer, A. A. 1962. *The Vertebrate Body*. 3d ed. Saunders, Philadelphia.

Walker, W. F., Jr. 1967. *Dissection of the Frog*. W. H. Freeman and Company, San Francisco

Wiggers, C. J. 1957. The Heart. *Scientific American,* May (Offprint No. 62). *Scientific American* Offprints are available from W. H. Freeman and Company, 660 Market Street, San Francisco 94104, and 58 Kings Road, Reading RG1 3AA, England. Please order by number.

Wright, A. H., and A. A. Wright. 1949. *Handbook of Frogs and Toads of the United States and Canada.* 3d ed. Comstock, Ithaca, New York.

Young, J. A. 1962. *The Life of Vertebrates*. 2d ed. Oxford University Press, New York.

EXERCISE 8

Movement of Materials Across Cell Membranes

If a cell is to perform its functions it must maintain a steady state in the midst of an ever-changing environment. This constancy is maintained by the regulation of movement of materials into and out of the cell. To achieve this control, cells are bounded by a delicate membrane that can distinguish between different substances, slowing down the movement of some while allowing others to pass through. Since not all substances penetrate the membrane equally well, the membrane is said to be **differentially permeable**.

The external and internal environment of cells is an aqueous solution of dissolved inorganic and organic molecules. Movement of these molecules, both in the solution and through the cell membrane, involves a physical process called diffusion—a spontaneous process by which molecules move from a region in which they are highly concentrated to a region in which their concentration is lower.

At a temperature of absolute zero—equivalent to −273°C, −459°F, or 0° on the Kelvin (K) temperature scale—molecules have no motion. At temperatures above absolute zero molecules are in constant motion. As a result of this motion the molecules possess kinetic energy in the form of heat, with increased molecular motion resulting in increased temperature. Molecules in motion move in a random pattern. Furthermore, individual molecules move independently of one another.

An example of diffusion occurs when a colored dye (the **solute**, that substance being dissolved) is dissolved in a volume of water (the **solvent**, that substance used to dissolve the solute). If a drop of this aqueous dye solution is gently placed on the surface of water, it will be diffused throughout the water after a period of time. Ultimately, the dye molecules become equally dispersed as a result of the random movements and kinetic energy of the solute and solvent molecules.

This type of diffusion is called **passive diffusion** because movement is caused by the internal kinetic energy of the diffusing molecules. No outside form of energy is required for passive diffusion to occur. In contrast, there is **active transport**, in which solute molecules may move **against a concentration gradient**. For example, it has been shown that human erythrocytes have almost 30 times more potassium than the surrounding blood plasma. Thus these cells are accumulating potassium by an active process requiring metabolic energy.

A special kind of diffusion is the phenomenon of osmosis. Simply defined in biological systems, **osmosis** is the diffusion of water through a differentially permeable membrane from a region in which it is highly concentrated to a region in which its concentration is lower. More often, however, osmosis is defined in terms of the effects that solutes have on the thermodynamic activity of water (i.e., the activity of the water molecules due to the kinetic energy of motion). For example, the addition of a solute to water tends to decrease the activity of the water. In other words, as more water molecules are displaced by solute molecules, the activity of the water goes down. Thus, in thermodynamic terms, water diffuses across membranes from a region in which the thermodynamic activity of water is high (low solute concentration) to one in which the thermodynamic activity is low (high solute concentration). This is diagramatically shown below:

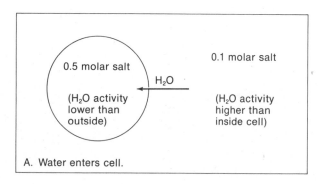

A. Water enters cell.

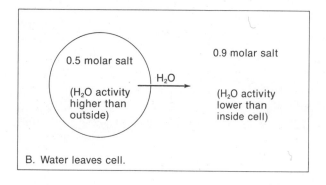

B. Water leaves cell.

Diffusion and osmosis, while basically phenomena resulting from the thermal activity of molecules, are affected by a number of other factors such as temperature, the physical nature of the medium in which diffusion is occurring, the molecular weight of the diffusing substance, the lipid solubility of the solute, and so forth. In this study you will become familiar with diffusion and osmosis and examine some of the factors regulating these processes.

A. BROWNIAN MOVEMENT

Brownian movement is a phenomenon where minute particles of solid matter—or bacteria and other small organisms—in a liquid suspension move in an erratic, random course. Robert Brown, an army surgeon and botanist, who first noticed this motion in 1827, thought it was due to some living activity until he discovered that boiling failed to stop the movement. He hypothesized that the motion was due to bombardment of the fluid by molecules. Since then this view has been fully confirmed. The energy that keeps a particle that is larger than a molecule in motion is the kinetic energy of the molecules of the surrounding liquid medium. The molecules of the dispersion medium strike the larger particles on all sides and impart to them a movement similar to the kinetic motion of the molecules themselves.

To observe Brownian movement, add a small drop of dilute India ink, or an aqueous suspension of carmine, to a slide. Add a cover slip and observe the motion of the particles with the high power of your microscope. It may be necessary to reduce the light to provide more contrast.

Examine a culture of *Closterium*, a crescent-shaped, unicellular alga. A striking feature of this alga is the presence of a small spherical vacuole at the extremities of the cell. Examine these vacuoles with the high power of your microscope. These vacuoles contain minute, insoluble crystals of calcium sulfate. The movement of these crystals is the same Brownian movement you observed with India ink.

B. DIFFUSION

The kinetic energy of molecules causes them to move from regions in which they are highly concentrated to regions in which their concentration

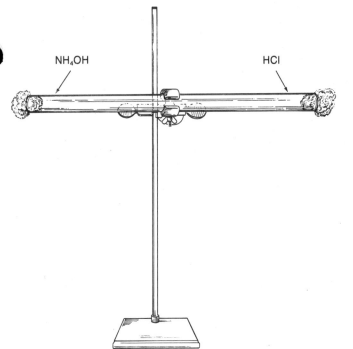

FIGURE 8-1
Apparatus for studying gas diffusion.

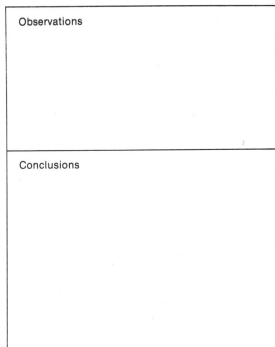

is lower until they become uniformly distributed throughout the space available to them. Thus, a gas set free in a room is in time equally distributed throughout the room; a crystal of salt put into a glass of water dissolves, and the molecules of which it is composed become uniformly distributed throughout the water. This movement of gases and dissolved substances from a region of high concentration to one of low concentration is termed diffusion. Diffusion can thus be defined as the net movement of the molecules of a substance, under their own kinetic energy, from a region of greater to one of lesser molecular activity of that particular substance. The rate at which particles will diffuse depends upon several factors, such as their size, their concentration, and the permeability of the dispersion medium.

1. Diffusion of a Gas in a Gas

A striking demonstration of diffusion is the movement of gases through air. An apparatus to demonstrate gas diffusion consists of a length of glass tubing clamped to a ring stand, cotton, and solutions of ammonium hydroxide (NH_4OH) and hydrochloric acid (HCl). Saturate a piece of cotton on one side with NH_4OH. Saturate another piece of cotton with HCl. (CAUTION: *Handle chemicals carefully.* Open only *one* reagent bottle at a time). Place the pieces of cotton *simultaneously* into opposite ends of the glass tubing as shown in Fig. 8-1.

Ammonium hydroxide and hydrochloric acid react to form ammonium chloride and water. The equation for this reaction is as follows:

$$NH_4OH + HCl \rightarrow NH_4Cl + H_2O$$

The ammonium chloride appears as a white precipitate. What do you expect will happen when these two gases meet?

_____

_____

_____

The molecular weight of the ammonium (NH_4^+) and chlorine (Cl^-) ions is 18 and 35.5, respectively. At which end of the tube does the reaction form?

_____

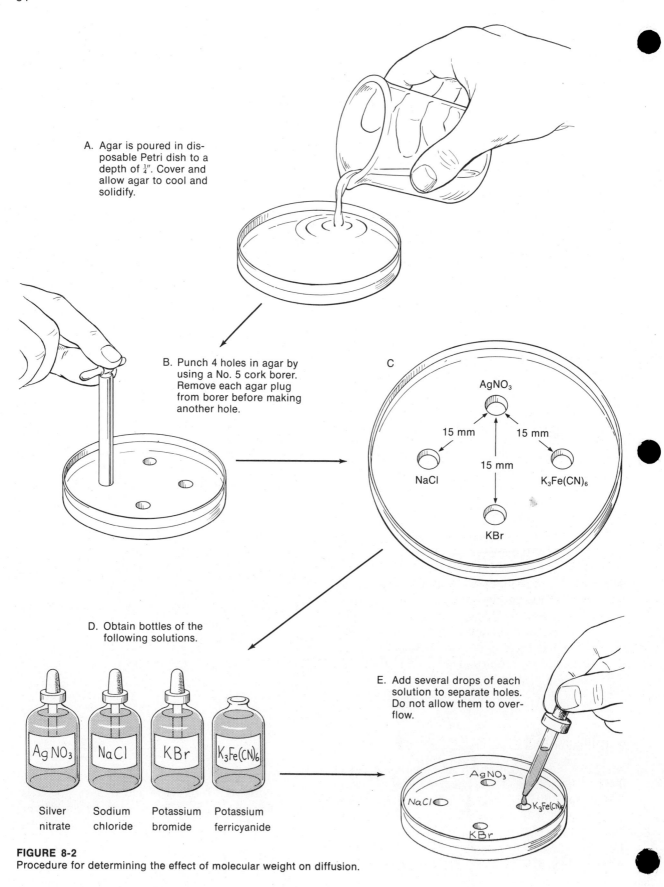

FIGURE 8-2
Procedure for determining the effect of molecular weight on diffusion.

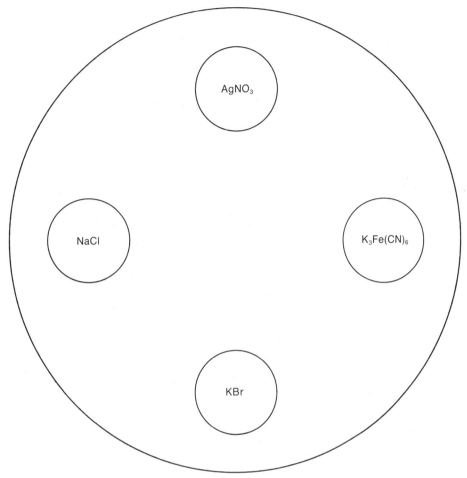

FIGURE 8-3
Effects of molecular weight on diffusion.

Suggest a relationship between molecular weight and the rate of diffusion in a gas?

_____

_____

Record your observations by completing Fig. 8-1.

2. Diffusion of a Liquid in a Solid

Your instructor will give you a disposable Petri dish containing agar, a substance obtained from certain sea weeds (Fig. 8-2A). Using a No. 5 cork borer, punch four holes in the agar as shown in Fig. 8-2B and C. Fill each of the holes uniformly with a small amount of 1 N solutions of sodium chloride (NaCl), potassium bromide (KBr), potassium ferricyanide [$K_3Fe(CN)_6$], or silver nitrate ($AgNO_3$). Periodically examine the Petri dishes and record your observations in Fig. 8-3. The approximate weights of each of the migrating groups formed when these substances are placed in solution are as follows: chloride anion, Cl−35; bromide anion, Br−80; ferricyanide anion, $Fe(CN)_6$ −212; nitrate anion, NO_3−112.

From this study what can you conclude about the relationship between the rate of diffusion of a molecule and its molecular weight?

_____

_____

_____

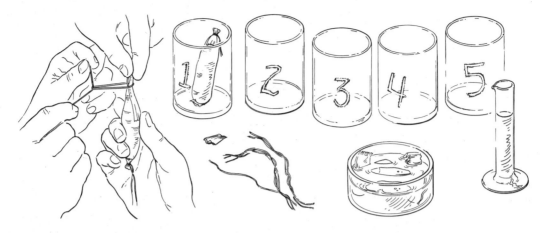

FIGURE 8-4
Preparation of dialysis bags. (From *Plants in Perspective* by Eldon H. Newcomb, Gerald C. Gerloff, and William F. Whittingham. W. H. Freeman and Company, Copyright © 1964.)

C. MEASURING THE RATE OF OSMOSIS

Each group of students will be given five dialysis bags. These will function as artificial, differentially permeable membranes. Each of these bags should be tied at one end by folding the end over and tying with a thread (Fig. 8-4). Fill these bags as follows:

Bag 1. 15 ml of tap water
Bag 2. 15 ml of 20% sucrose solution
Bag 3. 15 ml of 40% sucrose solution
Bag 4. 15 ml of 60% sucrose solution
Bag 5. 15 ml of tap water

After each bag has been filled, remove the air by gently squeezing below to bring the liquid to the top of the bag. Press the sides of the bag together so that air does not reenter. Fold the end of the bag over about $\frac{3}{4}$ inch and tie securely with a thread. Weigh each bag separately to the nearest 0.5 g and record the weights in Table 8-1, at "0" time.

TABLE 8-1
Osmosis data.

Time (minutes)	Weight of bags (grams)				
	1	2	3	4	5
0					
15					
30					
45					
60					
75					

Place Bags 1, 2, 3, and 4 in separate beakers of water, and place Bag 5 in a beaker of 60% sucrose solution.

At 15-minute intervals—that is, after 15, 30, 45, 60, and 75 minutes—remove the bags from the beakers; carefully wipe off all excess water; and again weigh each bag separately. Record the data in Table 8-1. Plot the changes in weight of each bag against time in Fig. 8-5. What relationships are there between the concentration of the solution and the rate of osmosis?

_____
_____
_____
_____

D. FACTORS AFFECTING MOVEMENT OF MATERIALS ACROSS CELL MEMBRANES

Red blood cells can be conveniently used to illustrate some of the factors that affect the osmotic relations of the living cell. The cell membrane of red blood cells is freely permeable to water, but differentially permeable to most other substances. The red blood cell increases in volume as substances such as water and dissolved solutes enter it from the outside, but it does not increase indefinitely. Instead, it reaches a limit of size, and then the cell membrane ruptures and the hemoglobin diffuses out of the cell, leaving only the

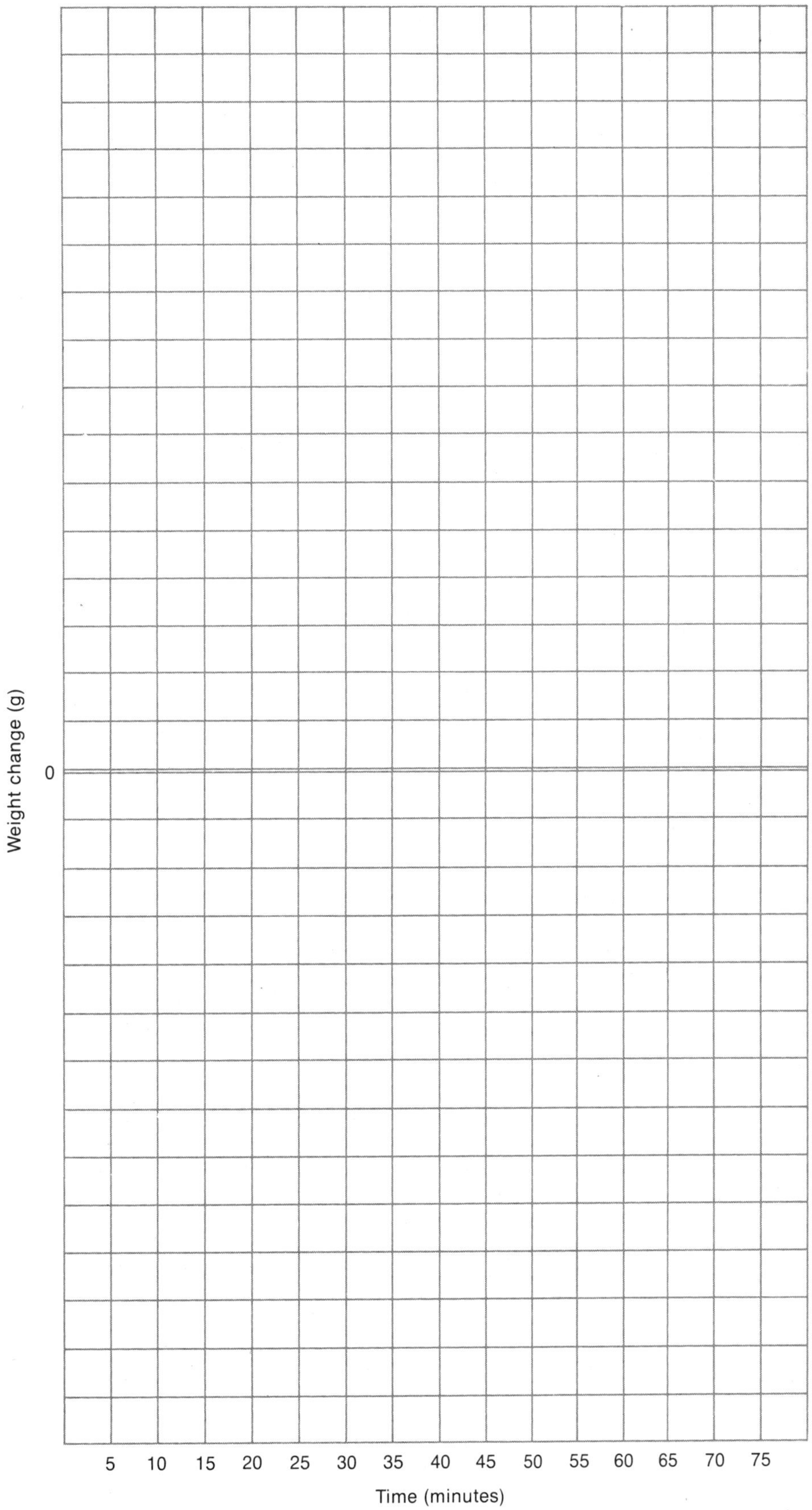

FIGURE 8-5
Plot of osmosis data.

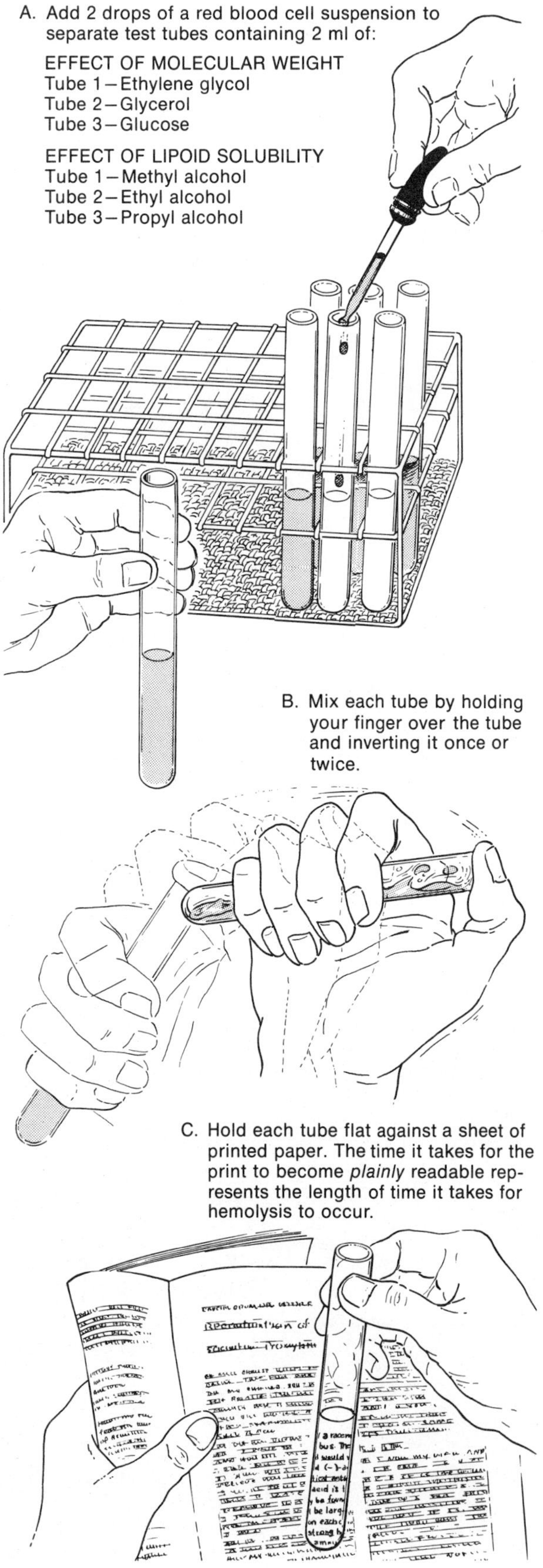

A. Add 2 drops of a red blood cell suspension to separate test tubes containing 2 ml of:

EFFECT OF MOLECULAR WEIGHT
Tube 1—Ethylene glycol
Tube 2—Glycerol
Tube 3—Glucose

EFFECT OF LIPOID SOLUBILITY
Tube 1—Methyl alcohol
Tube 2—Ethyl alcohol
Tube 3—Propyl alcohol

B. Mix each tube by holding your finger over the tube and inverting it once or twice.

C. Hold each tube flat against a sheet of printed paper. The time it takes for the print to become *plainly* readable represents the length of time it takes for hemolysis to occur.

FIGURE 8-6
Procedure for determining hemolysis time.

cell membrane (*ghost*) behind. This phenomenon is called **hemolysis**.

Dilute suspensions of red blood cells transmit very little light; but when these cells are hemolyzed, the solution becomes transparent. *The rate at which a dilute red blood cell suspension becomes transparent in solutions containing different kinds of molecules can be used as a measure of the rate at which such molecules enter the cell.* In this part of the exercise you will determine the effect of molecular weight and lipoid solubility on the permeability of cell membranes.

1. Effect of Molecular Weight

At "0" time add 2 drops of a red blood cell suspension to each of three test tubes containing 2 ml of solutions of ethylene glycol, glycerol, and glucose, respectively (Fig. 8-6A). Note: Test one solution at a time!) Mix each of the tubes by holding your finger over the tube and inverting it once or twice (Fig. 8-6B). Now hold each tube flat against a sheet of printed paper (Fig. 8-6C). The time required for the print to become plainly readable through the dilute suspension of blood cells represents the time it takes for hemolysis to occur. A stopwatch or the second hand of your watch should be used to make these measurements.

Determine the hemolysis time for each of these solutions and record your data in Table 8-2. Repeat this procedure for glycerol and glucose. Plot these data in Fig. 8-7. The molecular weights of these compounds are shown in Table 8-2. Which of the solutions appeared to enter the red blood cells the fastest and which ones entered the slowest?

___

___

TABLE 8-2
Data on effect of molecular weight on movement of materials across cell membranes.

Solution	Molecular weight	Hemolysis time
Ethylene glycol	62	
Glycerol	92	
Glucose	180	

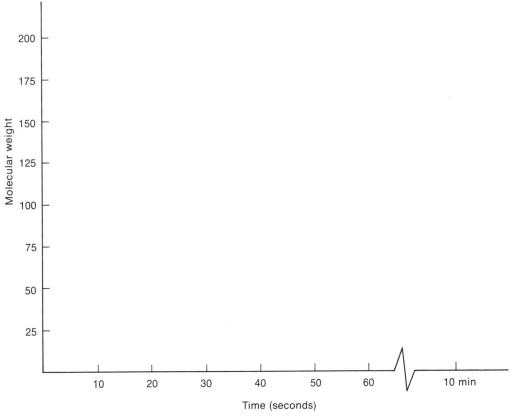

FIGURE 8-7
Effect of molecular weight on hemolysis.

How do your results compare with those of the other students in your class?

_____

_____

Why might you expect variation in hemolysis time as determined from one student to the next?

_____

_____

On the basis of your data what can you conclude about the effect of molecular weight on the permeability of cell membranes?

_____

_____

2. Effect of Lipoid Solubility on the Permeability of Cell Membranes

Nonpolar compounds are those in which electrons are equally shared by the two atoms forming a bond, as in the paraffins, olefins, and cyclic compounds. These compounds are readily soluble in fats or fat solvents, but have little solubility in water. The ratio of the solubility of a compound in lipoid substances (oil or fat) to its solubility in water is called the **partition coefficient**. Increasing lipoid solubility generally parallels increasing length of carbon chain and higher molecular weight; all three factors have been found to affect the rate at which compounds penetrate cells.

Add 2 drops of a suspension of red blood cells to each of three test tubes containing 2 ml of methyl alcohol, ethyl alcohol, and propyl alcohol, respectively (Fig. 8-6A). Mix each of the tubes and then, as before, hold them flat against a sheet of printed

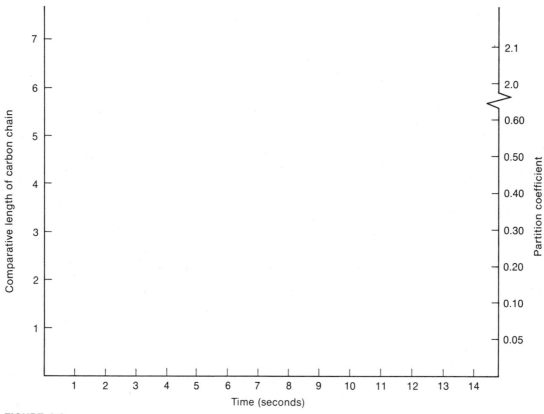

FIGURE 8-8
Effect of lipoid solubility on the permeability of cell membranes.

paper and determine the hemolysis time for each of the solutions (Fig. 8-6B and C).

Record your data in Table 8-3. Plot hemolysis time against partition coefficient and comparative length of carbon chain in Fig. 8-8. From these data, what can you conclude about the effect of the length of the carbon chain on the permeability of red cell membranes?

_____

_____

Based on these data and your knowledge of the factors determining the size of the partition coefficient, list these compounds in order of increasing partition coefficient.

_____

_____

Verify your answer by looking up the partition coefficients for these compounds in a chemistry handbook and recording them in Table 8-3.

TABLE 8-3
Effect of partition coefficient on the ability of materials to enter cells.

Compound	Formula	Partition coefficient	Hemolysis time (seconds)
3 M methyl alcohol	CH_3OH		
3 M ethyl alcohol	C_2H_5OH		
3 M propyl alcohol	C_3H_7OH		

REFERENCES

Barry, J. M. and E. M. Barry. 1969. *An Introduction to the Structure of Biological Molecules.* Prentice-Hall, Englewood Cliffs, New Jersey.

Bradshaw, L. J. 1966. Physical Concepts of Importance to Living Systems. In *Introduction to Molecular Biological Techniques.* Ch. 5. Prentice-Hall, Englewood Cliffs, New Jersey.

Butler, J. A. V. 1959. *Inside the Living Cell.* Basic Books, New York.

Edelman, I. S. 1961. Transport Through Biological Membranes. *Ann. Rev. Physiol.* **23**:37–70.

Giese, A. C. 1968. *Cell Physiology.* 3d ed. Saunders, Philadelphia.

Glynn, I. M. 1956. Sodium and Potassium Movement in Human Red Cells. *J. Physiol.*, **134**:278–310.

Holter, H. 1961. How Things Get into Cells. *Scientific American,* September (Offprint No. 96). *Scientific American* Offprints are available from W. H. Freeman and Company, 660 Market Street, San Francisco 94104, and 58 Kings Road, Reading RG1 3AA, England. Please order by number.

Loewy, A. G. and P. Siekevitz. 1969. *Cell Structure and Function.* 2d ed. Holt, Rinehart & Winston, New York.

McElroy, W. D. 1970. *Cellular Physiology and Biochemistry.* 3d ed. Prentice-Hall, Englewood Cliffs, New Jersey.

Novikoff, A. B. and E. Holtzman. 1970. *Cells and Organelles.* Holt, Rinehart & Winston, New York.

Robertson, J. D. 1962. The Membrane of the Living Cell. *Scientific American,* April (Offprint No. 151).

Rosenberg, E. 1971. *Cell and Molecular Biology.* Holt, Rinehart & Winston, New York.

Rustad, R. C. 1961. Pinocytosis. *Scientific American,* April.

Solomon, A. K. 1960. Pores in the Cell Membrane. *Scientific American,* December (Offprint No. 76).

Solomon, A. K. 1962. Pumps in the Living Cell. *Scientific American,* August (Offprint No. 131).

Young, J. H. and L. Brubaker. 1963. A Technique for Demonstrating Diffusion in a Gel. *Turtox News* **41**:274–276.

EXERCISE 9

Biologically Important Molecules: Proteins

The bulk of the dry matter of cells consists of carbon, oxygen, and hydrogen organized into large molecules of four main types: proteins, carbohydrates, lipids (fats), and nucleic acids. These molecules often combine to make even larger molecules of importance to the cell.

Proteins are large molecules that range in molecular weight from about 5,000 (insulin) to 40 million (tobacco mosiac virus protein). They are constructed of long chains of nitrogen-containing molecules known as **amino acids**, which are linked by **peptide bonds** and often conjugate with carbohydrates, lipids, nucleic acids, and heme groups. Protein molecules exhibit an unlimited variety of sizes, configurations, and kinds of physical properties because of:

- the large number of amino acids that enter into making a single protein molecule (thousands in many cases)
- the almost infinite number of combinations of different amino acids that can be formed
- the reactivity of side groups of the individual amino acids

For example, a single complex organism, such as man, is composed of thousands, possibly hundreds of thousands, of different proteins. Each protein has a special function, and by its unique chemical nature each is specifically suited to that function.

Proteins have no competitors in the diversity of roles they play in biological systems. They make up a significant portion of the structure of cells and form a large part of most cellular membranes and organelles (for example, mitochondria, ribosomes, spindle fibers, and chromosomes). In addition to their structural role, proteins act as biological catalysts (enzymes), regulate cellular and tissue functions (hormones), and fight disease (antibodies).

Because of their biological significance, it becomes important to know something about the chemistry of proteins. In this exercise you will become familiar with several tests used to identify

TABLE 9-1
Qualitative chemical reactions of amino acids and proteins.

Reagents tested	Test	Observations
0.1% egg albumin	Ninhydrin	
Distilled H_2O		
1.0% egg albumin	Sakaguchi	
Distilled H_2O		
0.1% arginine		
10 mg/ml histidine	Pauly	
10 mg/ml tyrosine		
10 mg/ml glycine		
Distilled H_2O		

the presence of protein molecules and to determine the amino acid composition of casein, a milk protein.

A. QUALITATIVE TESTS TO DETECT PROTEINS

The free terminal amino (-NH_2) and carboxyl (-COOH) groups of proteins undergo the same kinds of chemical reactions as those of the amino and carboxyl groups of free amino acids. For example, the terminal amino acid molecules of derivatives. This reaction is widely used for the derivatives; this reaction is widely used for the detection and quantitative estimation of proteins in electrophoretic and chromatographic procedures used to separate complex mixtures of molecules.

1. Ninhydrin Reaction

The reaction of ninhydrin with amino acids is of particular importance for the detection and quantitative estimation of amino acids. Ninhydrin is a powerful oxidizing agent that removes the amino groups of amino acids, liberating ammonia, carbon dioxide, the corresponding aldehyde, and a reduced form of ninhydrin. The ammonia then reacts with an additional mole of ninhydrin and the reduced ninhydrin to form a purple substance.

Add 3.0 ml of distilled water to one test tube and 3.0 ml of 0.1% egg albumin solution to a second tube. Add solid sodium acetate to each test tube (scoop-type spatula loaded to a depth of 1 inch). Add 8 drops of ninhydrin to each tube. Heat for 3 minutes in a boiling water bath, and cool. Record your observations in Table 9-1.

2. Sakaguchi Test

Alkaline solutions of proteins that contain the amino acid arginine react with α-naphthol and sodium hypobromite to produce an intense red color. This color disappears rapidly unless stabilized by the addition of urea. Thus the Sakaguchi test is a useful tool for the detection of proteins containing arginine.

Pipet 3.0 ml of distilled water into test tube 1; 3.0 ml of a 1% albumin solution into test tube 2; and 3.0 ml of a 0.1% arginine solution into test tube 3. Add 1.0 ml of 10 N sodium hydroxide to each tube, followed by 1.0 ml of a 0.02% α-naphthol solution. Add 2 drops of sodium hypobromite to tube 1, followed immediately (within 10 seconds) by 1.0 ml of a 40% urea solution. Repeat this procedure on tubes 2 and 3. Record your observations in Table 9-1.

3. Pauly Test

When the amino acids tyrosine and/or histidine are present in a protein hydrolysate (product of enzymatic hydrolysis of protein), they react in alkaline solution with sulfanilic acid to give an intense red

TABLE 9-2
Protocol for quantitative determination of protein.

Tube No.	Serum albumin 2.5 mg/ml (ml)	0.5 M KCl (ml)	Biuret reagent (ml)	T(%)
1	0.0	5.0	3.0	100
2	1.0	4.0	3.0	
3	2.0	3.0	3.0	
4	3.0	2.0	3.0	
5	4.0	1.0	3.0	
6	5.0	0.0	3.0	
Unknown No. 1	5.0	0.0	3.0	
Unknown No. 2	5.0	0.0	3.0	

color. No other amino acids react. Thus, this test is useful as a confirmation of the presence of histidine and/or tyrosine.

Pipet 2.0 ml of histidine (10 mg/ml) into tube 1; 2.0 ml of tyrosine (10 mg/ml) into tube 2; 2.0 ml of glycine (10 mg/ml) into tube 3; and 2.0 ml of distilled water into tube 4. Add 1.0 ml of sulfanilic acid reagent and 1 ml of 5% sodium nitrite to each tube. Mix and let stand for 30 minutes. Add 3.0 ml of 20% sodium carbonate to each tube, and mix. Record your observations in Table 9-1.

B. QUANTITATIVE CHEMICAL DETERMINATION OF PROTEIN

In protein molecules, the successive amino acid molecules are bonded together between the carboxyl and amino groups of adjacent molecules to form long, unbranched molecules. Such bonds are commonly called **peptide bonds**. The structures resulting from the formation of peptide bonds are called **peptides** or polypeptides and the individual amino acids are called **residues.**

Biuret, a simple molecule prepared from urea, contains what may be regarded as two peptide bonds and thus is structurally similar to simple peptides. This molecule when treated with copper sulfate in alkaline solution (the biuret reaction) gives an intense purple color. The reaction is based on the formation of a purple-colored complex between copper ions and two or more peptide bonds. Proteins give a particularly strong biuret reaction because they contain a large number of peptide bonds. The biuret reaction may be used to quantitate the concentration of proteins because peptide bonds occur with approximately the same frequency per gram of material for most proteins.

In this experiment you will determine the concentration of unknown protein solutions by measuring colorimetrically the intensity of their color production in the biuret reaction as compared with the color produced by a known concentration of the protein serum albumin.

Prepare a set of five test tubes, containing increasing amounts of a standard solution of serum albumin and 0.5 M KCl, as shown in Table 9-2. Also prepare two test tubes containing 5 ml each of two unknown protein solutions.

Add the biuret reagent last and mix thoroughly by rotating the tubes between the palms. The color is fully developed in 20 minutes and is stable for at least an hour. While waiting for the color to develop, standardize the colorimeter using tube 1, which is a blank containing all reagents except protein. Set the colorimeter at a wavelength of 540 nm (see Appendix A for instructions on the use of the Bausch & Lomb Spectronic 20 Colorimeter). After zeroing the instrument, set tube 1 at 0.00 absorbance (100% transmittance) and read tubes 2 through 5 against it (see Appendix B for a discussion of colorimetry). Record your readings in Table 9-2.

Plot the percent transmittance against the serum albumin content in each tube to obtain a standard curve. Using this curve, determine the concentration of your unknown protein solutions.

Unknown No. 1 _____

Unknown No. 2 _____

C. CHROMATOGRAPHIC SEPARATION OF AMINO ACIDS

A variety of physical procedures are available for the isolation and purification of proteins and amino acids. One of the simplest and most elegant techniques available to the biologist for the separation of the components of a mixture of molecules is **chromatography**. This technique permits the resolution of mixtures of substances by separation of the components between two phases. In **column chromatography** these phases are a solid or **stationary phase**, such as calcium carbonate or sephadex, and a **mobile phase**, a liquid solvent. In paper chromatography, the paper acts as the stationary phase and the solvent as the mobile phase. **Thin-layer chromatography** (TLC) is a recent adaptation of chromatographic methods that is designed to increase the resolution of component separations. The name of the process is derived from the thin, uniform layer of an adsorbent, such as silica gel or alumina, spread on a flat glass plate and treated in a similar manner to a paper strip. The thinness of the adsorbent layer requires less material and results in a very rapid migration of the components in a mixture. (See Appendix C for a more detailed discussion of chromatography.)

In this study you will use thin-layer chromatography to separate and identify the amino acid residues of a hydrolysate of casein, a milk protein.

1. Obtain a *clean* glass slide. Holding the slide by the sides, immerse it as far as possible into a jar containing silica gel. Swish the slide back and forth several times. Stop, and then carefully lift the slide *straight up* (Fig. 9-1).

2. Allow the slide to air-dry for several minutes. (Note: The white, silica gel coat is very fragile. Do not damage the surface.)

3. Select the side of the slide with the smoothest surface. Then, by wiping with a paper towel, remove the silica gel from the other side. (Note: Avoid excessive handling of the slide, since your hands may contaminate it with amino acids. Touch it only at the edges.)

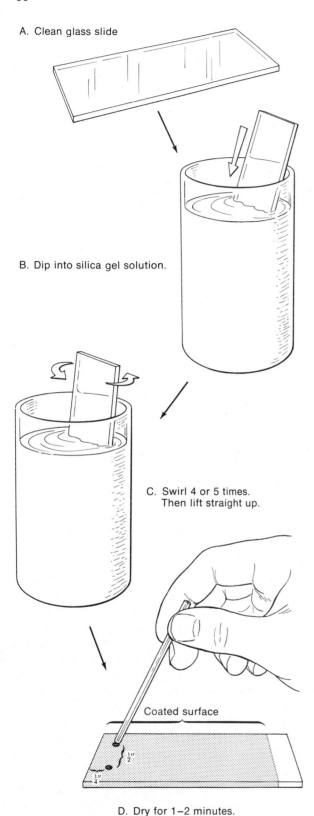

FIGURE 9-1
Preparation of plate for thin-layer chromatography.

BIOLOGICALLY IMPORTANT MOLECULES: PROTEINS

TABLE 9-3
Chromatography of amino acids.

Reagents	Distance spot moved (mm)	Color of spot	R_f value
Aspartic acid			
Glutamic acid			
Methionine			
Proline			
Tyrosine			
Histidine			
Alanine			
Lysine			
Slide No.			
Slide No.			
Slide No.			
Slide No.			
Slide No.			
Slide No.			
Slide No.			
Slide No.			

4. Lay the slide down with the coated side upward. "Spot" the coated surface at two points approximately ½ inch apart and ¼ inch from the bottom. To Spot 1, add a very small drop of casein hydrolysate, a milk protein solution. To Spot 2, add a known amino acid provided by your instructor. Other students in the class will be given other known amino acids to test.

5. Allow the spots to dry. Then carefully place the slide into a chromatography developing jar and cover. When the **solvent front** (i.e., the leading edge of the solvent) has moved to about ¼ or ½ inch from the top of the slide (30–45 minutes), remove the slide from the jar. Allow the slide to dry for 4–5 minutes.

6. In an area designated by your instructor, cautiously spray the surface of the slide with ninhydrin. *Do not inhale the fumes.* Allow the slide to dry for 2–3 minutes and then heat the slide as directed by your instructor for 2–3 minutes.

The amino acids in the hydrolysate will react with the ninhydrin and show up as colored spots on the slide. The ninhydrin test yields purple colors with most amino acids and a yellow spot with the amino acid proline. Record the color of the spots on your chromatogram in Table 9-3.

Roughly estimate the center of each amino acid spot and measure the distance it has travelled up the slide from its point of application. This distance, divided by the total distance travelled by the solvent from the origin line, is known as the R_f **value**. Two substances having the same R_f are probably identical and thus this value can be used to identify specific amino acids separated from a mixture.

Record your observations as well as those of other students who were given known amino acids other than your own in Table 9-3.

REFERENCES

Blackburn, S. 1968. *Amino Acid Determination.* Marcel Dekker, New York.

Downes, H. R. 1962. *The Chemistry of Living Cells.* 2d ed. Harper, New York.

Greenstein, J. P., and M. Winitz. 1961. *Chemistry of the Amino Acids*. 3 vols. Wiley, New York.

Haurowitz, F. 1963. *The Chemistry and Function of Proteins*. 2d ed. Academic Press, New York.

Karlson, P. 1965. *Introduction to Modern Biochemistry*. 2d ed. Academic Press, New York.

Kendrew, J. C. 1961. The Three-Dimensional Structure of a Protein Molecule. *Scientific American*, December (Offprint No. 121). *Scientific American* Offprints are available from W. H. Freeman and Company, 660 Market Street, San Francisco 94104, and 58 Kings Road, Reading RG1 3AA, England. Please order by number.

Loewy, A. G., and P. Siekevitz. 1969. *Cell Structure and Function*. 2d ed. Holt, Rinehart & Winston, New York.

Mahler, H. R., and E. H. Cordes. 1966. *Biological Chemistry*. Harper & Row, New York.

McElroy, W. D. 1970. *Cell Physiology and Biochemistry*. 3d ed. Prentice-Hall, Englewood Cliffs, New Jersey.

Meister, A. 1965. *Biochemistry of the Amino Acids*. 2d ed. 2 vols. Academic Press, New York.

Neurath, H., Ed. 1963–1965. *The Proteins*. 4 vols. Academic Press, New York.

Perutz, M. 1962. *Proteins and Nucleic Acids*. Elsevier, New York.

Randerath, K. 1963. *Thin-Layer Chromatography*. Academic Press, New York.

Stein, W. H., and S. Moore. 1961. The Chemical Structure of Proteins. *Scientific American*, February (Offprint No. 80).

Steiner, R. F., and H. Edelhoch. 1965. *Molecules and Life*. Van Nostrand, Princeton, New Jersey.

White, A., P. Handler, and E. L. Smith. 1968. *Principles of Biochemistry*. 4th ed. McGraw-Hill, New York.

EXERCISE 10

Biologically Important Molecules: Carbohydrates and Lipids

A. CARBOHYDRATES

Carbohydrates, formed when a green plant captures light energy in the process of photosynthesis, play a key role in making this captured energy available to the cell. They also supply important carbon "skeletons" (carbon atoms linked together) for the synthesis of other biologically important molecules.

The term "carbohydrate" means *hydrate of carbon*. This name is used because it includes many compounds that contain atoms of hydrogen and oxygen in the same proportion occurring in water—two hydrogens to one oxygen. Thus, a carbohydrate can be described by the general formula $C(H_2O)n$; n represents the number of $C(H_2O)$ units that make up carbohydrates ranging from relatively simple molecules, called sugars, to the complex molecules of starch and cellulose.

Most carbohydrates have a basic unit of six carbon atoms (sugars), which are linked together in various ways. On the basis of the number of these 6-carbon compounds that are linked together to make longer units, the carbohydrates are divided into three classes: **monosaccharides**, which consist of a single 6-carbon molecule (for example, glucose); **disaccharides**, which consist of two single sugar molecules linked together (sucrose is made up of glucose linked to fructose); and **polysaccharides**, which consist of three or more sugar molecules linked together (starch and glycogen are long chains of glucose molecules). Starch and glycogen are storage forms of carbohydrates; starch is usually produced in plants, and glycogen in animals.

Carbohydrates may be identified by color reactions with specific reagents. These tests can be used to determine the amount as well as the kind of carbohydrate by measuring the variations in color obtained with different concentrations of reagent. Most of the procedures require heating the carbohydrate and reagent together in a hot water bath. It is usually best to carry out a procedure on all the carbohydrates in the same water bath at the same time so that a direct comparison can be made of the relative behaviour of all of the carbohydrates in a given test.

In this study you will become familiar with some of the more common tests for detecting the presence of specific types of carbohydrates. You will be asked to identify the carbohydrate composition of an "unknown" solution containing one or more carbohydrates.

The unknown will be a solution containing carbohydrates, each present at a concentration of 6%. Only the carbohydrates you test will be found in the unknown. A portion of each unknown solution (about one-half) should be diluted to give a 1% solution, which will be used in most of the tests. *The unknowns should always be run simultaneously with the known carbohydrates.*

1. Tests for Reducing Sugars

A carbohydrate with a free or a potentially free aldehyde ($-C{\overset{O}{\underset{OH}{\diagdown\!\!\!\diagup}}}$) or ketone ($\diagup\!\!\!\!\diagdown C=O$) group is a reducing sugar and, in a solution of sufficiently high pH, can reduce weak oxidizing agents such as cupric, silver, or ferricyanide ions. For example, Cu^{++} ions react with glucose to form a colored precipitate of cuprous oxide. The color of the precipitate will range from green to reddish brown, depending on the quantity of the reducing sugar present.

$$\text{glucose} + Cu(OH)_2 \xrightarrow{\text{heat}} \underset{\text{colored}}{Cu_2O} + H_2O + \text{oxidized glucose}$$

a. Benedict's Test

Benedict's solution contains sodium bicarbonate, sodium citrate, and copper sulfate. After heating with a reducing sugar (such as glucose or fructose), the divalent copper ion (Cu^{++}) of copper sulfate ($CuSO_4$) is reduced to the monovalent copper ion (Cu^+) of cuprous oxide (Cu_2O), which forms a precipitate.

Place 5.0 ml of Benedict's reagent in a test tube and add 8 drops of a 1% solution of the carbohydrate to be tested. Heat the contents for 2 minutes over a Bunsen burner and allow to cool to room temperature. Alternatively, place the test tube in a boiling water bath for 3 minutes and allow to cool to room temperature. When several sugars are tested at one time, the latter method is preferred.

Record the color and amount of precipitate formed for each of the carbohydrates tested in Table 10-1.

TABLE 10-1
Qualitative carbohydrate tests.

Carbo-hydrates	Tests		
	Benedict's	Barfoed's	Phenylhydrazine
Glucose			
Fructose			
Galactose			
Mannose			
Xylose			
Lactose			
Maltose			
Sucrose			
Starch			
Unknown			

b. Barfoed's Test.

Barfoed's test is used to distinguish between monosaccharides and disaccharides. The reagent is similar to Benedict's reagent except that it is slightly acidic, having a pH of about 4.5. At this pH, disaccharides *will not* reduce Cu^{++} to Cu_2O, whereas monosaccharides *will* reduce the Cu^{++} when heated for 2 minutes. Longer heating of disaccharides may lead to some reduction because of the formation of monosaccharides by hydrolysis, so it is essential that all the sugars be treated in exactly the same way and that the time of appearance of a precipitate be noted.

For each carbohydrate to be tested, put 5 ml of Barfoed's reagent in a test tube and add 0.5 ml (10 drops) of a 1% solution of the carbohydrate. Mix the solutions well. Place the test tubes in a boiling water bath and heat for 2 minutes. A positive test for monosaccharides is the appearance of a red precipitate of Cu_2O within 1 or 2 minutes. Record the results in Table 10-1, noting the time of appearance of the precipitate.

c. Reactions with Phenylhydrazine

Reducing sugars react with phenylhydrazine to form phenylhydrazones. This reaction takes place

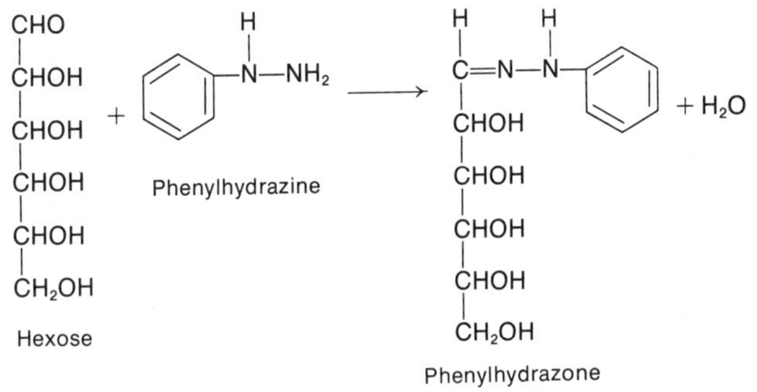

BIOLOGICALLY IMPORTANT MOLECULES: CARBOHYDRATES AND LIPIDS

FIGURE 10-1
Reaction of phenylhydrazine with hexose.

at room temperature. All of the common reducing sugars, except mannose, form phenylhydrazones that are soluble in H_2O. Mannose forms an insoluble phenylhydrazone (Fig. 10-1). Thus, this test is specific for the detection of mannose.

For each carbohydrate put about ¼ inch of solid phenylhydrazine reagent into a test tube. (CAUTION: *Phenylhydrazine is poisonous. Do not spill or get on skin.*) To each test tube add 2 ml of a 1% solution of one of the carbohydrates to be tested. Mix until the solid has dissolved and allow to stand at room temperature for 10–15 minutes. The formation of a white crystalline precipitate indicates mannose. Record your results in Table 10-1.

B. UNKNOWN CARBOHYDRATES

Using the tests in Part A, determine the carbohydrate content of the "unknown" provided by your instructor. Confirm your results with your instructor.

C. LIPIDS

Lipids are a diverse group of fatty or oily substances that are classified together because they are insoluble in water and soluble in the so-called fat solvents (for example, ether, acetone, and carbon tetrachloride). The simplest lipids (butter, coconut oil, and animal and plant fats) are composed of carbon, hydrogen, and oxygen atoms, and upon hydrolysis yield glycerol and fatty acids (Fig. 10-2). They have a higher proportion of carbon hydrogen bonds than carbohydrates and consequently release a larger amount of energy upon oxidation than other organic substances.

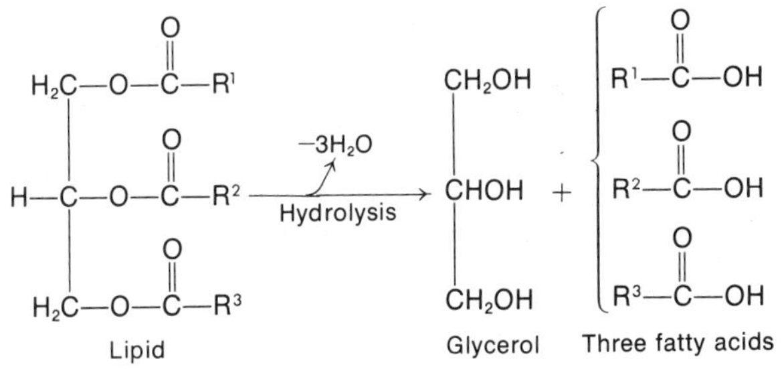

FIGURE 10-2
Hydrolysis of a lipid.

Fats, for example, release about twice the calories as an equivalent amount of carbohydrates. These molecules also form a large part of most cellular membranes, in which they control the movements of other lipids in and out of cells.

In this study you will attempt to separate a lipid extract into its component fatty acids using thin-layer chromatography (see Appendix C for a discussion of the principles of chromatography).

1. Obtain a *clean* glass slide. Holding the slide by the sides, immerse it as far as possible into a jar containing silica gel. Swish the slide back and forth several times. Stop, and then carefully lift the slide *straight up*.

2. Allow the slide to air-dry for several minutes. (Note: The white, silica gel coat is very fragile. Do not damage the surface.)

3. Select the side of the slide with the smoothest surface. Then, by wiping with a paper towel, remove the silica gel from the other side. (Note: Avoid excessive handling of the slide, since your hands may contaminate it with lipids. Touch it only at the edges.)

4. Lay the slide down with the coated side up. Spot the lipid extract on the coated surface about ½ inch from the bottom. Allow the spot to dry. Then

carefully place the slide into a chromatography developing jar and cover. When the solvent front (i.e., the leading edge of the solvent) has moved to about ¼ or ½ inch from the top of the slide (30–45 minutes), remove the slide from the jar. Allow the slide to dry for 4–5 minutes.

5. Place the slide into an iodine jar, cover, and leave until brownish-colored spots appear.

Only fatty acids migrate in the solvent used. Other components present in the extract either do not absorb on the silica gel (and will be found at the edge of the solvent front, i.e., triglycerides) or do not migrate and thus remain at the origin (e.g., phospholipids). Locate these components on your slide.

REFERENCES

Downes, H. R. 1962. *The Chemistry of Living Cells.* 2d ed. Harper, New York.

Karlson, P. 1965. *Introduction to Modern Biochemistry.* 2d ed. Academic Press, New York.

Loewy, A. G., and P. Siekevitz. 1969. *Cell Structure and Function.* 2d ed. Holt, Rinehart & Winston, New York.

Mahler, H. R., and E. H. Cordes. 1966. *Biological Chemistry.* Harper & Row, New York.

McElroy, W. D. 1970. *Cell Physiology and Biochemistry.* Prentice-Hall, Englewood Cliffs. New Jersey.

Pigman, W. W. 1957. *The Carbohydrates: Chemistry, Biochemistry, Physiology.* Academic Press, New York.

Randerath, K. 1963. *Thin-Layer Chromatography.* Academic Press, New York.

Stacey, M., and S. A. Barker. 1962. *Carbohydrates of Living Tissues.* Van Nostrand, Princeton, New Jersey.

Steiner, R. F., and H. Edelhoch. 1965. *Molecules and Life.* Van Nostrand, Princeton, New Jersey.

White, A., P. Handler, and E. L. Smith. 1968. *Principles of Biochemistry.* 4th ed. McGraw-Hill, New York.

EXERCISE 11

Photosynthesis

The living world, with few exceptions, operates at the expense of the energy captured by the photosynthetic machinery of green plants. From the products of photosynthesis, and from a small number of inorganic compounds available in the environment, living organisms are able to build up the numerous complex molecules that contribute to their cellular structure, or, in other ways, are essential to their existence. Furthermore, the ultimate source of the energy expended by living organisms is the converted energy of sunlight that is trapped within the newly synthesized organic molecules during photosynthesis.

Classically, photosynthesis is defined as:

$$6\ CO_2 + 6\ H_2O \xrightarrow[\text{Chlorophyll}]{\text{Light}} C_6H_{12}O_6 + 6\ O_2$$

Carbon dioxide — Water — Sugar — Oxygen

This equation suggests that carbohydrate synthesis is the central feature in this process. Photosynthesis, however, is not a single step reaction, as might be indicated by this equation. It is a complex process involving the interaction of many compounds. The large number of individual reactions can be divided into two groups: (1) those in which light is required (light, or photochemical, reactions) and (2) the so-called dark, or biosynthetic, reactions, so named because they do not require light for the reactions to proceed.

In the **photochemical (light) reactions**, radiant energy is used for two purposes. First, light is used to split water molecules, which then produce oxygen and hydrogen. These work to generate what has been called "reducing power." Molecules rich in reducing power can readily transfer electrons to other oxidized molecules. $NADPH_2$ (the reduced form of nicotinamide adenine dinucleotide phosphate, NADP), is the form of the reduced molecule produced in photosynthesis.

Second, light energy absorbed by chlorophyll is converted to chemical energy, which is stored in the molecule ATP (adenosine triphosphate). This conversion occurs in the chloroplast and consists of the transport of electrons from "excited" chlorophyll through a series of acceptor molecules (including the cytochromes) that constitute an electron transport system. Thus mitochondria are not the

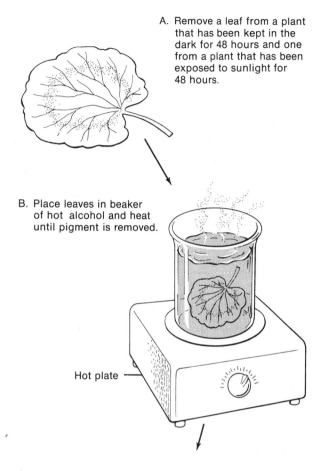

A. Remove a leaf from a plant that has been kept in the dark for 48 hours and one from a plant that has been exposed to sunlight for 48 hours.

B. Place leaves in beaker of hot alcohol and heat until pigment is removed.

Hot plate

C. Place leaves in dish containing iodine for several minutes.

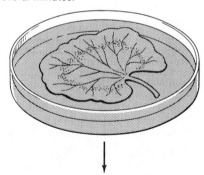

D. Remove leaves. If starch is present, leaves will become deep bluish black.

FIGURE 11-1
Procedure for determining the necessity of light for photosynthesis.

only cytoplasmic structures capable of generating ATP. This phenomenon of light-dependent generation of ATP has been called **photophosphorylation** to differentiate it from **oxidative phosphorylation**, which occurs in mitochondria.

Thus, the light reactions result in the formation of $NADPH_2$, ATP, and the release of oxygen. In the "dark" reactions, $NADPH_2$ and ATP are then used to reduce CO_2 to carbohydrate.

In the following parts of this exercise you will examine the role of light, carbon dioxide, and chlorophyll in the photosynthetic process.

A. ROLE OF LIGHT

1. Necessity of Light for Photosynthesis

There is no practical method in introductory courses to determine precisely the amount of sugar or oxygen produced during photosynthesis. However, it has been shown experimentally that during photosynthesis much of the sugar produced in the leaves is rapidly condensed into starch. Although starch is not a direct product of photosynthesis, the fact that it is a condensation product of the glucose produced during photosynthesis allows us to use its presence as indirect evidence for photosynthetic activity.

Your instructor placed several geranium plants in the dark 48 hours ago. Remove one of the plants from its cabinet and test one of the leaves for the presence of starch by the following procedure (Fig. 11-1A–D):

a. Remove the pigment by putting the leaf in hot alcohol.

b. Pour iodine over the leaf in a Petri dish. If starch is present the leaf will turn a deep bluish-black.

Similarly, test for the presence of starch in plants that were exposed to light for 48 hours.

Based upon your observations, what conclusions can be made about the necessity of light for photosynthesis?

_____

_____

_____

_____

PHOTOSYNTHESIS

TABLE 11-1
Effects of light intensity on photosynthetic activity.

Distance from light (cm)	Volume of oxygen (ml)	Average bubble count
10		
30		
50		

2. Effect of Light Intensity on the Rate of Photosynthesis

The intensity of sunlight striking the earth's surface varies from hour to hour, from day to day, and from season to season. Since oxygen is a by-product of photosynthesis, oxygen liberation may be used in designing an experiment to measure the effect of variations in light intensity on photosynthesis. In this study, light intensity is varied by placing an Elodea sprig at varying distances from a constant light source.

a. Method I

(1) Select a 1-ml pipet graduated in hundredths. Turn it upside down and place a short piece of rubber tubing over the delivery end (Fig. 11-2).

(2) Select a "healthy looking" sprig of Elodea about 15 cm in length. Insert it upside down into a large test tube filled with "pond" or "spring water" prepared as a 0.25% solution of sodium bicarbonate ($NaHCO_3$). Before completely submerging the plant, cut 2–3 mm of stem from the end opposite the growing point with a sharp razor blade, being careful not to crush the stem. If there are any leaves within a few millimeters of the cut end, remove them with a forceps.

(3) Swab the rubber tubing on the pipet with cotton moistened with 70% ethyl alcohol. Allow to dry. Suck the "spring water" into the pipet until it is full. Hold your finger over the end of the rubber tubing to prevent the water column from dropping and attach a clamp over the tubing as shown in Fig. 11-2B.

(4) Next, position the pipet gently over the cut end of the Elodea and clamp in place on a ring stand, as shown in Fig. 11-2C and D. Keep the pipet and Elodea below the level of the water. Obtain a reflector containing a 250-watt bulb and a container of cool water, and set them in the position shown in Fig. 11-2D. Why is the water container used in this setup?

_____

_____

With the Elodea plant at a distance of 50 cm, turn on the lamp and allow the setup to equilibrate for 7–10 minutes. Why?

_____

_____

Determine the total amount of oxygen given off by the plant during a 10-minute period, *by determining the amount of water in the pipet that is displaced during this time*. Figure 11-3 illustrates how this is done.

Determine the amount of oxygen produced at distances of 50, 30, and 10 cm from the light source. Enter your results in Table 11-1, and plot your data in Fig. 11-4.

b. Method II

Set up materials as in Method I. It is not necessary here, however, to use a graduated pipet. Any glass tubing that will fit closely over the cut end of the Elodea will do. In this exercise, the changes in photosynthetic rate will be measured as changes in the rate of oxygen produced *as bubbles*. Set the tube at the 50-cm distance and allow the system to equilibrate for 7–10 minutes. Then determine the rate of photosynthesis by counting the bubbles produced each minute for a 5-minute period. Calculate the *average* number of bubbles per minute and record your data in Table 11-1. Move the tube to the 30-cm distance, allow the system to equilibrate, and calculate the average number of bubbles. Repeat this procedure for the 10-cm distance. Plot your data in Fig. 11-4.

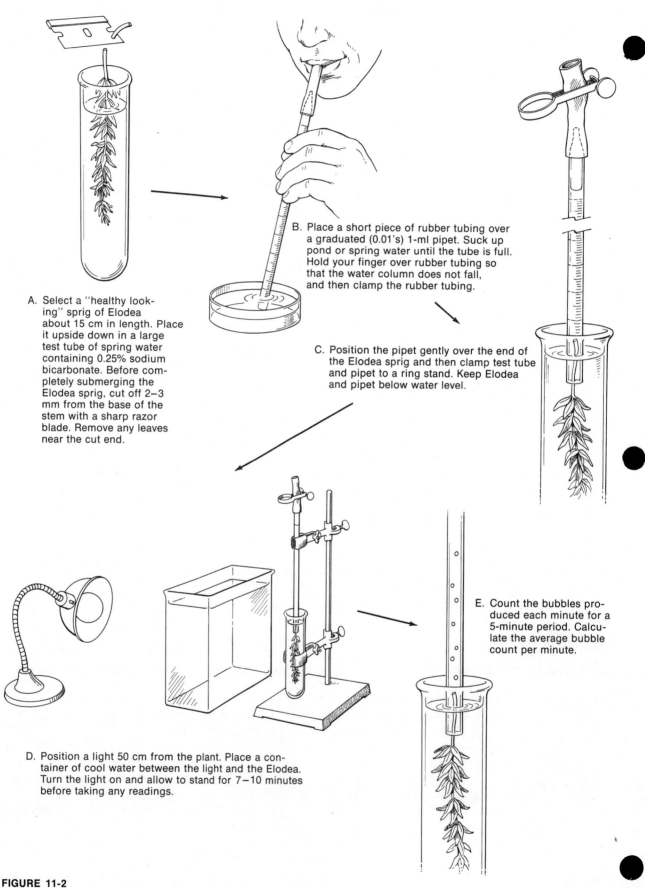

FIGURE 11-2
Procedure for determining the effect of light intensity on photosynthesis.

PHOTOSYNTHESIS

As light intensity increases, does the rate of photosynthesis (as measured in oxygen production) increase along with it? If not, what does this suggest?

_____
_____
_____

B. ROLE OF CARBON DIOXIDE

1. Necessity of CO_2 for Photosynthesis

To determine the necessity of CO_2 in photosynthesis, the apparatus in Fig. 11-5D will be used.

Select a leaf from a geranium plant that has been in the dark for 24 hours. Test the leaf for the presence of starch by immersing it in hot alcohol until it loses its green color, and then place it in a Petri dish containing iodine (Fig. 11-5A and B). Return the plant to the dark while performing the starch test. If a strong positive starch test occurs, select another plant and test the leaves until a negative or very weak starch test is obtained. Why is this step necessary?

_____
_____
_____

Select another leaf from the plant giving the negative starch test and place the leaf in the jar as shown in Fig. 11-5D.

Place a 250-watt shielded lamp near the setup, but not close enough to heat the jar. (Better results may be obtained if this setup can be placed in direct sunlight.)

What is the "control" for this experiment?

_____
_____

Run the control simultaneously with the "experimental" setup. This experiment should be allowed to run for about 24 hours, after which time you should remove the leaf and test it for photosynthetic activity in terms of starch production.

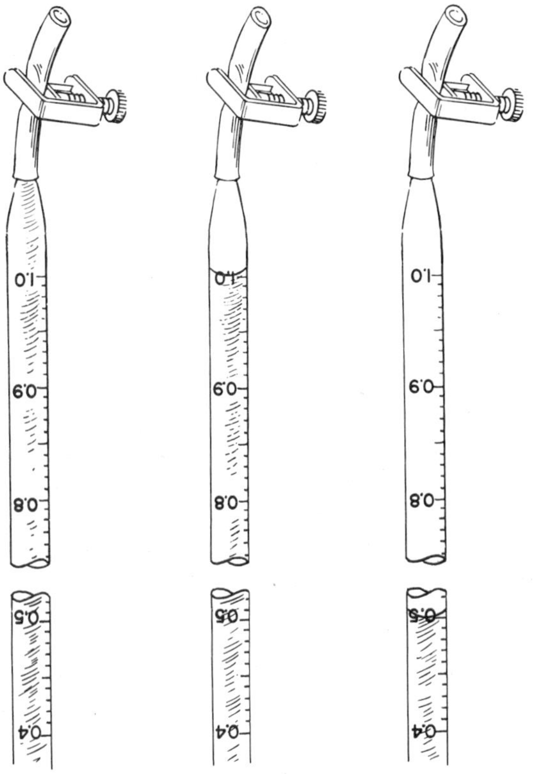

FIGURE 11-3
Method of measuring oxygen evolved during photosynthesis.

At beginning of equilibration

At start of timing (initial reading, 1.0 ml)

At end of timing (final reading) Amount of oxygen released, 0.5 ml

After setting up your experiment, examine the demonstration arranged by your instructor (Fig. 11-6).

Why is there a beaker of $Ba(OH)_2$ in the bell jar?

_____
_____

What "control" would be needed for this experiment?

_____
_____

To save time, your instructor has tested the leaves of the "experimental" and "control" plants for photosynthetic activity. Under which condition is the starch test negative?

_____
_____

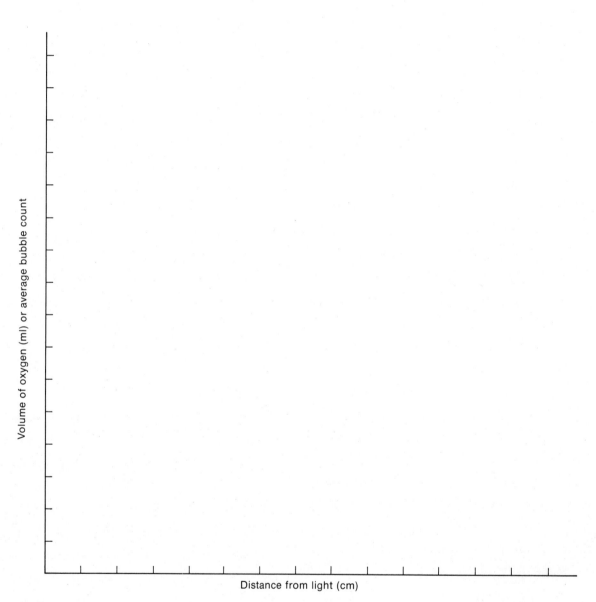

FIGURE 11-4
Effect of light intensity on photosynthetic rate.

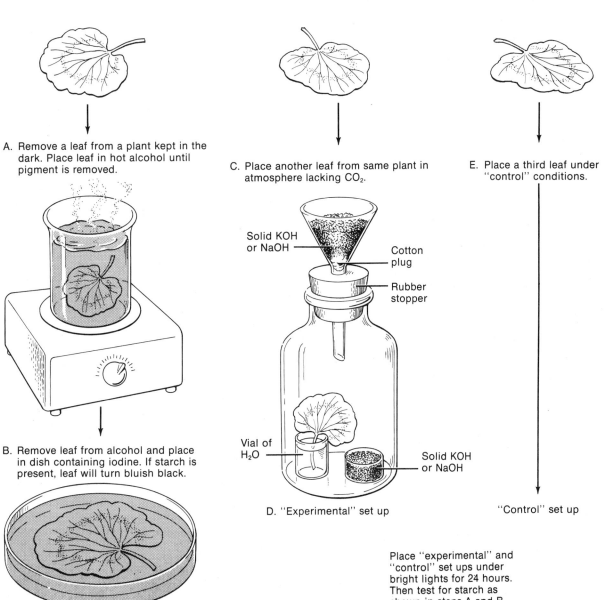

FIGURE 11-5
Procedure for determining the necessity of CO_2 for photosynthesis.

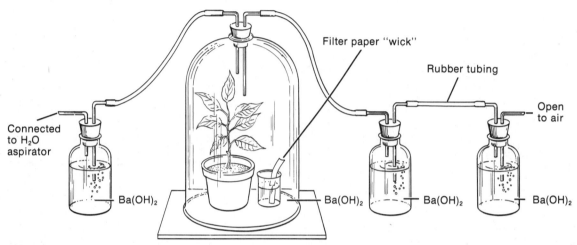

FIGURE 11-6
Necessity of CO_2 for photosynthesis.

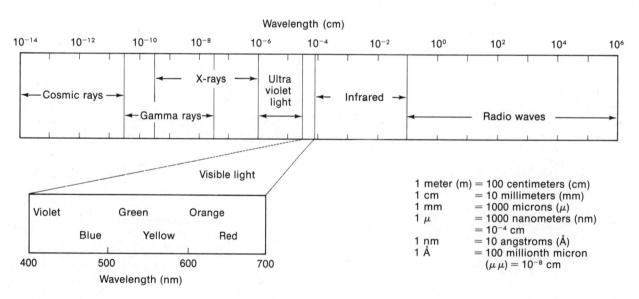

FIGURE 11-7
The electromagnetic spectrum.

PHOTOSYNTHESIS

If the results of *your* experiment do not agree with those of your instructor, suggest reasons for this difference.

_____

_____

_____

2. Uptake of CO_2 by Aquatic Plants

That CO_2 is used during photosynthesis can be demonstrated by placing an Elodea plant in a test tube containing a chemical indicator that will change color in the presence or absence of CO_2. Phenol red is a chemical indicator that is red in an alkaline solution and yellow in an acid solution. Using this information, devise and run an adequately controlled experiment showing (1) that CO_2 is taken up by Elodea and (2) that light affects the plant's ability to take up CO_2.

C. ROLE OF CHLOROPLAST PIGMENTS

Light is a type of electromagnetic radiation. Visible light is but a small part of the spectrum of electromagnetic radiation, comprising the band of wavelengths between 400 nm (nanometers) and 700 nm (Fig. 11-7).

However, since only absorbed light can transfer its energy, it is apparent that the colored components of plant cells may act to absorb visible light. Those substances that have the ability to absorb light are called **pigments**. In this part of the exercise you will determine the necessity for chlorophyll in photosynthesis, the nature of the green color of plants, and the absorption spectrum of chloroplast pigments.

1. Necessity of Chlorophyll for Photosynthesis

a. Select a leaf from a *Coleus* and a silver-leafed geranium plant. In Row 1 of Table 11-2, draw an outline of each leaf, showing the distribution of the different pigments. The obvious pigments will be the "green" chlorophyll and the "red" anthocyanin.

b. Place the leaves in a beaker of cold water for several minutes. Remove the leaves and record your observations (by drawings or written comments) in Row 2 of Table 11-2.

c. Transfer the leaves to a beaker of boiling water for several minutes. Remove, and record any changes observed in Table 11-2. Account for the differences observed between Rows 2 and 3 in Table 11-2.

_____

_____

_____

d. Next, place the leaves in hot alcohol. (CAUTION: *Use a hot water bath to heat the alcohol.*) After several minutes the leaves will become whitish in color. At this point transfer them to a Petri dish containing iodine. Swirl the dish gently. Outline the distribution of starch in each leaf in Row 4 of Table 11-2. How do these experiments demonstrate the necessity for chlorophyll in photosynthesis?

_____

_____

2. Pigment Composition of Chlorophyll

Complex mixtures of chemical substances may be separated by chromatography. The separation of the mixture is based on differences in the solubilities of the constituents of the mixture in different solvents. (See Appendix C for a discussion of the principles of chromatography.) In this study you will use the techniques of paper and thin-layer chromatography to analyze the pigment composition of chlorophyll.

In paper chromatography, filter paper is usually used to separate the mixture. In thin-layer chromatography, a gellike material is thinly applied to a support such as glass, aluminum, plastic, or cardboard. The gel can be prepared in several ways. In this study the powdered, adsorbent material is dissolved in a solvent to make a slurry. Glass slides are then dipped into the slurry and removed. Then the solvent is allowed to evaporate.

A unique advantage of using thin-layer chromatography is the speed at which separation

TABLE 11-2
The role of chlorophyll in photosynthesis.

Row	Treatment	Coleus	Silver-leafed geranium
1	None		
2	Cold H_2O for several minutes		
3	Boiling H_2O for several minutes		
4	Hot alcohol for several minutes. Place in dish with iodine.		

PHOTOSYNTHESIS

occurs. Whereas paper chromatography can require up to 24 hours to separate a complex chemical mixture, the same separation using thin-layer methods can be done in an hour.

a. Procedure I. Paper Chromatography

(1) Prepare a "chlorophyll" extract by grinding two or three spinach leaves in 5 ml of acetone (Fig. 11-8A). Adding a small quantity of quartz sand will make the grinding easier.

(2) Using a small paint brush, apply a narrow strip of chlorophyll extract to the filter paper (Fig. 11-8B). Dry thoroughly by blowing on the paper or waving it in the air. Apply the extract five or six more times. Let it dry thoroughly after each application.

(3) Place the strip in a test tube containing benzene-petroleum ether as shown in Fig. 11-8. Examine the chromatogram for the next several minutes. How long does it take for the solvent to reach the top of the paper?

_____

Describe any separation that occurs.

_____

_____

_____

b. Procedure II. Thin-Layer Chromatography

(1) Place 20 g of silica gel into a 250-ml beaker. Slowly add 50 ml of chloroform-methanol to the powder. Stir vigorously to form a slurry (Fig. 11-9).

(2) Transfer the slurry to a Coplin jar. Cover. (When not used for a long period of time, the cover should be greased to prevent evaporation.)

(3) Place two microscope slides back to back. Holding the slides at the top, insert them into the slurry to a depth of about 1 cm from the top.

(4) Remove the slides and allow to air-dry for several minutes. Separate the slides and place them on paper toweling, coated side up.

(5) In preparation for spotting the extract on the plate, place about $\frac{1}{2}$ cm of the developing solvent (isooctane: acetone: diethyl ether, 2:1:1) into a Coplin jar. Cover to allow the interior to

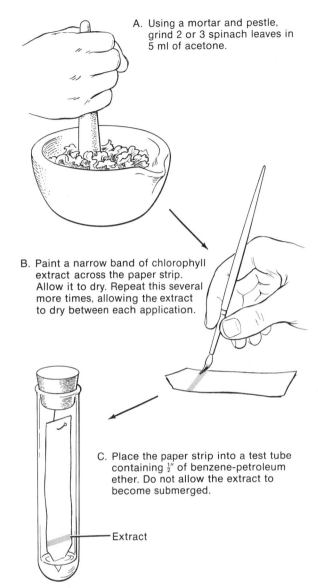

A. Using a mortar and pestle, grind 2 or 3 spinach leaves in 5 ml of acetone.

B. Paint a narrow band of chlorophyll extract across the paper strip. Allow it to dry. Repeat this several more times, allowing the extract to dry between each application.

C. Place the paper strip into a test tube containing $\frac{1}{2}$" of benzene-petroleum ether. Do not allow the extract to become submerged.

FIGURE 11-8
Procedure for separating chlorophyll pigments using paper chromatography.

become saturated with the fumes of the solvent (Fig. 11-10).

(6) Dip the narrow edge of a clean microscope slide into the previously prepared chlorophyll extract. Apply this edge to the thin-layer slide, about 1 cm from the end. Make two or three applications, keeping the band as narrow as possible. (Note: Do not break the surface of the powdered coat when applying the extract.)

(7) Insert the "spotted" slide into the Coplin jar containing the solvent. Cover the jar.

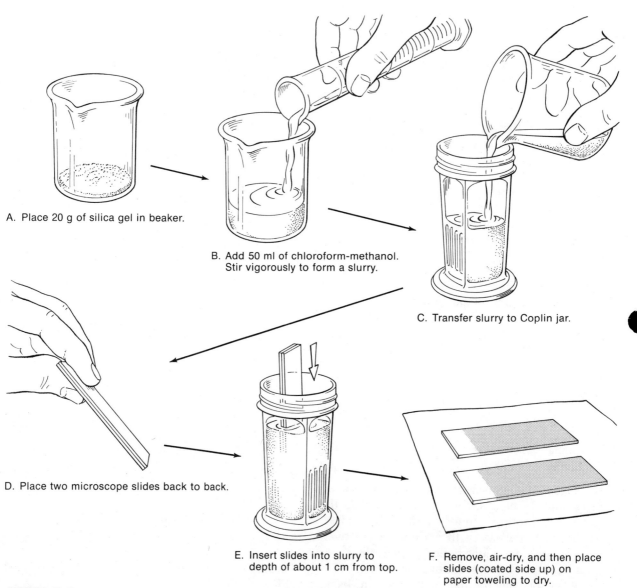

FIGURE 11-9
Procedure for preparing thin-layer chromatographic plates.

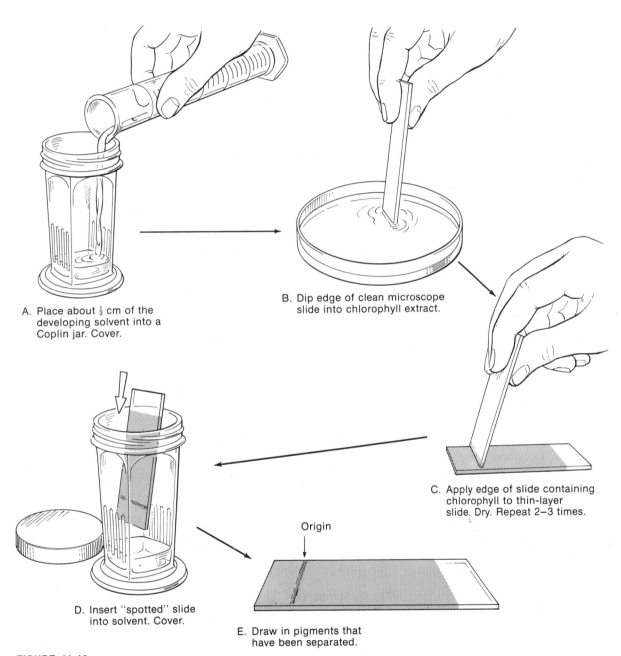

FIGURE 11-10
Procedure for thin-layer chromatography of chloroplast pigments.

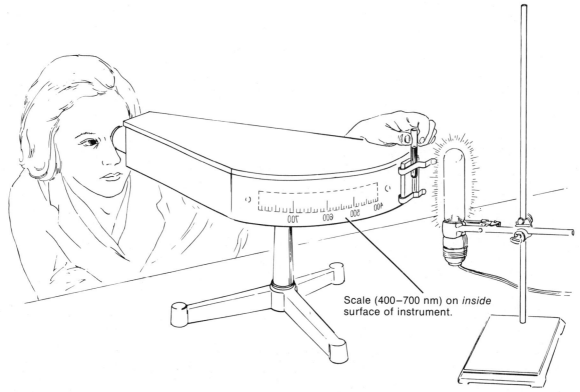

FIGURE 11-11
Spectroscopic determination of the absorption spectrum of chloroplast pigments.

Approximately how long does it take to separate the pigments, using the thin-layer as compared to the paper chromatographic method?

___

What advantages does this procedure have over paper chromatography?

___

___

3. Absorption Spectra of Chlorophyll

The portion of the visible spectrum that is absorbed by the chloroplast pigments can be determined by using an instrument that disperses visible light into its component colors or wavelengths (Fig. 11-11). The disappearance from the spectrum of various colors, or wavelengths, as a result of passing light through a pigment solution, indicates that those wavelengths were absorbed by the pigments. A plot of the ability of a substance to absorb light against the wavelength of light is called an **absorption spectrum**. The absorption spectrum for a hypothetical substance is shown in Fig. 11-12.

What wavelengths are strongly absorbed by this hypothetical substance?

___

What wavelengths are weakly absorbed?

___

In this study you will attempt to determine the absorption spectrum of a chloroplast extract using two different methods.

PHOTOSYNTHESIS

a. Method I. Spectroscopic Determination

Pipet a sample of the extract available in the laboratory into a small test tube. Your instructor will help you use the instrument for determining the absorption spectra for your extract. In the particular instrument used here, two spectra will be projected onto a scale (400 nm–700 nm) on the back, inside surface of the instrument. The upper reference spectrum shows the various colors (wavelengths) of light. The lower sample spectrum results from passage of the light through the sample. In the space below indicate the wavelengths of light that are absorbed by the chlorophyll extract.

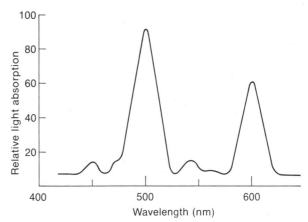

FIGURE 11-12
A sample absorption spectrum.

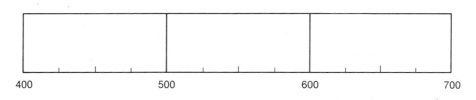

Nanometers (nm)

b. Method II. Spectrophotometric Determination

In this method an instrument called a spectrophotometer will be used to more accurately determine the absorption spectrum of the chloroplast pigments. (Refer to Appendices A and B for a description of the theory and mechanics of using this instrument.)

(1) Beginning at a wavelength of 400 nm, standardize the instrument with the acetone-ethanol solvent used to extract the chloroplast pigments. Why is this solvent used to standardize the spectrophotometer?

_____

_____

(2) Place the tube containing the extract into the sample holder and determine the percent transmittance.

(3) Remove the sample and reset the wavelength control to 425 nm. **Restandardize** the instrument to 0% and 100% transmittance and then determine the percent transmittance of the sample.

(4) Repeat the above procedure at 25-nm intervals. It will be necessary, however, to insert an accessory red filter and red sensitive phototube for determinations above 625 nm.

(5) Plot your data in Fig. 11-13. At what wavelength(s) does the chlorophyll extract absorb maximally?

_____

_____

In those plants having chlorophyll as the predominant pigment, why do the leaves appear green?

_____

_____

Since the chloroplast pigments consist of both chlorophylls and carotenoids, you cannot tell from the absorption spectrum which pigments

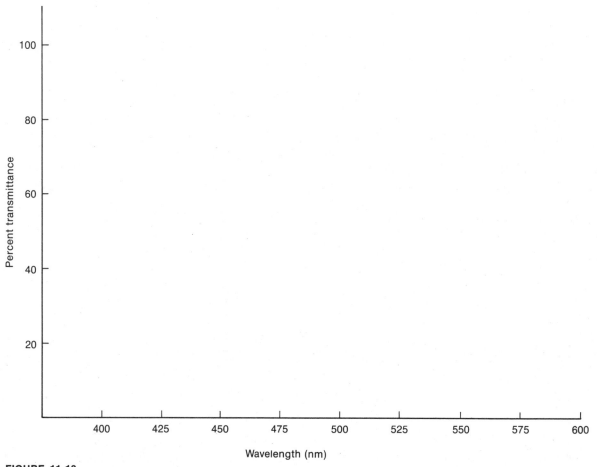

FIGURE 11-13
Absorption spectrum of chloroplast extract.

are absorbing which wavelengths. How could you determine this?

_____

_____

_____

_____

REFERENCES

Arditti, J., and A. Dunn. 1969. *Experimental Plant Physiology*. Holt, Rinehart & Winston, New York.

Arnon, D. I. 1959. Conversion of Light into Chemical Energy in Photosynthesis. *Nature* (London) **184**: 10–21:

Arnon, D. I. 1960. The Role of Light in Photosynthesis. *Scientific American*, November (Offprint No. 75). *Scientific American* Offprints are available from W. H. Freeman and Company, 660 Market Street, San Francisco 94104, and 58 Kings Road, Reading RG1 3AA, England. Please order by number.

Blass, U., J. M. Anderson, and M. Calvin. 1959. Biosynthesis and Possible Functional Relationships Among the Carotenoids; and Between Chlorophyll-a and Chlorophyll-b. *Plant Physiol.* **34**:329–333.

Bulteman, V., H. Ruppel, and H. T. Witt. 1964. Intermediary Reactions in the Water-Splitting Part of Photosynthesis. *Nature* (London) **204**:646–648.

Feldman, S. 1962. Paper Chromatography. In *Techniques and Investigations in the Life Sciences*. Ch. 4. Holt, Rinehart & Winston, New York.

Gabrielsen, E. K. 1948. Influence of Light of Different Wavelengths on Photosynthesis in Foliage Leaves. *Physiol. Plantarum* **1**:113–123.

Heftmann, E., Ed. 1967. *Chromatography.* 2d ed. Reinhold, New York.

Jensen, A., and O. Aasmundrud. 1963. Paper Chromatographic Characterization of Chlorophylls. *Acta Chem. Scand.* **17**:907–992.

Lenhoff, E. S. 1966. Separation Techniques. In *Tools of Biology.* Ch. 5. Macmillan, New York.

Levine, R. P. 1969. The Mechanism of Photosynthesis. *Scientific American,* December (Offprint, No. 1163).

Price, C. A. 1970. *Molecular Approaches to Plant Physiology.* McGraw-Hill, New York.

Rabinowitch, E. I. 1956. *Photosynthesis and Related Processes.* Vol. 2. Part 2. Interscience, New York.

Rabinowitch, E. I., and Govindjee. 1965. The Role of Chlorophyll in Photosynthesis. *Scientific American,* July (Offprint No. 1016).

San Pietro, A., and C. C. Black. 1965. Enzymology of Energy Conversion in Photosynthesis. *Ann. Rev. Plant Physiol.* **16**:155–174.

Van Norman, R. W. 1971. Chromatography. In *Experimental Biology.* 2d ed. Ch. 11. Prentice-Hall, Englewood Cliffs, New Jersey.

Wilkins, M. B., Ed. 1969. *Physiology of Plant Growth and Development.* McGraw-Hill, Maidenhead, Berkshire, England.

EXERCISE 12

Carbohydrate Metabolism

In the broadest sense, metabolism includes all those events that occur in living cells. One of the more important results of these events is the formation of new cells. Concomitant with new cell formation is the active synthesis and utilization of carbohydrates, proteins, lipids, and other complex organic materials. In this exercise we will confine our studies to carbohydrate metabolism and, specifically, to the hydrolysis and synthesis of starch.

Starch is a long-chain polysaccharide composed of glucose molecules linked together as shown in Fig. 12-1. Starches are frequently stored by plants and animals for use as a potentially available energy source. The breakdown of starch into its component "sugar units" is accomplished by hydrolysis. The important sugar unit we are concerned with is glucose because (1) it is a key compound in cellular metabolism (recall that it is synthesized from CO_2 and H_2O by photosynthesizing plants), and (2) it contains energy (found in the chemical bonds holding the molecule together) that can be released to do the work of the cell.

The hydrolysis of starch is brought about through the activity of organic catalysts called **enzymes**. Enzymes (1) are effective in minute amounts (often too small to be detected by ordinary chemical tests) because they are not used up in the reactions they catalyze, (2) are specific in the reactions they affect, and (3) do not affect the direction of the reaction but hasten the attainment of equilibrium of a reaction.

The hydrolysis of starch is accomplished by **amylases**, which cleave the starch molecule into smaller and smaller subunits until maltose, a reducing sugar, is obtained. This sequence is diagrammatically shown in Fig. 12-2.

Maltose is enzymatically converted to glucose, which may then be shunted into the glycolytic and Krebs cycles, where it is further broken down into carbon dioxide, water, and energy. This process, called cellular respiration, will be considered in another exercise.

In the experiments that follow, the rate of hydrolysis of starch by salivary amylase will be measured colorimetrically under various conditions

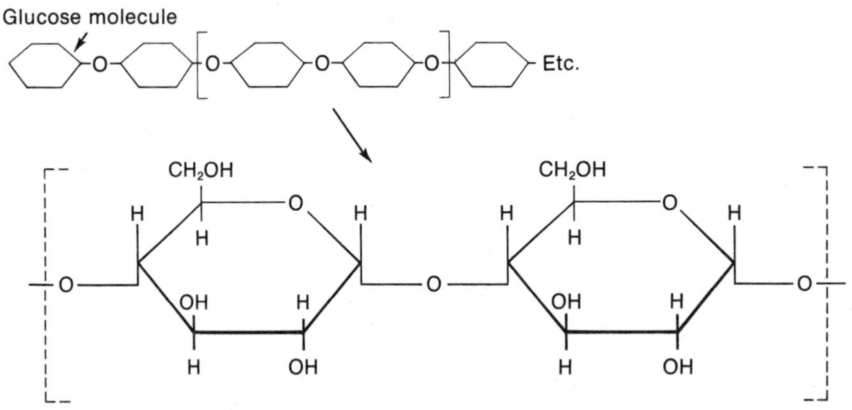

FIGURE 12-1
Structure of a starch molecule.

of temperature, pH, and enzyme and substrate concentrations. In these experiments you will first test for the presence of starch by adding several drops of iodine to the sample. If starch is present, the solution will turn a deep blue-black. If a starch solution is tested with iodine, 15–20 minutes after the addition of salivary amylase, the sample will not show any starch but will take on only the color of the added iodine. As hydrolysis occurs over a period of time, the amount of starch in the sample will be reduced. This will be reflected in the coloration of the sample upon the addition of iodine, so that if the sample is tested at various time intervals after adding the enzyme, a graded series of colors from deep blue to red to iodine color will be seen. Thus you may qualitatively measure starch hydrolysis by simply noting the color changes over a period of time. In this exercise you will quantitatively measure the rate of hydrolysis by measuring the amount of light that the sample absorbs when placed in a colorimeter (see Appendix B for a discussion of the principles of colorimetry).

A. EFFECT OF TEMPERATURE ON THE ACTIVITY OF SALIVARY AMYLASE

Collect 10 ml of saliva in a clean test tube and dilute with an equal amount of tap water. (You can stimulate the flow of saliva by chewing a small piece of paraffin.) Mix the saliva and water thoroughly and filter through a double layer of cheesecloth into a small beaker. This is your stock enzyme solution. If additional enzyme is needed, dilute whatever amount is collected with an equal volume of water.

Into each of five Erlenmeyer flasks (125-ml) pour 50 ml of the starch solution. Immerse these containers in large beakers of water or water baths adjusted to temperatures of 5°, 15°, 30°, 45°, and 70°C. This may be accomplished by adding ice water or hot water to the beakers. The temperatures in the beakers should be maintained throughout the experiment and should not vary more than ±3°C.

While waiting for the temperature to equilibrate in the beakers, prepare a test-tube rack containing five rows of 10 to 12 colorimeter tubes each. Label the first tube in each row to correspond with the five temperature values used.

Turn on the colorimeter (see Appendix A for instructions in the use of the Bausch & Lomb Spectronic 20 colorimeter) and allow the instrument to warm up for 5 minutes. Check the meter needle to make sure it records 0% transmittance at a wavelength of 560 nm (nanometers).

Prepare an "iodine control" tube by adding 3 drops of iodine to 5 ml of water in another colorimeter tube. Place the iodine control in the colorimeter, close the cover, and adjust the light control until the needle records 100% transmittance. Why is this step required?

_____

_____

_____

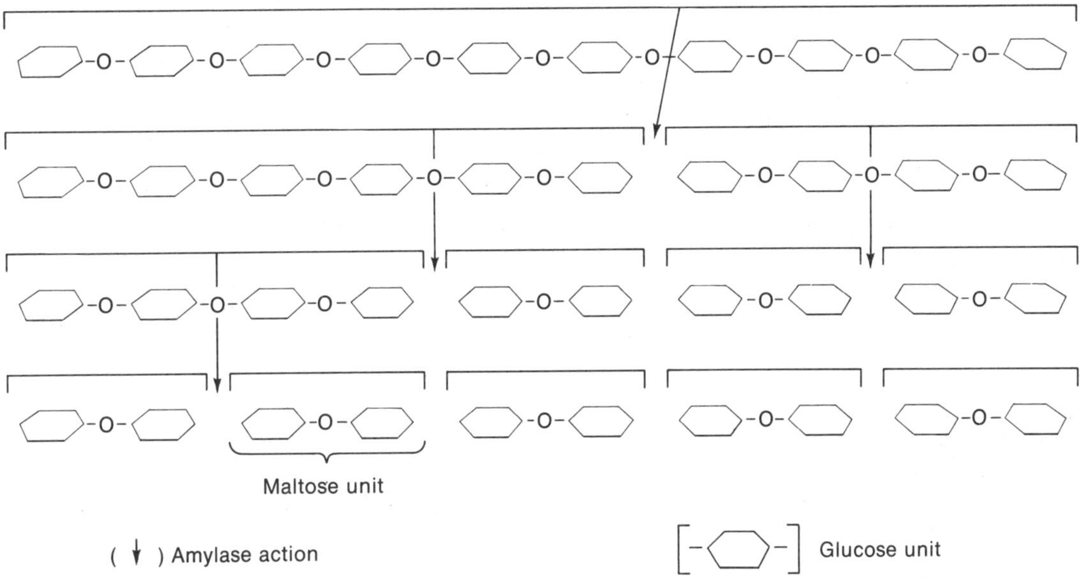

CARBOHYDRATE METABOLISM

FIGURE 12-2
Action of amylase on starch.

Is this iodine control used here really adequate? If not, how would you improve on it?

_____

_____

_____

_____

When the starch solution in the flasks has reached the temperature of the water bath (verify this by placing a thermometer *in the starch solution*), add 1 ml of enzyme solution to each flask, starting with that at 5°C. Note the time you add the enzyme to each flask. Mix each flask thoroughly immediately upon adding the enzyme, and replace it in the water bath.

Two minutes after the addition of the enzyme, quickly transfer 5 ml of the contents of each flask to the corresponding first colorimeter tube in each series. *Five milliliters will fill about one-half of the colorimeter tube.* Immediately test for the presence of starch by adding 3 drops of iodine to each colorimeter tube. *Mix thoroughly.* (At the 70°C temperature, cool the tube under the tap *before* adding iodine.) Repeat this procedure *at 2-minute intervals* for the remaining tubes of each series. Be prepared to test for starch at longer, or shorter, intervals of time if necessary.

In all probability, even if the starch has been completely digested, you may not obtain the 100% transmittance given by the iodine control. Why?

_____

_____

_____

For these experiments a reading of 90% or better will indicate that hydrolysis is complete. Establish the length of time required to reach complete hydrolysis at each temperature. Record this time in the appropriate space in Table 12-1. Plot your

TABLE 12-1
Effect of temperature on starch hydrolysis.

	Temperature (°C)				
Hydrolysis time (minutes)	5	15	30	45	70

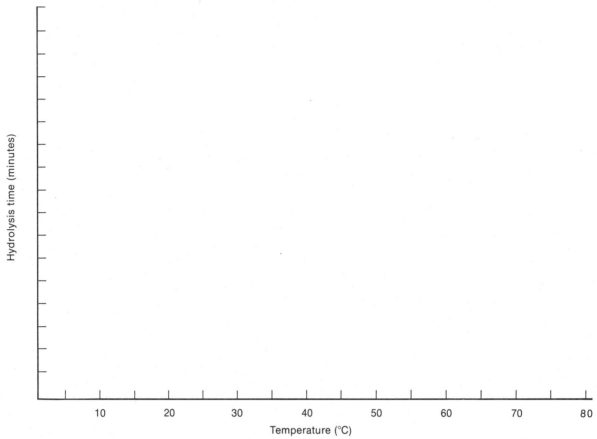

FIGURE 12-3
Effect of temperature on the activity of salivary amylase.

data in Fig. 12-3. What is the effect of temperature on the activity of salivary amylase?

_____
_____
_____
_____
_____

B. EFFECT OF pH ON THE ACTIVITY OF SALIVARY AMYLASE

Prepare a test-tube rack containing four rows of 10 to 12 colorimeter tubes each. Label the first tube in each row to correspond to pH values of 5, 6, 7, and 8.5, respectively. To each of four Erlenmeyer flasks (125-ml) add 25 ml of buffer solutions at a pH of 5.0, 6.0, 7.0, and 8.5, respectively. Into each flask pipet 0.5 ml of enzyme solution. *Mix thoroughly* by swirling each flask. Beginning with the tube of lowest pH, add 25 ml of starch solution to each flask. (Note the time!) *Thoroughly mix* the contents immediately following the addition of the starch.

Two minutes after adding the starch to the first flask, quickly transfer approximately 5 ml (recall that 5 ml fills the colorimeter tube about half full) from each flask to the first tube in each corresponding series. Test for the presence of starch, using 3 drops of iodine. Record colorimeter readings as quickly as possible after the addition of iodine.

Repeat this procedure at *2-minute intervals* for the remaining tubes of each series.

Establish the length of time required for complete hydrolysis of the starch (colorimeter readings of 90% or greater indicate complete hydrolysis). Record this time in Table 12-2, and plot your data

CARBOHYDRATE METABOLISM

TABLE 12-2
Effect of pH on enzyme activity.

Hydrolysis time (minutes)	pH			
	5	6	7	8.5

in Fig. 12-4. What is the optimum pH for the activity of salivary amylase?

_____

Are your results consistent with your knowledge of where this enzyme functions in the body? Explain.

_____

_____

_____

_____

C. EFFECT OF SUBSTRATE CONCENTRATION ON THE ACTIVITY OF SALIVARY AMYLASE

In this study you will determine the effect of altering the amount of substrate (starch) available to the enzyme.

Prepare a test-tube rack containing four rows of 10 colorimeter tubes each. Label the first tube in each row 1:2, 1:4, 1:8, and 1:16, corresponding to their respective dilutions of starch.

Label four Erlenmeyer flasks to correspond to the dilutions indicated, and add 25 ml of distilled water to each flask. To the first flask (dilution of 1:2) add 25 ml of starch solution. *Mix thoroughly* and transfer 25 ml of this mixture to the second flask (dilution 1:4). *Mix thoroughly* and transfer 25 ml of the mixture to the third flask (dilution 1:8). *Mix thoroughly* and transfer 25 ml of the mixture to the last flask (dilution 1:16). *Mix thoroughly.* Remove and discard 25 ml of the last mixture.

Using a 1-ml pipet, add 0.1 ml of enzyme solution to each flask and *mix thoroughly* by swirling each flask. Note the time.

Two minutes after adding the enzyme, transfer a sample from each flask so that the colorimeter tube is one-third full. Test for the presence of starch. Establish the time required for hydrolysis to occur in each case and record this value in Table 12-3. Plot your data in Fig. 12-5. Interpret your results in terms of the amount of substrate available to a constant amount of enzyme.

_____

_____

_____

D. EFFECT OF ENZYME CONCENTRATION ON THE ACTIVITY OF SALIVARY AMYLASE

Prepare a test-tube rack containing four rows of 10 colorimeter tubes each. Label the first tube in each row 1:5, 1:25, 1:125 and 1:625 to correspond

TABLE 12-3
Effect of substrate concentration on the activity of salivary amylase.

Hydrolysis time (minutes)	Substrate concentration			
	1:2	1:4	1:8	1:16

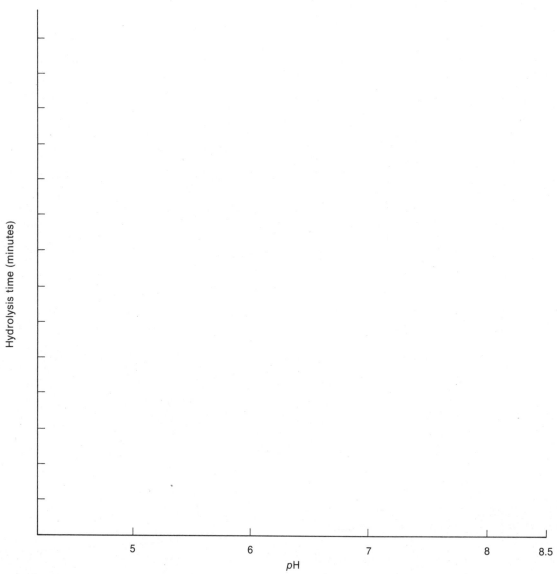

FIGURE 12-4
Effect of pH on enzyme activity.

to the enzyme dilutions that will be used in this experiment.

Next, set up four Erlenmeyer flasks (125-ml), each containing 20 ml of distilled water. Label each flask to correspond to the dilution series above. Add 5 ml of the enzyme solution to the first flask (1:5 dilution). *Mix thoroughly* and transfer 5 ml to the second flask (1:25 dilution). Repeat this procedure for the third and fourth flasks (1:125 and 1:625 dilutions, respectively), *thoroughly mixing each flask before each transfer.*

Next add 25 ml of starch solution to each flask, starting with the first flask. Note the time.

Two minutes after adding the starch, remove approximately 5 ml of solution from each flask (about one-half of the colorimeter tube), transfer it to the first colorimeter tube in each series, and immediately test for the presence of starch.

Repeat this procedure at *2-minute intervals.*

Colorimetrically determine the hydrolysis time for each concentration series and enter these values in Table 12-4.

CARBOHYDRATE METABOLISM

TABLE 12-4
Effect of enzyme concentration on the activity of salivary amylase.

Hydrolysis time (minutes)	Enzyme concentration			
	1:5	1:25	1:125	1:625

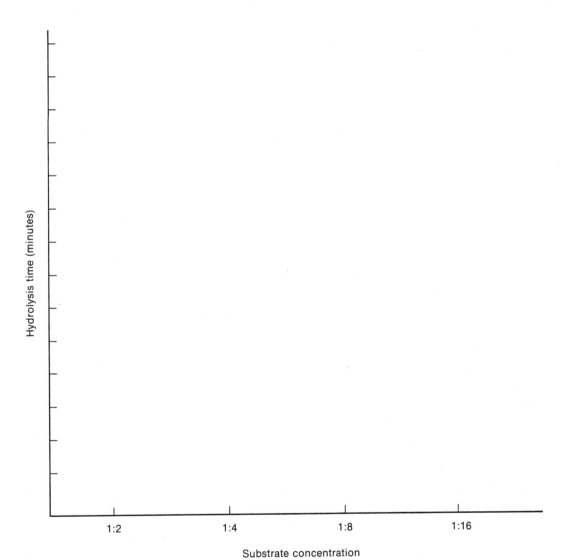

FIGURE 12-5
Effect of substrate concentration on the activity of salivary amylase.

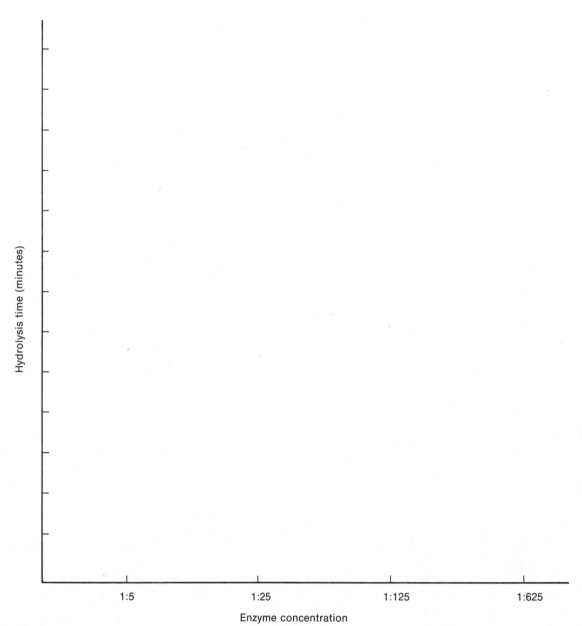

FIGURE 12-6
Effect of enzyme concentration on the activity of salivary amylase.

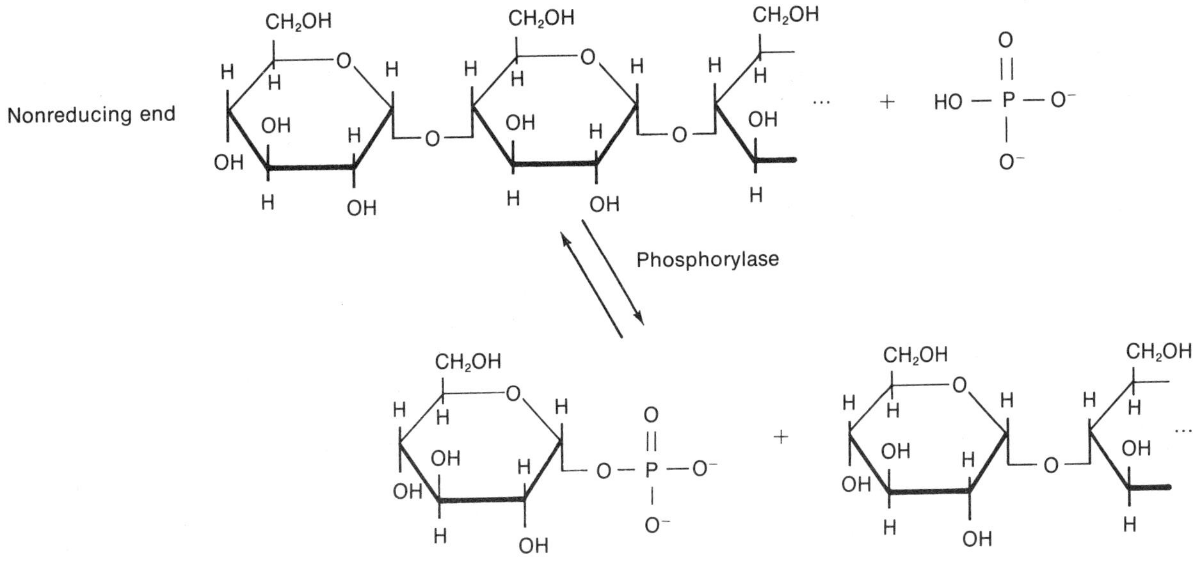

CARBOHYDRATE METABOLISM

FIGURE 12-7
Action of phosphorylase. (From *Experimental Biochemistry* edited by John M. Clark, Jr. W. H. Freeman and Company. Copyright © 1964.)

Plot your data in Fig. 12-6. Interpret your results.

_____
_____
_____
_____
_____

E. ENZYMATIC SYNTHESIS OF STARCH

In the first part of this exercise you examined the hydrolysis of starch by a class of enzymes called amylases. However, there are other enzymes that also function in this reaction. These enzymes, called **phosphorylases,** degrade glycogen or starch more rapidly than any of the known amylases.

The activity of the two classes of enzymes differs in a number of ways. The differences in their rate have already been mentioned. A second difference is that amylases cleave this polysaccharide into maltose units that require a second enzyme, called **maltase,** to complete the conversion to glucose. Phosphorylases cleave polysaccharides into glucose-phosphate units, which are further hydrolyzed to glucose and phosphoric acid through the activity of the enzyme **phosphatase.**

The most important difference in these enzymes, however, lies in the reversibility of the phosphorylase reaction, which goes readily in either direction; the amylase reaction is almost irreversible (Fig. 12-7). Explain the difference in the reversibility of these two reactions.

_____
_____
_____
_____
_____

In this experiment you will attempt to show that the phosphorylase reaction is reversible if the following conditions are met: (1) phosphorylase activity requires the presence of inorganic phosphate; (2) the reaction proceeds from glucose-1-phosphate and not glucose; (3) there must be present a small amount of primer or starter polysaccharide (this enzyme adds glucose units to the ends of preexisting polysaccharide chains).

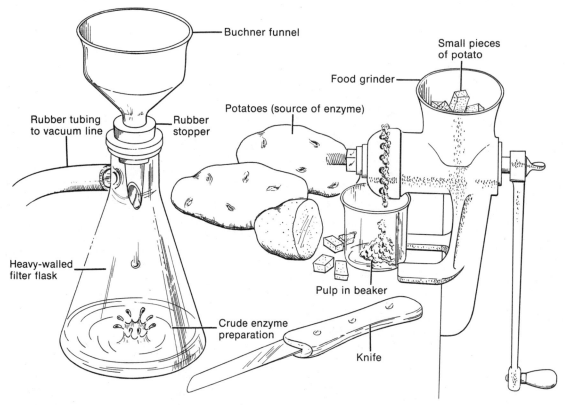

FIGURE 12-8
Preparing the enzyme solution for synthesizing starch. (Adapted from *Plants in Action* by Leonard Machlis and John G. Torrey. W. H. Freeman and Company. Copyright © 1956.)

1. Preparation of Enzyme

Phosphorylases are known to occur in many plant and animal tissues. An easily available and inexpensive source of this enzyme is potatoes. Before preparing the active enzyme, however, set up and number a series of test tubes, as shown in Table 12-5. If enzyme is called for, it is to be added immediately after it is prepared, since this particular enzyme denatures rapidly. Peel and cube a small potato, and place the cubes immediately into a beaker containing 30–40 ml of 0.01 N neutralized potassium cyanide (KCN). (CAUTION: *KCN is a poison!*) It is used here to inhibit the activity of potato phosphatase. Why?

Transfer the cubes to a food grinder and grind to pulp (Fig. 12-8). Filter the pulp through a Buchner funnel and use the extract immediately.

2. Enzymatic Synthesis of Starch by Phosphorylase

Place the enzyme preparation into the tubes called for in Table 12-5. Mix the reagents thoroughly. Immediately after adding the enzyme, remove a small sample from each tube and test it with a drop of iodine on a spot plate. Repeat the iodine test at 3-minute intervals thereafter. The reaction should go to completion within 30 minutes. Record your observations in Table 12-5. What is the role of KH_2PO_4 (potassium dihydrogen phosphate)?

CARBOHYDRATE METABOLISM

TABLE 12-5
Enzymatic synthesis of starch by phosphorylases.

Tube No.	Reagents	Observations
1	3 ml of 0.01 M glucose, a small drop of 0.2% starch solution, and 3 ml of the enzyme preparation	
2	3 ml of 0.01 M glucose-1-phosphate, a small drop of 0.2% starch solution, and 3 ml of the enzyme preparation	
3	3 ml of 0.01 M glucose-1-phosphate and 3 ml of the enzyme preparation	
4	3 ml of 0.01 M glucose-1-phosphate, a small drop of 0.2% starch solution, and 3 ml of the boiled enzyme preparation	
5	3 ml of 0.01 M glucose-1-phosphate, 1 ml of 0.2 KH_2PO_4, a small drop of 0.2% starch solution, and 3 ml of the enzyme preparation	
6	3 ml of 0.2% starch solution, 1 ml of 0.2 M KH_2PO_4, and 3 ml of the enzyme preparation	
7	3 ml of 0.2% starch solution, 1 ml of 0.2 M KH_2PO_4, and 3 ml of boiled enzyme preparation	

What role does each of the test tubes play in this experiment?

_____

_____

_____

_____

What would have happened if the sodium cyanide were left out of the enzyme preparation?

_____

_____

REFERENCES

Awapara, J. 1968. *Introduction to Biological Chemistry.* Prentice-Hall, Englewood Cliffs, New Jersey.

Barry, J. M., and E. M. Barry. 1969. *An Introduction to the Structure of Biological Molecules.* Prentice-Hall, Englewood Cliffs, New Jersey.

Downes, H. R. 1962. *The Chemistry of Living Cells.* 2d ed. Harper, New York.

Florkin, M., and E. Stotz, Eds. 1963. *Comprehensive Biochemistry.* Vol. 5. Carbohydrates. American Elsevier, New York.

Fruton, J. S., and S. Simmonds. 1959. *General Biochemistry.* 2d ed. Wiley, New York.

Giese, A. C. 1968. *Cell Physiology.* 3d ed. Saunders, Philadelphia.

Hassid, W. Z. 1960. Biosynthesis of Complex Saccharides. In *Metabolic Pathways.* D. M. Greenberg, Ed. Vol 1. Academic Press, New York.

Lehninger, A. L. 1965. *Bioenergetics.* Benjamin, New York.

Lehninger, A. L. 1960. Energy Transformation in the Cell. *Scientific American,* May (Offprint No. 69). *Scientific American* Offprints are available from W. H. Freeman and Company, 660 Market Street, San Francisco 94104, and 58 Kings Road, Reading RG1 3AA, England. Please order by number.

Leloir, L. F., C. E. Cardini, and E. Cabib. 1960. Utilization of Free Energy for the Biosynthesis of Saccharides. In *Comparative Biochemistry.* M. Florkin and H. S. Mason, Eds. Vol. 2. Academic Press, New York.

Loewy, A. G., and P. Siekevitz. 1969. *Cell Structure and Function* 2d ed. Holt, Rinehart & Winston, New York.

McElroy, W. D. 1960. *Cellular Physiology and Biochemistry.* Prentice-Hall, Englewood Cliffs, New Jersey.

Nielands, J. B., P. K. Stumpf, and R. Y. Stanier. 1958. *Outlines of Enzyme Chemistry.* 2d ed. Wiley, New York.

Pigman, W., Ed. 1957. *The Carbohydrates. Chemistry, Biochemistry, Physiology.* Academic Press, New York.

Soskin, S., and R. Levine. 1946. *Carbohydrate Metabolism.* University of Chicago Press, Chicago.

White, A., P. Handler, and E. L. Smith. 1968. *Principles of Biochemistry.* 4th ed. McGraw-Hill, New York.

EXERCISE 13

Cellular Respiration

Some of the energy of sunlight, through the process of photosynthesis, is converted to chemical energy in glucose molecules. During respiration most of the energy in the glucose molecule is released and becomes available to carry out the various processes essential to maintaining the "living" state. The glucose molecule is broken down and its energy released in three stages: glycolysis, the Krebs cycle, and the electron transport chain.

During glycolysis the 6-carbon sugar molecule is converted, through a complex series of enzyme-controlled reactions, to two 3-carbon molecules of pyruvic acid. In the presence of molecular oxygen, pyruvic acid enters another series of enzyme-mediated reactions called the Krebs cycle. These latter reactions occur in the mitochondria.

Before entering the Kreb's cycle, pyruvic acid is "activated," losing a molecule of carbon dioxide and picking up a compound called coenzyme A. This new complex is called acetyl-CoA. In the first step of the Krebs cycle, coenzyme A is removed and the remaining 2-carbon acetyl group combines with oxaloacetic acid, a 4-carbon acid, to produce citric acid. During subsequent steps the 6-carbon citric acid is broken down to produce carbon dioxide, water, and another molecule of oxaloacetic acid, which combines with acetyl-CoA to start the cycle over.

More important, however, energy in the form of ATP is generated during the Krebs cycle. This is accomplished as a result of a series of oxidation-reduction reactions occurring in the electron transport chain. During this process hydrogen atoms, removed from Krebs-cycle intermediates, are split into positively charged protons and negatively charged, "high energy" electrons. The electrons are transported through a series of molecules that are alternately oxidized (lose electrons) and reduced (gain electrons). During these oxidation-reduction reactions some of the energy of the electrons is incorporated into ATP. Finally, at the end of the chain the free protons combine with the now "low energy" electrons and oxygen to form water.

Through this process of cellular respiration the cell is able to capture about 55% of the chemical energy within the glucose molecule. In its energy conversion the cell shows a greater efficiency than an automobile motor, which is able to convert only about 25% of the energy in gasoline into energy used to propel the car. The use of glucose by the cell may be simply outlined as follows:

changes occurring in the respiration chamber.

In this study you will determine respiratory rate in terms of oxygen intake.

1. Following the diagram in Fig. 13-1A, fill the respiration test tube half full with germinating peas.

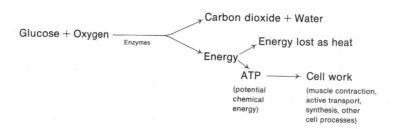

There are several ways in which respiration taking place in an organism can be demonstrated. These include measurements of energy given off in the form of heat, the amount of glucose used, the amount of oxygen consumed, and the amount of carbon dioxide released. In the following parts of the exercise you will study respiration using several of these approaches.

A. AEROBIC RESPIRATION

Many of the methods used in the study of respiration depend upon the measurement of changes in the volume or pressure of gas. Any variation in the volume or pressure within a closed system in which an organism is respiring represents the net difference between oxygen consumption (which would decrease pressure and volume in the closed container) and carbon dioxide production (which would increase pressure and volume). However, if the carbon dioxide produced is absorbed in some way, any changes could be attributed to oxygen consumption.

A simple respirometer that can be used to detect changes in gas pressure and volume is shown in Fig. 13-1. This equipment consists of two vessels that can be closed to the outside. Respiring material is placed in one of the containers, along with potassium hydroxide (KOH), an agent used to absorb carbon dioxide. Because gas volume is influenced by such physical factors as atmospheric pressure and temperature, the second container (identical except for the living material) is employed as a compensation chamber. Any changes in the volume of gas in the compensation chamber must be taken into consideration when evaluating

2. Place a loose wad of cotton over the peas. Place about ½ inch of potassium hydroxide (KOH) pellets over the cotton. The cotton keeps the KOH from contacting the living seeds; it should not be packed tightly. KOH is a substance that will remove the CO_2 from the atmosphere of the tube as fast as it is given off by respiring peas. Why is it necessary to remove the CO_2 from the tube?

_____

_____

_____

3. Prepare the compensation tube in the same way, except use glass beads in place of the peas. Why is it necessary to place the inert glass beads into the compensation tube?

_____

_____

4. Insert a rubber stopper, with attached capillary tubing, firmly into each tube (Fig. 13-1B).

5. Place the tubes in a vertical position by clamping to a ring stand (Fig. 13-1C). Using an eyedropper, add enough dye to the end of each capillary tube so that about ½ inch of the dye will be drawn into the tube (Fig. 13-1D).

6. After allowing 2–3 minutes for the gas pressures to reach equilibrium, note the position of the outer end of the dye column on the millimeter scale (Fig. 13-1E). Record this initial reading in Table 13-1. (Note: Since the respirometer is very sensitive to volume changes due to heat, keep it away from any heat sources such as lamps, hot

CELLULAR RESPIRATION

plates, etc.) Then attach pinch clamps to the rubber tubing on each test tube.

7. Take readings of the location of the column at 1-minute intervals for the next 5 minutes. Record your data in Table 13-1. (Note: If the movement of the dye is fairly rapid, you must be prepared to take your readings at shorter intervals—20 or 30 seconds—or the column may reach the bent portion of the tube before you have enough readings to complete your data.) The dye column can be returned to the outer end of the tube by opening the pinch clamp and tilting the capillary tube.

Why did the dye move toward the respiration chamber and not away?

___

___

Under what circumstances might the dye move away from the chamber?

___

___

8. Repeat the procedure to determine the effects of temperature on respiration. Students should select different temperatures in order to reflect a variety of temperature effects.

9. Plot the data tabulated in Table 13-1 in Figs. 13-2 and 13-3. Identify each line on the graph with the appropriate label.

B. BIOLOGICAL OXIDATION

As previously described, oxidation-reduction reactions are important in releasing the energy in glucose and incorporating this energy into ATP molecules. Iron-containing compounds called **cytochromes** take part in these reactions, which are carried out in the mitochondria. In these reactions oxidation takes place in the presence of oxygen.

Cells can also accomplish oxidation in the absence of oxygen by removing hydrogen (and electrons) from an oxidizable substrate. Oxidation proceeds as shown in reaction (1):

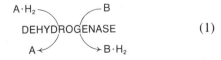

(1)

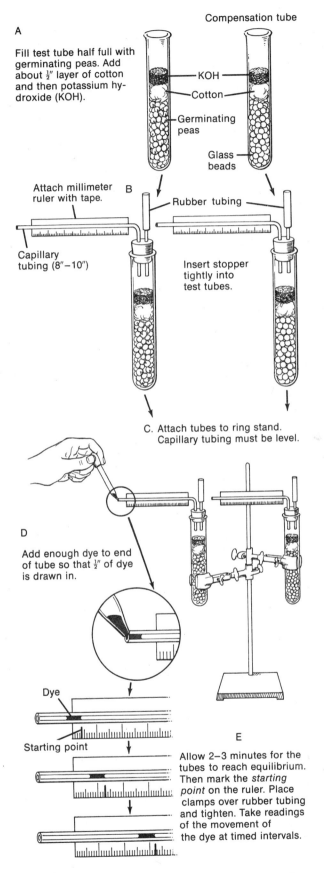

FIGURE 13-1
Procedure for measuring oxygen consumption in germinating peas.

TABLE 13-1
Respiration data.

Time (minutes)	Respirometer readings (in mm)					
	Germinating peas (Room temp: ____°C)			Germinating peas (____°C)		
	Respiration tube (1)	Compensation tube (2)	Corrected data (1 minus 2)	Respiration tube (1)	Compensation tube (2)	Corrected data (1 minus 2)
0						
1						
2						
3						
4						
5						

"$A \cdot H_2$," the hydrogen donor, is called the **reductant**; "B," the hydrogen acceptor is called the **oxidant**. Thus every oxidation must be accompanied by a simultaneous reduction. The energy required for the removal of hydrogens in oxidation reactions is supplied by the accompanying reduction. In order for this type of reaction to proceed, specific enzymes called **dehydrogenases** are required.

Dehydrogenases, called **oxidases**, use oxygen as a hydrogen acceptor. Most oxidases are conjugated enzymes whose prosthetic groups contain metals, such as copper, iron, or zinc, and a riboflavin-containing complex. The transfer of hydrogen from the substrate (the reductant) to oxygen by the oxidase usually results in the formation of hydrogen peroxide (H_2O_2). See reaction (2) below. Although H_2O_2 is toxic to the tissues, its toxic effect is prevented by two important enzymes: **peroxidase** and **catalase**.

1. Peroxidase

This enzyme combines with H_2O_2, activating it to donate its oxygen to act as a hydrogen acceptor for another substrate molecule, as shown in reaction (2):

$$A \cdot H_2 \xrightarrow{\text{OXIDASE}} O_2 \xrightarrow{\text{PEROXIDASE}} B \cdot H_2 \quad (2)$$
$$A \leftarrow H_2O_2 \rightarrow B$$

guaiacol (reduced) + H_2O_2 $\xrightarrow{\text{peroxidase}}$ guaiacol (oxidized colored complex) + $2H_2O$ (3)

Since guaiacol, a phenolic compound, is oxidized to a colored product by H_2O_2 in the presence of peroxidase, reaction (3) can be used as a measure of enzyme activity. The rate of enzyme activity can be determined by measuring the amount of light the sample absorbs when placed in a colorimeter.

With a paring knife, peel a turnip and cut it into small cubes; place these in a homogenizer and grind for 30 seconds. Or place the tubes into a mortar containing silica sand, and grind to pulp with a pestle (Fig. 13-4). Filter the homogenate through a

CELLULAR RESPIRATION

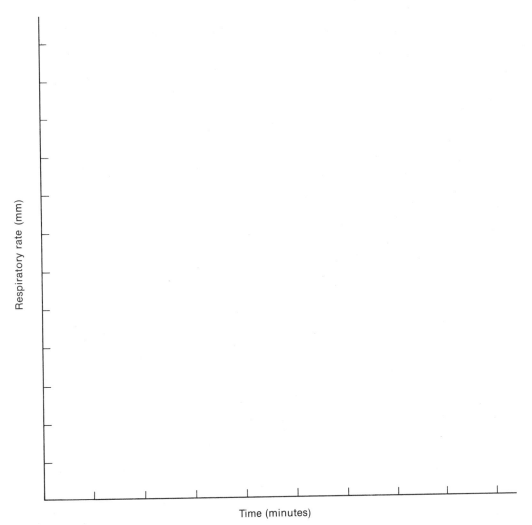

FIGURE 13-2
Respiratory rate of germinating peas at room temperature.

double layer of cheesecloth into a beaker. Squeeze out as much of the liquid as you can. Dilute 1 ml of the juice to 200 ml with distilled water.

Add 0.01 ml of liquid guaiacol, together with 0.2 ml of 0.9% H_2O_2 and 4.7 ml of distilled water, to a test tube. Put 1.0 ml of diluted turnip extract and 4 ml of water into a colorimeter tube. Pour the guaiacol-H_2O_2 solution into the colorimeter tube. To mix, pour the mixture quickly back into the empty test tube and then back into the colorimeter tube. Wipe the colorimeter tube clean and immediately place it in the Spectronic 20 colorimeter. The wavelength to be used is 500 nanometers (nm). What solution should you use to standardize the colorimeter for this test?

Start the stopwatch and read the percent transmittance with the wavelength set at 500 nm. Take readings every 20 seconds and record the data in Table 13-2. Plot these readings against time in Fig. 13-5. Repeat the determination, using one-half and then twice as much extract. Also measure the effect of extract that, immediately before assay, has been placed in a bath of boiling water for several minutes and then cooled. Also, test the effect of adding 0.5 ml of 0.01 M sodium cyanide (CAUTION: Poison!) to the reaction mixture.

From these data, what can you conclude about peroxidase activity in turnip tissue?

TABLE 13-2
Peroxidase activity in turnip tissue.

Time (seconds)	Percent transmittance				
	1 ml turnip extract	0.5 ml turnip extract	2.0 ml turnip extract	1 ml boiled turnip extract	1 ml extract +0.5 ml NaCN
20					
40					
60					
80					
100					
120					

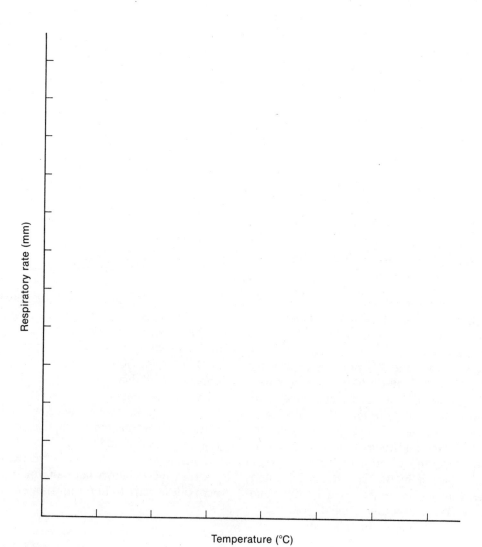

FIGURE 13-3
Effect of temperature on respiratory rate.

CELLULAR RESPIRATION

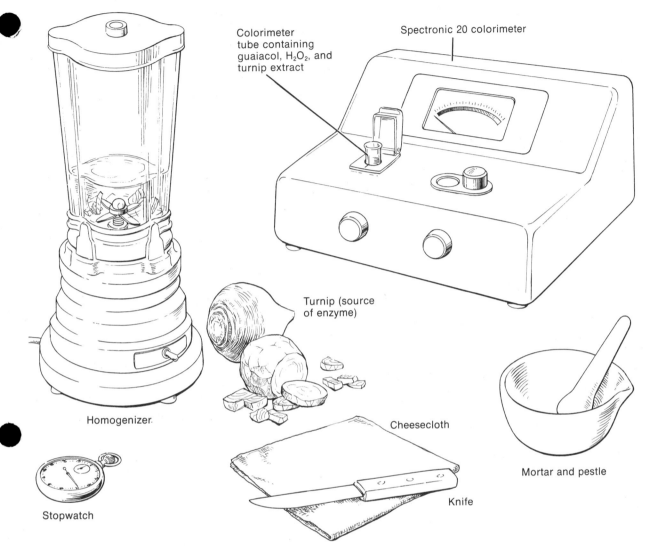

FIGURE 13-4
Apparatus and procedure for extracting the enzyme peroxidase.

Why is peroxidase activity affected by heating?

_____

_____

What is the mechanism by which peroxidase activity is affected by such compounds as sodium cyanide?

_____

_____

_____

_____

2. Catalase

Catalase contains an iron-containing prosthetic group identical with that of blood hemoglobin. This enzyme specifically catalyzes the breakdown of H_2O_2 into water and oxygen. The catalytic action of this enzyme is the most rapid enzyme-catalyzed process known (one molecule of catalase can catalyze the decomposition of 44,000 molecules of peroxide per second).

The assay for catalase to be used in this experiment is based upon the measurement of oxygen liberated by the enzyme action. However, the volume of oxygen released is not measured as such. Rather, the buoyant effect of liberated oxygen on paper discs saturated with the enzyme solution is measured, indirectly, by recording the time

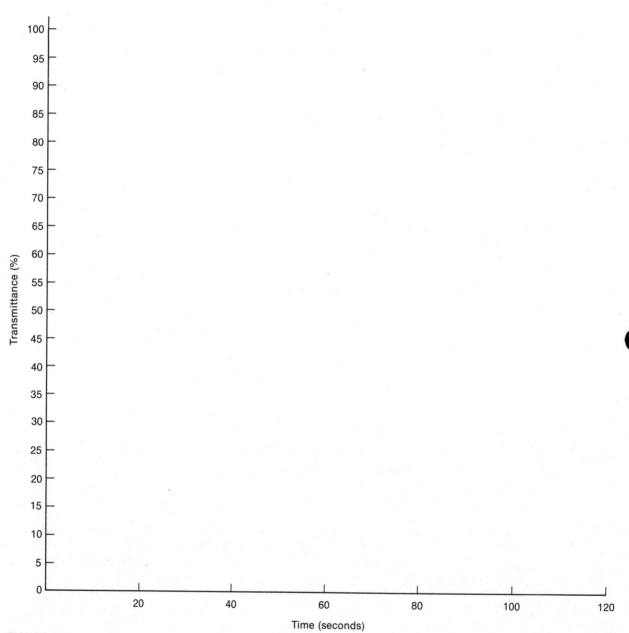

FIGURE 13-5
Peroxidase activity in turnip tissue.

CELLULAR RESPIRATION

for the disc to rise to the surface of a solution of H_2O_2.

a. Preparation of a Catalase Standard Curve

Add 10 ml of 3% H_2O_2 to a clean 100-ml beaker. Dip paper discs ($\frac{1}{2}$ inch in diameter) held by sharply pointed forceps into a 0.01% catalase solution until they are saturated with enzyme (Fig. 13-6). Then immediately drop the disc into the beaker of peroxide solution. With a stopwatch record the time that elapses between the moment the disc touches the surface of the H_2O_2 (as it falls toward the bottom of the beaker) and the time when it again reaches the surface. Record the time to the nearest one-half second. Note the bubbles of oxygen that form about the disc as it floats on the solution. Repeat the experiment with the following dilutions of enzyme: 0.00250, 0.00100, 0.00025, and 0.00010% catalase. Record the data in Table 13-3. Using these data, plot the concentration of catalase against disc flotation time in Fig. 13-7. This will represent the standard curve from which you can calculate the concentration of catalase in various tissues and body fluids.

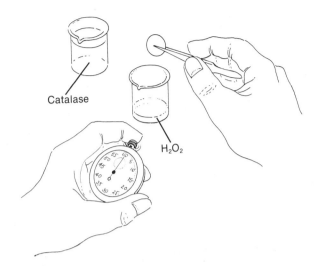

FIGURE 13-6
Measuring disc flotation times of various catalase concentrations.

TABLE 13-3
Disc flotation times at various catalase concentrations.

Catalase concentration (%)	Disc flotation time (seconds)
0.01000	
0.00250	
0.00100	
0.00025	
0.00010	

b. Determination of Catalase Activity in Blood

Prick your finger and collect enough blood to fill a capillary tube. Spin down the blood sample in a hematocrit centrifuge. Break the tube at about the level of the surface of the packed cells, and blow the packed cells into a colorimeter test tube containing 1 ml of distilled water. The volume of cells will approximate 0.02 ml, since you have diluted them about 50 times. Assay the plasma and diluted red blood cells for catalase activity, as described above.

A control should be carried out, using discs soaked with distilled water or buffer. The time required for the control disc to rise is used as a reference point for the detection of catalase activity. Beyond this time the disc flotation is not related to the reaction times nor the presence of the enzyme.

Calculate the concentration of catalase present in the plasma and red blood cells, and record these values (in milligrams of catalase per gram of original material) in Table 13-4.

Also assay the activity of catalase in several unknown tissues provided by your instructor. Record the data in Table 13-4.

TABLE 13-4
Catalase activity of tissues.

Sample tested	Disc flotation time (seconds)	Catalase concentration (mg/g tissue)
Control (distilled water)		
Red blood cells		
Plasma		
Tissue No. 1		
Tissue No. 2		
Tissue No. 3		
Tissue No. 4		
Tissue No. 5		

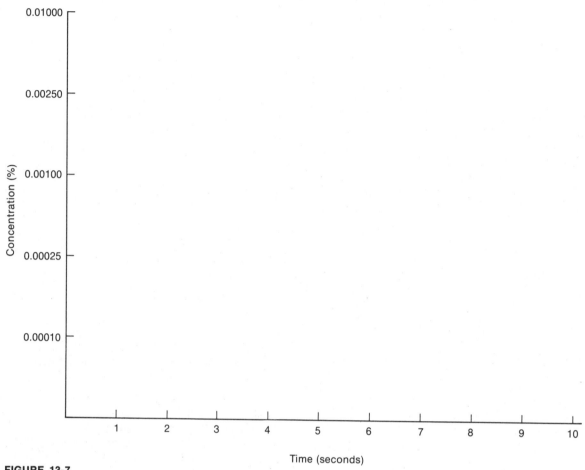

FIGURE 13-7
Standard curve for catalase.

3. Enzyme Inhibition

With a capillary pipet, saturate a sodium-azide-treated disc with the 0.01% standard catalase solution. Wait about 1 minute and test for catalase activity. Repeat the experiment with discs saturated with sodium cyanide, sodium fluoride, and mercuric chloride. Record your observations in Table 13-5.

From these studies what can you conclude about the effect of these substances on the activity of catalase?

TABLE 13-5
Enzyme inhibition.

Samples tested	Disc flotation time (seconds)
Control	
0.01% catalase	
0.01% catalase + sodium azide	
0.01% catalase + sodium cyanide	
0.01% catalase + sodium fluoride	
0.01% catalase + mercuric chloride	

REFERENCES

Arditti, J., and A. Dunn. 1969. *Experimental Plant Physiology.* Holt, Rinehart & Winston, New York.

Baldwin E. 1967. *Dynamic Aspects of Biochemistry.* 5th ed. Cambridge University Press, New York.

Clark, J. M., Jr. 1964. *Experimental Biochemistry.* W. H. Freeman and Company, San Francisco.

Fruton, J. S., and S. Simmonds. 1959. *General Biochemistry.* 2d ed. Wiley, New York.

Goldsby, R. A. 1967. *Cells and Energy.* Macmillan, New York.

Hayashi, O. 1962. *The Oxygenases.* McGraw-Hill, New York.

Lehninger, A. L. 1965. *Bioenergetics.* Benjamin, New York.

Prosser, C. L., and F. A. Brown. 1961. *Comparative Animal Physiology.* 2d. ed. Saunders, Philadelphia.

Schmidt-Nielsen, K. 1964. *Animal Physiology.* 2d ed. Prentice-Hall, Englewood Cliffs, New Jersey.

White, A., P. Handler, and E. L. Smith. 1968. *Principles of Biochemistry.* 4th ed. McGraw-Hill, New York.

Zeuthen, E. 1953. Oxygen Uptake as Related to Body Size in Organisms. *Quart. Rev. Biol.* **28**:1–12.

EXERCISE 14

Transport in Biological Systems

The movement of water, metabolic wastes, food, hormones, minerals, gases, and many other materials within the organism is extremely important to higher plants and animals. Almost every metabolic activity of the organism depends ultimately upon the exchange of materials between cells and between the organism and its environment.

Nutrient molecules must be absorbed, body wastes removed, and oxygen obtained to support cellular respiration. A variety of substances, including such diversified materials as chemical regulators (hormones) and defensive materials (antibodies), must be exchanged if the organism is to function coherently. Basically, exchange depends upon physical processes such as diffusion, osmosis, and pinocytosis and biochemical processes such as active transport. Since these phenomena can affect rapid movement over short distances, very small organisms (i.e., protozoa) or organisms with most of their cells in contact with the environment (sponges) rely almost entirely on them to achieve exchange (Fig. 14-1A).

Larger and more complex organisms require supplementary systems. With only a small fraction of their cellular mass in direct contact with the environment, exchanges based solely on diffusion would not occur rapidly enough to meet their needs. In larger organisms, cellular exchanges take place in an internal environment (the extracellular fluid system), and a means must be developed for the transport of these fluids throughout the organism. Exchange between the organism and the environment then takes place through the body wall or some specialized part of it. These mechanisms constitute the transport system of plants and the circulatory systems of animals.

A. TRANSPORT IN ANIMALS

Circulatory systems in animals vary enormously in complexity. In the simplest cases movement of extracellular fluids is achieved by the compressive action of body wall muscles or by ciliary activity

A. Direct exchange between cells and environment

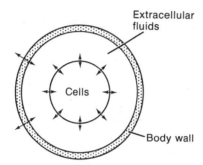

B. Circulation of extracellular fluids in the absence of a defined circulatory system

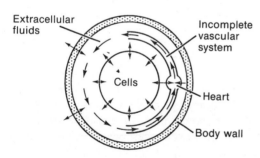

C. Open circulatory system

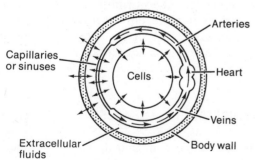

D. Closed circulatory system

FIGURE 14-1
Types of exchange and circulation systems.

(Fig. 14-1B). Usually, however, some type of pump-like heart is formed. The pump may consist simply of a group of pulsatile vessels, as is the case in the earthworm, or it may take the form of a simple membranous chamber, as in insects. In many molluscs and vertebrates the heart is a complex, chambered structure. In any event, the evolution of a specialized heart is normally associated with the development of vessels that direct the flow of extracellular fluids to specific areas of the body. In insects and some other invertebrates such vessels form incomplete systems, since the fluids they carry are emptied into irregular cavities and return to the heart through the body spaces (Fig. 14-1C). Such circulatory systems are said to be "open" in type, and there is no distinction between the fluids within and outside the heart and vessels.

In vertebrates the circulatory system is completely enclosed in a continuous series of vessels. The blood carried in these vessels is distinctly different from the extracellular fluid in many respects (Fig. 14-1D). Blood flow takes place within this tubular system, and exchange between the blood and extracellular fluids occurs through the smallest vessels (capillaries) and irregular cavities (sinuses).

In this exercise you will investigate certain aspects of capillary circulation and heart action. Before beginning these experiments see Appendix D for a discussion of the proper use of live animals in the laboratory.

B. PERIPHERAL CIRCULATION

Select a large, lightly pigmented frog, and anesthetize it until it is limp by immersion in tricaine methane sulphonate solution. Remove the animal and place it ventral side up on a cork board. Extend and secure the limbs to the board with pins, spreading the forelimbs as far apart as possible (Fig. 14-2A).

With a sharp scalpel make an incision through the skin just lateral to the ventral midline, extending it as indicated in Fig. 14-2A. Make lateral extensions of this incision and reflect the skin fold thus formed, tucking it securely under the animal's body. With forceps, lift the abdominal wall and force the points of a pair of closed scissors through it. *Do not force the points into the underlying viscera.* Extend the scissor points to make an anterior-posterior hole in the body wall.

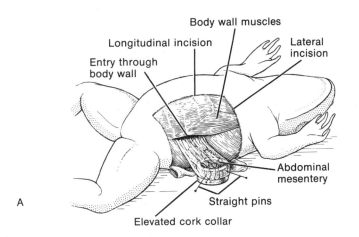
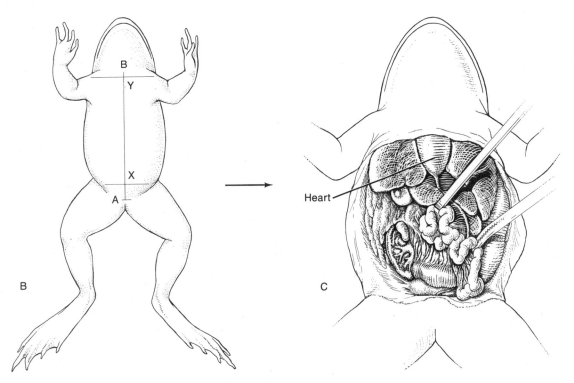

FIGURE 14-2
Dissection of frog to expose (A) the abdominal mesentery and (C) the heart.

With forceps, *gently* draw a fold of abdominal mesentery through the incision. The cork board will be provided with a hole and an elevated cork collar. Draw the fold over the hole, being careful not to stretch the mesentery. Secure the fold to the collar with pins.

Place paper toweling moistened with tap water around the body. Be sure that the exposed mesentery is kept damp with Ringer's solution at all times. Should the animal show signs of recovery, moisten the paper toweling with anesthetic.

Observe the mesentery with the low-power lens of a compound microscope or with a stereoscopic microscope at a magnification of no less than 40×.

1. Blood Flow Through Arteries, Veins, and Capillaries

Observe the blood vessels in the mesentery. Note their relative diameters and wall thicknesses. Observe the direction and velocity of flow through the vessels. How might you distinguish between arteries, veins, and capillaries?

_____

_____

_____

Pick out a single, fairly large artery and observe the red blood cells moving through it. Describe any differences in flow velocity in the central and peripheral regions of the vessel.

_____

_____

_____

What might cause such differences?

_____

_____

What consequences will this have, and how is the circulatory system adapted to compensate for this?

_____

_____

_____

_____

Examine a bed of capillaries, using a higher magnification if necessary. Compare the internal diameters of several capillaries with those of the red blood cells passing through them. What is the relation between the diameters of the red blood cells and the capillaries?

_____

_____

What significance might this relation have with regard to exchange processes?

_____

_____

_____

Does the red blood cell appear to be a rigid or flexible structure?

_____

What advantages and disadvantages does this have?

_____

_____

Do all the capillaries remain open at all times, or do some close periodically?

_____

_____

What might be the functional significance of this?

_____

_____

2. Factors Influencing Blood Flow

The flow of blood is influenced by a variety of factors, most of which operate to ensure that blood

TRANSPORT IN BIOLOGICAL SYSTEMS

is delivered at adequate rates to each section of the body. In the following sections you will observe the effects of temperature and of the hormones adrenalin and acetylcholine.

With an eyedropper apply chilled Ringer's solution to the mesentery. Record the sequence of changes that occur.

_____

_____

_____

_____

After all changes have ceased, bring the mesentery back to its original temperature by washing it with Ringer's solution at room temperature. Next apply Ringer's solution heated to 40°C. (Note: *Do not use solutions warmer than* 45°–50°C.) Record in detail the sequence of heat-induced changes.

_____

_____

_____

_____

After washing the mesentery with Ringer's solution at room temperature, use a pin to place *one* drop of 1:50,000 adrenalin in Ringer's solution on the area under observation. Note and record the changes.

_____

_____

_____

Wash the mesentery thoroughly with Ringer's solution. Then, as before, place 1 drop of 1:50,000 acetycholine in Ringer's solution on the observed area, noting and recording the changes.

_____

_____

_____

_____

What do your observations suggest regarding possible modes of control of blood flow through various parts of the body?

_____

_____

_____

Carefully replace the mesentery into the abdominal cavity before proceeding with Part C.

C. HEART ACTION AND CONTROL

With a sharp scissors make a midline cut from A to B as shown in Fig. 14-2B. Extend the incision forward to the shoulder girdle. Lift upward and forward with each cut to avoid damaging the underlying organs. Cut through the bones of the shoulder girdle. Then at points X and Y cut laterally and pull back the skin to expose the internal organs (Fig. 14-2B and C). Use clean filter paper to sponge up any blood in the area.

Note that the heart is enclosed in a membrane, the **pericardium**. Observe the volume of the heart during its diastolic (expansion) and systolic (contraction) phases. Carefully slit the pericardium, and note any changes in volume during **diastole** and **systole**. There are two anterior chambers and a posterior third chamber; the anterior chambers are the **atria**. The single posterior chamber is the **ventricle**. Note any differences in the texture of its wall relative to those of the atria. Observe the groove marking the boundary between the atria and ventricle. This is the **atrioventricular groove**. Observe the **truncus arteriosus**, a thick-walled tube that arises dorsally from the ventricle and divides into two **aortic arches**. With finely nosed forceps, lift the ventricle by its apical tip and reflect it forward. Note the narrow V-shaped **sinus venosus**, which opens into the right atrium.

Finally, observe the main veins, which return blood to the heart. The right and left **superior venae cavae**, or caval veins, enter the sinus venosus from the anterior part of the body. A single **inferior vena cava** or caval vein empties into the posterior part of the sinus venosus.

Carefully observe the sequence of contraction of all parts of the heart and the veins and truncus arteriosus. List the order in which contraction occurs.

_____

_____

With the second hand of a watch, or with a stopwatch, determine the number of contractions per minute made by the atria, ventricle, truncus arteriosus, and sinus venosus.

      left atrium   _____/minute
      right atrium   _____/minute
      ventricle   _____/minute
      truncus   _____/minute
      sinus   _____/minute

Are the rates equal or different? What may you conclude from this?

_____

_____

_____

_____

Observe and record the rate of contraction of the heart: _____/minute. Decerebrate the frog as outlined in Appendix D, Part C. After 5 minutes again observe and record the rate of contraction: _____/minute. Destroy the midbrain and hindbrain areas by pithing the animal as described in Appendix D, Part B. After 5 minutes, observe and record the rate of contraction: _____/minute.

Finally, insert a dissecting needle into the part of the vertebral column containing the spinal cord, and complete the destruction of the central nervous system by forcing the needle down as far as it will go. After a recovery period of 5 minutes, observe and record the rate of contraction of the heart: _____/minute.

Do your observations suggest that the central nervous system plays a role in governing heart action?

_____

_____

If so, which part of the central nervous system appears to be most important?

_____

Did the heart cease beating with complete destruction of the central nervous system?

_____

What conclusions are suggested by this?

_____

_____

_____

Chill several glass or metal rods in a beaker of ice water and apply the rods to the ventricle. When it has cooled, record the contraction rates of the atria, ventricle, and sinus venosus in Table 14-1. Restore the heart to room temperature with Ringer's solution. Repeat, cooling the atria and then determining the contraction rates as before. Again warm the heart. Finally, cool the sinus venosus and determine the rate of contraction of each chamber. (It will be necessary to tip up the ventricle to do this.)

Do your observations suggest that any part of the heart has more influence over heart rate than the other parts?

_____

How do you account for this?

_____

_____

_____

TABLE 14-1
Effects of differential cooling on heart contraction.

Portion cooled	Rate (contractions/minute)		
	Ventricle	Atria	Sinus venosus
Ventricle			
Atria			
Sinus venosus			

TRANSPORT IN BIOLOGICAL SYSTEMS

To examine further this aspect of heart control, remove the heart from the body. To do this, lift the ventricular apex and cut through the anterior and posterior vena cavae and the right and left aortae. Be careful not to stretch the heart, and avoid any damage to the chambers, particularly the sinus venosus. In addition, be careful not to puncture the gall bladder, a small, dark green sac lying close to the heart near the liver.

Place the heart in a watch glass containing Ringer's solution at room temperature. After 2–3 minutes the heart should begin to contract again. Does this support your earlier conclusions? Explain.

_____

_____

_____

_____

Determine the rate of contraction of the whole isolated heart: _____/minute. Carefully separate the sinus venosus from the right atrium, without damage to either. After 2–3 minutes, determine the contraction rate of the sinus venosus and of the atrialventricular portion: sinus venosus: _____/minute; atrialventricular: _____/minute. As a last step, separate the atria from the ventricle. It may be necessary to wait a few minutes for the ventricle to resume contractions. If it shows no signs of contraction after 5 or 10 minutes, stimulate it by touching it with a dissecting needle. After it resumes beating, determine the contraction rate of each segment; sinus venosus: _____/minute; atria: _____/minute, ventricle: _____/minute. Summarize your conclusions regarding the control of heart activity in the frog.

_____

_____

_____

_____

_____

D. TRANSPORT IN PLANTS

Water is one of the basic raw materials of photosynthesis. It is the major component of plant tissues, making up 90% of the plant body. Water is the substance in which most materials enter and leave the cells of plants and it is the solvent for the various biochemical reactions that occur in living cells.

The amount of water used by plants is far greater than that used by animals of comparable weight. The reason for this is that a large amount of the water used by the animal is recirculated in the form of blood plasma or tissue fluid. In plants, over 90% of the water taken in by the root system is evaporated into the air as water vapor. This process, which largely occurs through the leaves, is called **transpiration**. Consequently, plants not only have developed extensive and efficient transport systems but have evolved numerous morphological adaptations to conserve water. In this study you will examine some of the modifications of plants and plant parts to obtain and retain water and become familiar with some of the factors that regulate transpiration.

1. Environmental Adaptations of Vascular Plants

This exercise is limited to an examination of those morphological adaptations that enable higher plants to survive in a given habitat. Specific attention will be paid to structural changes that have evolved with respect to the ability of plants to obtain or retain water.

There are many types of habitats found in nature with respect to water supply. These can be conveniently divided into xeric, mesic, and hydric habitats. The plants that are adapted for living in these habitats are called xerophytes, mesophytes, and hydrophytes, respectively.

Xerophytes include a number of species that live in habitats where the supply of water is physically or physiologically deficient (xeric habitat). **Mesophytes** inhabit regions of average or optimum water conditions (mesic habitat) and include the majority of wild and cultivated plants of the temperate regions. **Hydrophytes** form an extensive flora living on the surface of water or submerged at various depths (hydric habitat).

In this study representative examples of hydrophytes and xerophytes will be studied, since the structural adaptations are more obvious in these groups. Following your examination of specimens characteristic of these groups, you will be asked to classify unknown specimens.

a. Hydrophytic Adaptations

The chief structural modifications exhibited by hydrophytes are an increase in leaf surface, the presence of air chambers, and a reduction in protective, supportive, and conductive tissues.

(1) *Dissected Leaves.* Examine the hydric habitat at the demonstration table. Note that many of the plants have finely divided leaves. How may this condition be advantageous to the plant?

_____

_____

(2) *Air Chambers.* The leaves and stems of many submerged plants have chambers that are filled with gases. Examine prepared slides of cross sections of the leaf of *Potamogeton*. Locate these large air spaces, which are separated from each other by partitions of photosynthetic tissue. List several ways in which these air spaces may be of benefit in a hydric habitat.

_____

_____

(3) *Supporting Tissues.* Remove one of the hydrophytic plants from the water and note how flaccid it becomes. This condition is due to the marked reduction of thick-walled supporting tissues or cells. Confirm this by examining prepared slides of *Potamogeton* stems and leaves. Why is it that submerged hydrophytes do not need large amounts of supporting tissues?

_____

(4) *Vascular Tissues.* Since aquatic plants are submerged in, or floating on, a nutrient solution, the structures necessary for absorption and transport of mineral nutrients and water are greatly reduced, and in some cases, absent. The greatest reduction occurs in the xylem. The phloem, although reduced in amount, is fairly well developed. Examine prepared slides of *Potamogeton* stems and leaves. Locate the vascular tissue and note the absence of xylem. By what process does water enter aquatic plants?

_____

_____

_____

(5) *Protective Tissues.* The epidermis of aerial plants has become modified to prevent or reduce desiccation. Under normal conditions aquatic plants do not lose water through the epidermis, so the epidermis in hydrophytes is not a protective tissue. In these plants nutrients and gases may be absorbed directly from the water. The cuticle overlying the epidermis is extremely thin and may be lacking. The epidermal cells usually contain chloroplasts and may form a considerable part of the photosynthetic tissue. Where would you expect to find guard cells in submerged hydrophytes?

_____

Where would they be located in floating hydrophytes?

_____

Examine slides of *Potamogeton* and locate the modifications indicated above.

b. Xerophytic Adaptations

The deficiency of water that characterizes xeric habitats may be the result of various environmental conditions—for example, intense light, heat, high velocity winds, and others. Many of the adaptations of xerophytes that have evolved prevent desiccation of the plant when exposed to one, or any combination of the above factors.

(1) *Stomata.* The stomates function in the exchange of carbon dioxide and oxygen between the plant and the environment. When the stomates are open, however, water may also leave the plant, which is much to the detriment of the plant as a whole. As a consequence it becomes important to xerophytic plants to reduce the rate of transpiration. One way this is accomplished is by having the stomates situated below the level of the surrounding epidermal cells. Examine prepared slides of the cross section of a pine leaf and locate sunken stomates. How may this position of the stomates effect a reduction in the rate of transpiration?

_____

(2) *Protective Tissues.* In contrast to the hydrophytes, the epidermis of xerophytes commonly has a thick layer of cutin. In addition, the walls of the epidermal cells may be highly lignified. How would this serve to cut down on the amount of water lost from the plant?

_____

_____

Examine slides of pine leaf and locate the modifications mentioned here.

(3) *Supporting Tissues.* Xerophytes generally have a large proportion of supportive tissues that not only prevent water loss but help support the stem or leaf. Examine a slide of pine leaf and locate a layer of thick-walled tissue just below the epidermis. Why is it important that stems and leaves of aerial plants be supported in some manner?

_____

(4) *Leaf Rolling.* During drying conditions, the leaves of many xerophytes—notably the xerophytic grasses—roll up tightly. Since the stomates in these plants are more numerous on the upper surface, what effect would this have on the rate of transpiration? Explain.

_____

_____

_____

_____

Examine a prepared slide of a corn leaf. In the upper epidermis locate large, bulbous cells. These are bulliform cells (motor cells), which function in the rolling of the leaves in dry weather.

(5) *Water Storage.* Some xerophytes possess large amounts of water-storage tissue. These plants are called fleshy xerophytes. Examine the specimens of this group that are on demonstration. In some plants the leaves may be fleshy (leaf succulents). In others, the stem is fleshy and therefore the plants are called stem succulents. Note the absence, or greatly reduced number, of leaves on the stem succulents. What is the primary photosynthetic organ in these plants.

_____

Cut a thin cross section of a leaf succulent (for example, *Aloe*) and mount it in a drop of water on a slide. Examine microscopically. Note the large amount of water-storage tissue that makes up the bulk of the leaf. What other xerophytic characteristics are present that you are able to observe microscopically?

_____

_____

_____

_____

_____

c. Unknown Specimens

Examine slides of the species of plants listed in Table 14-2. Enter the information asked for in the table and then decide in what habitat (xeric or hydric) each plant would be found. Your decision should be based upon the information obtained from the examination of the slide and characteristics cited in this study.

2. Effect of Environmental Factors on Transpiration Rate

a. Cut a branch from a geranium or *Coleus* plant and insert it into a rubber stopper as shown in Fig. 14-3. Place the branch and stopper into the potometer flask that was previously filled with distilled water. Insert the stopper slowly to avoid creating bubbles. If this is done properly, water will be forced out of the end of the capillary tubing. However, when the pressure on the stopper is released the fluid in the capillary tube will tend to recede. If this should occur, fill a 5-ml syringe with water and insert the needle into the rubber coupling between the flask and the capillary tubing. Slowly inject water until it comes out of the end of the tubing. If the apparatus has been properly set up, the water in the tube will begin to recede slowly. The rate that the meniscus moves along the tubing

TABLE 14-2
Adaptations of vascular plants to the environment.

Species	Air chambers	Supporting tissues	Protective tissues	Leaf modifications (stomata, motor cells, other)	Vascular tissue	Water storage tissue	Habitat and reason for choice (xeric, hydric)
Potamogeton							
Yucca							
Myriophyllum							
Ammophila							
Acorus							
Typha							
Pinus							

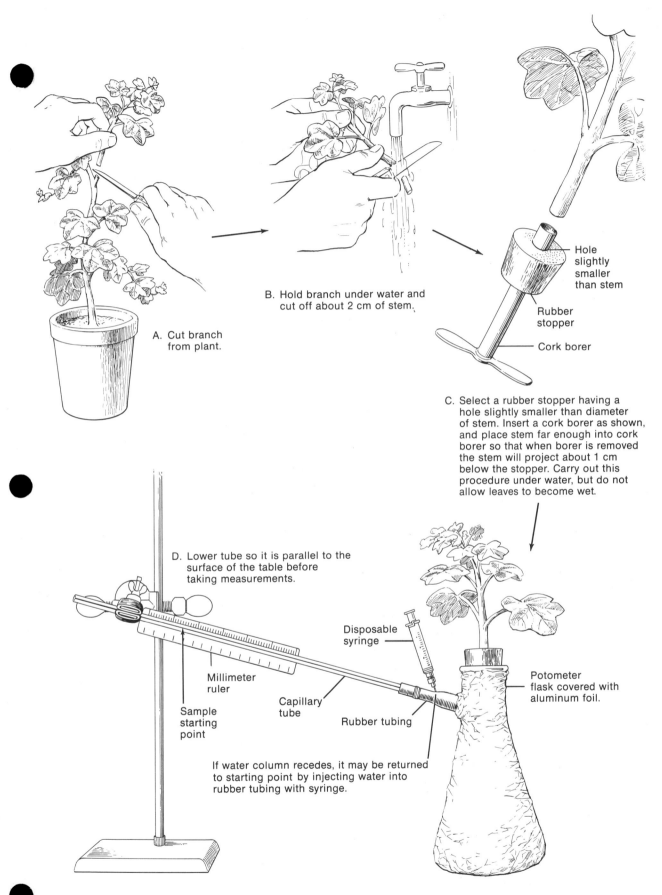

FIGURE 14-3
Procedure for determining the rate of transpiration.

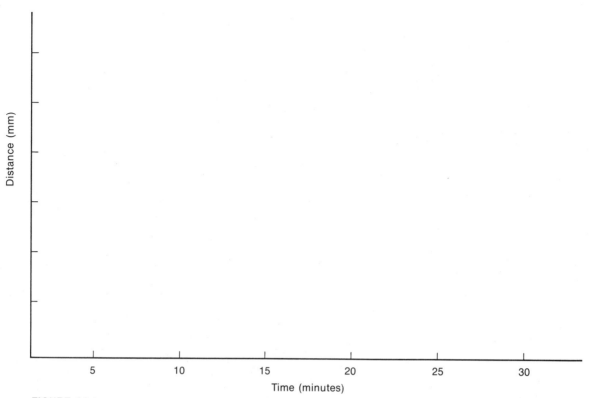

FIGURE 14-4
Comparison of transpiration rates under various environmental conditions.

is a measure of the rate of water uptake by the branch and may be used as a measure of the rate of transpiration. Determine the transpiration rate by recording the distance the meniscus moves each minute for a period of 10 minutes. If the meniscus goes beyond the graduated scale at the right, it may be returned to the "zero" mark by injecting water into the rubber coupling as previously described. Record your results in Table 14-3. Plot your data in Fig. 14-4.

b. Design and perform experiments to show the effects (if any) of light intensity, air movement, and humidity and the role of the leaves in the process of transpiration. Record the experimental conditions and your results in Table 14-3. Plot your data in Fig. 14-4. What would be the control for these various experiments?

What is the reason for covering the potometer flask with aluminum foil?

All aerial parts of plants may lose water by transpiration. In most herbaceous and woody plants, a large proportion of the water is lost through openings called **stomates**. Discuss the effects of the various environmental factors as they affect transpiration in terms of their effects on the activity of the stomates.

Why is important that ornamental evergreen plants be thoroughly watered before winter sets in?

TRANSPORT IN BIOLOGICAL SYSTEMS

TABLE 14-3
Potometer data.

Experimental conditions	Time (minutes)									
	1	2	3	4	5	6	7	8	9	10
	Distance (mm)									
Control										
Light intensity										
Air movement										
Humidity										
Effect of leaves										

In what way is transpiration a beneficial process to the plant?

3. Movement of Minerals

The study of relationships between structure and function in plants and animals has been aided by the use of radioactive substances called **radioisotopes** (see Appendix E). Radioisotopes can be used to trace the pathways of substances during their movement throughout an organism. When used in this way they are called **tracers**. For example, biologists using radioactive carbon dioxide ($C^{14}O_2$) have been able to identify the various intermediate compounds that carbon becomes associated with in the photosynthetic process.

In this part of the exercise a radioactive compound, uranyl nitrate, will be used to determine if the minerals that are absorbed by roots are distributed throughout the plant.

Carefully remove two young bean seedlings from their container (Fig. 14-5A). Try not to damage the roots. Wash the roots with tap water (Fig. 14-5B). Place one of the plants into a flask containing a solution of uranyl nitrate so that the roots are completely submerged (Fig. 14-5C). Label this flask "experimental." The second plant should be set up as your control (Fig. 14-5D). What solution should be put in the control flask?

Cover both flasks with aluminum foil and place them on a window ledge in the light for one to two hours. Next lift the plants out of the flasks just high enough so that the root can be cut off at the base of the stem. Allow the roots to fall back into the flasks (Fig. 14-5E). Return the containers to your instructor for disposal. Place each plant on a separate X-ray film pack. Spread the leaves so they do not overlap and tape the plants to the film packs (Fig. 14-5F). Mark each film pack with your name and the date, and indicate whether it is the "experimental" or the "control" plant. Now place each film pack between two pieces of blotting paper (or paper towels). Add a heavy weight (two or three heavy books) and allow to remain this way for 10–15 days. At the end of this time (they may be left for a longer period) develop each film according to the directions given by your instructor. Discard the plants and blotting paper in the container provided for this purpose.

The radioactive uranyl nitrate is emitting high energy particles. When these strike the emulsion of the film, latent images are formed, which, when treated with photographic developing solution, are reduced to metallic silver. This produces a black spot on the film. In this way you will be able

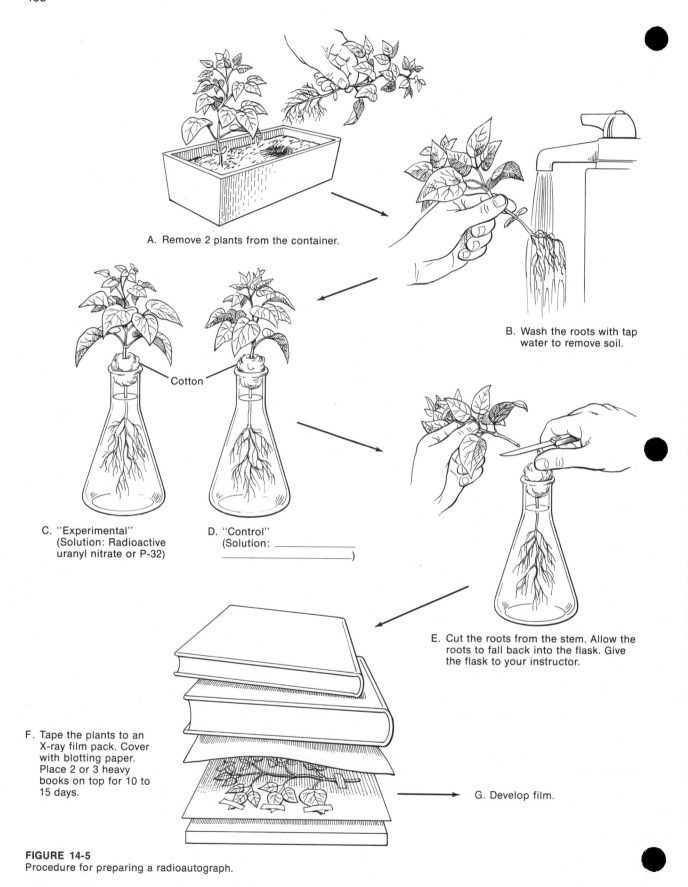

FIGURE 14-5
Procedure for preparing a radioautograph.

to show the distribution of the radioactive uranyl nitrate in the plant. This procedure is called **radioautography**. The picture you obtain is a **radioautograph**.

REFERENCES

Biddulph, O. 1959. Translocation of Inorganic Solutes. In *Plant Physiology.* F. C. Steward, Ed. Academic Press, New York.

Biddulph, S., and O. Biddulph. 1959. The Circulatory System of Plants. *Scientific American,* February (Offprint No. 53). *Scientific American* Offprints are available from W. H. Freeman and Company, 660 Market Street, San Francisco 94104, and 58 Kings Road, Reading RG1 3AA, England. Please order by number.

Crafts, A. S. 1961. *Translocation in Plants.* Holt, Rinehart & Winston, New York.

Galston, A. W. 1964. *The Life of the Green Plant.* 2d ed. Prentice-Hall, Englewood Cliffs, New Jersey.

Koontz, H., and O. Biddulph. 1957. Factors Affecting Absorption and Translocation of Foliar Applied Phosphorus. *Plant Physiol.* 32:463–470.

Kramer, P. J. 1959. Transpiration and the Water Economy of Plants. In *Plant Physiology.* Vol. 2. Academic Press, New York.

Leopold, A. C. 1964. *Plant Growth and Development.* McGraw-Hill, New York.

Macey, R. I. 1968. *Human Physiology.* Prentice-Hall, Englewood Cliffs, New Jersey.

Moore, J. A., Ed. 1964. *Physiology of the Amphibia.* Academic Press, New York.

Morrison, T. F., F. D. Cornett, and J. E. Tether. 1959. *Human Physiology.* Holt, Rinehart & Winston, New York.

Ray, P. M. 1963. *The Living Plant.* Holt, Rinehart & Winston, New York.

Stout, P. R., and D. R. Hoaglund. 1939. Upward and Lateral Movement of Salt in Certain Plants as Indicated by Radioactive Isotopes of Potassium, Sodium and Phosphorus Absorbed by Roots. *Am. J. Botany* 26:320–324.

Swanson, C. A., and J. B. Whitney. 1953. Studies on the Translocation of Foliar Applied P^{32} and Other Radioisotopes in Bean Plants. *Am. J. Botany* 40:816–823.

Telfer, W., and D. Kennedy. 1965. *The Biology of Organisms.* Wiley, New York.

Zimmerman, M. H. 1963. How Sap Moves in Trees. *Scientific American,* March (Offprint No. 154).

Zweifach, B. W. 1959. The Microcirculation of the Blood. *Scientific American,* January (Offprint No. 64).

EXERCISE 15

Biological Coordination: Plants

As organisms become more complex and individual cells become more specialized, the various activities within the organism must be coordinated. Furthermore, organisms need effective ways to respond to physical and chemical changes in their environment. Thus, all organisms must become adapted to receive various internal and external stimuli and to respond to these stimuli in a coordinated manner.

Although higher plants do not have nervous systems, the coordination of many activities is a necessity if the plant is to grow and develop. There are two types of mechanisms that bring about coordination in plants. One, consisting of a system of chemical messengers, directs the cells of the plant to carry out different functions. This group includes the various plant hormones and other growth regulators. A second coordinating system consists of various physical forces, for example electrical gradients, gas-exchange gradients, and metabolic or pressure differences throughout the plant. The end product of these controls is a highly complex plant body having localized areas of growth, special regions of photosynthetic activity, and the ability to translocate and accumulate the necessary raw materials for growth and differentiation into reproduction or other functions to assure the continuation of the species.

A. COORDINATION IN PLANTS

1. Phototaxis in *Euglena*

The ability to respond to a stimulus (a chemical or physical change in the environment) is a striking characteristic of living organisms. The response of the organism to any stimulus serves to coordinate the activities of the organism so that equilibrium with the environment is maintained. In this study you will examine the effects of various wavelengths of light in orientating movements of the green alga, *Euglena*.

Following the directions outlined in Fig. 15-1, expose a tube of *Euglena* to various wavelengths of light. The *Euglena* will tend to concentrate and

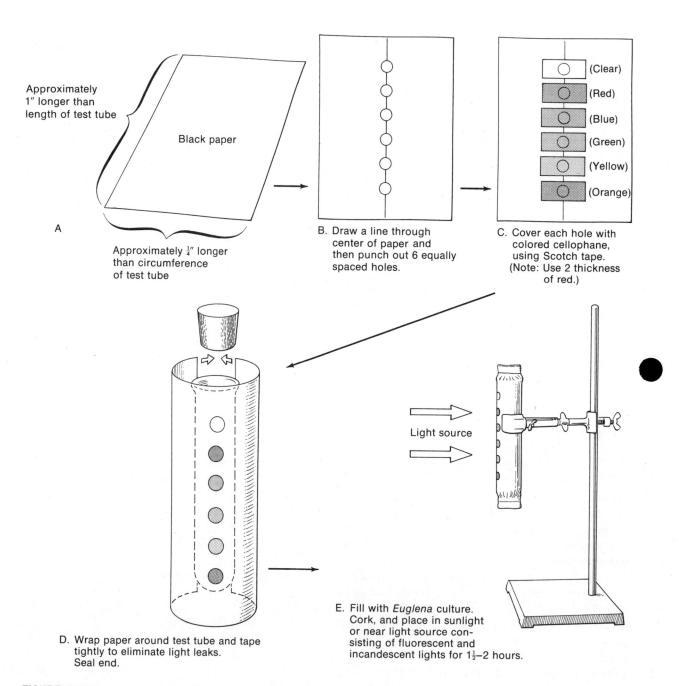

FIGURE 15-1
Procedure for studying the effect of wavelengths of light on *Euglena*.

BIOLOGICAL COORDINATION: PLANTS

adhere to the sides of the tube, depending upon the wavelength to which they are most sensitive. At the conclusion of the study, carefully decant off the liquid in the tube and remove the black paper sleeve. Record your observations in Fig. 15-2 by shading the appropriate circles to represent the density of *Euglena* found associated with each colored filter. What wavelength of light is most effective in attracting *Euglena*?

_____

Prepare a wet mount of *Euglena* and locate a red-pigmented "eye-spot" near the flagella. Suggest a role for this structure in the response just studied.

_____

_____

_____

Of what survival value is the response observed to *Euglena*?

_____

_____

_____

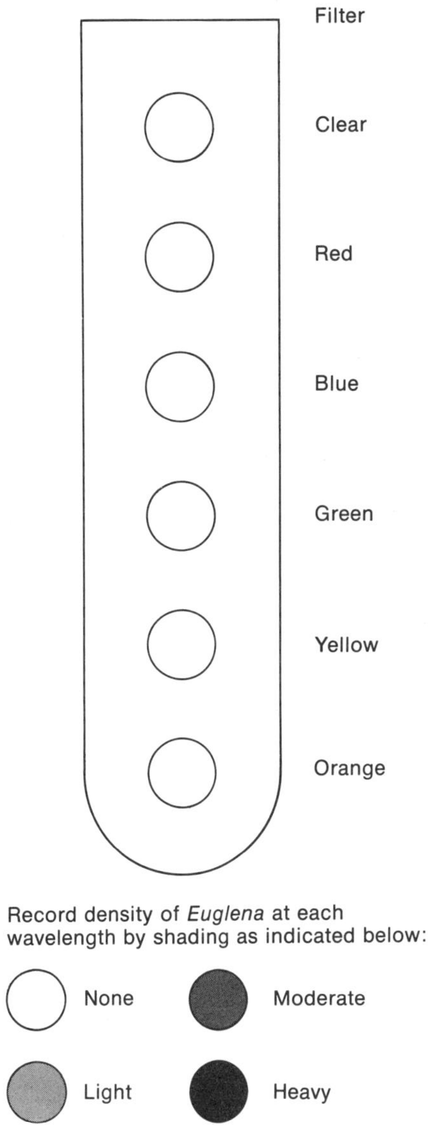

2. Geotropism

The effect of gravity in determining the direction of growth in various plant parts is called **geotropism**. Stems generally grow in a direction opposite the force of gravity. Is this negative or positive geotropism?

_____

Characterize the response of roots in regard to their reaction to gravity.

_____

In the following parts of this study, special attention will be given to the geotropic response of *Coleus* or *Iresine* shoots. For convenience, each team will be assigned one aspect of the problem. Since the response of the plant usually requires several hours, set the experiment aside for examination during the next laboratory period. In Table 15-1 fill in the expected results and then compare

FIGURE 15-2
Effect of wavelength of light on the orientating movements of *Euglena*

Record density of *Euglena* at each wavelength by shading as indicated below:

- None
- Light
- Moderate
- Heavy

these with the actual results.

a. What is the effect of gravity on stem orientation? Remove three *Coleus* or *Iresine* branches, keeping the surface moist during the cutting operation. Set them up as indicated in Fig. 15-3. What is the "control" in this experiment?

_____

_____

_____

TABLE 15-1
Geotropism data.

Results	Stem position		
	Vertical	Horizontal	Inverted
Expected			
Observed			
Conclusions			

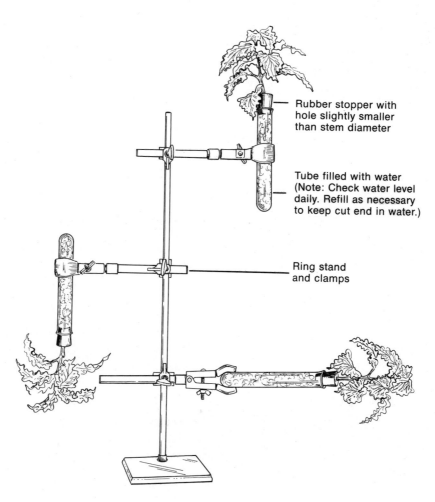

FIGURE 15-3
Experimental setup to show the effect of gravity on growth.

b. Does defoliation affect the geotropic response of *Coleus* or *Iresine* shoots? Design and perform an experiment to answer this question. Complete Table 15-2.

c. Does the shoot apex exert any effect on the geotropic response of *Coleus* or *Iresine*? Design and perform an experiment to answer this question. Complete Table 15-3.

d. Will replacing the shoot apex with auxin have any effect on the geotropic response of *Coleus* or *Iresine*? Design and perform an experiment to answer this question. Complete Table 15-4.

3. Chemotropism

The response of nonmotile cells to chemical stimulation, in which growth occurs toward or away from the chemical stimulus, is called **chemotropism**. In this part of the exercise, the chemotropic response of lily pollen tubes to various materials (pistils, auxin, and gibberellic acid) will be studied. Petri dishes containing a nutrient medium will be used in these experiments. Pollen and the materials to be tested for chemotropic activity can be obtained from the instructor. A small piece of material to be tested is placed on the surface of the nutrient medium and the pollen positioned around it, using a camel-hair brush and a binocular microscope. For best results distribute the pollen in small clumps within 2 or 3 mm of the material to be tested. To test the chemotropic activity of liquids, cut a well in the medium with a cork borer, place the liquid in the well, and position the pollen around the rim.

a. Is chemotropic activity exhibited by the lily pistil? Obtain Petri dishes, pollen, and lily pistils from the instructor. Determine if chemotropic activity is exhibited by all parts of the pistil. Design and perform an experiment to answer this question. Test for chemotropic activity as described above. What controls should be used?

What criteria would be used to determine if positive or negative chemotropism is indicated?

Complete Table 15-5.

b. Is the chemotropic activity of pistils altered by heat? Design and perform an experiment to answer this question. Complete Table 15-6.

c. Is chemotropic activity exhibited by auxin or gibberellic acid? Design and perform an experiment to test the chemotropic activity of each of these known plant growth substances. It will be necessary to know the solvents for each of these substances so that adequate controls can be established.

If one (or both) of these substances shows chemotropic activity, determine its optimum concentration by testing with a series of concentrations — for example, 1, 10, 50, and 100 parts per million (ppm). Complete Table 15-7.

REFERENCES

Audus, L. J. 1959. *Plant Growth Substances.* 2d ed. Leonard Hill, London.

Galston, A. W. 1964. *The Life of the Green Plant.* 2d ed. Prentice-Hall, Englewood Cliffs, New Jersey.

Leopold, A. C. 1955. *Auxins and Plant Growth.* University of California Press, Berkeley, California.

Machlis, L., and J. G. Torrey. 1956. *Plants in Action: A Laboratory Manual of Plant Physiology.* W. H. Freeman and Company, San Francisco.

Rosen, W. G. September, 1962. Cellular Chemotropism and Chemotaxis. *Quart. Rev. Biol.* **37**:242–259.

Salisbury, F. B. 1957. Plant Growth Substances. *Scientific American,* April (Offprint No. 110). *Scientific American* Offprints are available from W. H. Freeman and Company, 660 Market Street, San Francisco 94104, and 58 Kings Road, Reading RG1 3AA, England. Please order by number.

Went, F. W. 1962. Plant Growth and Plant Hormones. In *This is Life.* W. H. Johnson and W. C. Steere, Eds. Holt, Rinehart & Winston, New York.

TABLE 15-2
Data on the effect of defoliation on the geotropic response.

Procedure:
  1. Experimental treatment
 	
  2. Control treatment

Expected results:
Observed results:
Conclusions:

BIOLOGICAL COORDINATION: PLANTS

TABLE 15-3
Data on the effect of the shoot apex on the geotropic response.

Procedure:
  1. Experimental treatment
 	
  2. Control treatment

Expected results:
Observed results:
Conclusions:

TABLE 15-4
Data on the effect of replacing the apex with auxin on the geotropic response.

Procedure:
1. Experimental treatment
2. Control treatment

Expected results:

Observed results:

Conclusions:

BIOLOGICAL COORDINATION: PLANTS

TABLE 15-5
Data on the chemotropic activity of lily pistils.

Procedure:
 1. Experimental treatment

 2. Control treatment

Results:

Conclusions:

TABLE 15-6
Data on the effect of heat on the chemotropic activity of lily pistils.

| Procedure: |
| 1. Experimental treatment |
| |
| 2. Control treatment |

| Results: |

| Conclusions: |

BIOLOGICAL COORDINATION: PLANTS

TABLE 15-7
Data on the chemotropic activity of auxin and gibberellic acid.

Procedure:
  1. Experimental treatment
 	
  2. Control treatment

Results:
Conclusions:

EXERCISE 16

Biological Coordination: Animals

The intact animal has functional capabilities over and above those characteristic of its constituent cells, tissues, organs, and organ systems. This observation provides the basis for the generalization that ". . . the organism is more than a simple sum of its parts." Each subunit of the body has a specific contribution to make, but the proper functioning of the whole organism depends upon the integration of a wide spectrum of individual processes into a coherent unit. Such integration in turn depends upon the exchange of information between parts of the body, and in this vital process the nervous system and endocrine organs play a central role. Although structurally quite different, both exert their effects in essentially similar fashion—through the production and release of molecules that modify cellular activity. The effects produced by the nervous system are normally quite localized, as the materials released at nerve endings (such as acetylcholine or norepinephrine) act on specific targets—other nerve cells, muscle cells, and secretory cells. The effects of the endocrine glands are more generalized, since their secretions (hormones) are distributed throughout the body by the circulatory system. Specificity in this instance is achieved by variations in the sensitivity of different types of cells to particular hormones. In a general sense the nervous system contributes to organismic coordination by providing rapid and highly specific but short-term control, whereas long-term modifications of physiological activity are normally the province of the endocrine organs.

The nervous system consists basically of circuits linking **sensory receptors** to **effector units** such as muscle or gland cells. These circuits, termed neuromotor units or **reflex arcs**, vary considerably in their complexity. For example, among very primitive animals **direct effector** systems are found, in which a nerve cell responds directly to stimulation and transmits directly to an effector (Fig. 16-1A). Usually, however, a specialized sensory organ detects stimuli and activates nerve

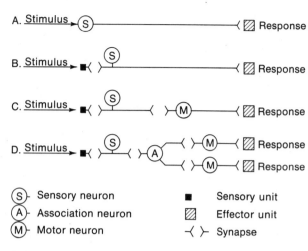

FIGURE 16-1
Typical reflex arcs.

cells. The nervous pathway from sense organ to effector unit may consist of a sensory-motor neuron (Fig. 16-1B) or separate sensory and motor nerve cells (Fig. 16-1C), or one or more association neurons may be interpolated to increase the number of circuits possible (Fig. 16-1D).

The nerve cells found in both primitive and highly evolved animals are surprisingly similar in their functional properties. However, very marked differences are found in the organization of the nervous systems that they form. Perhaps the most primitive type of nervous system is seen in coelenterates (e.g., *Hydra*, corals), in which the nerve cells form a simple, diffuse network. Among higher forms there is a marked tendency for the cell bodies of neurons concerned with the control of a given function to become grouped into masses (**ganglia** and **nuclei**). These are located in a central nerve cord or cords and are interlinked by association nerve cells. The anterior portions of such cords tend to be enlarged by the accumulation of nerve-cell masses and their association pathways to form a "brain." The major sense organs also accumulate in the anterior part of the body and form direct connections with the brain. Ultimately, the brain becomes the dominant integrative center. The term **cephalization** is used to describe this overall evolutionary process.

Since the nervous system usually responds much more rapidly than the other coordinating system, the endocrine system, this exercise will be concerned with the anatomy and physiology of the nervous system.

A. SPINAL CORD ANATOMY

Study a stained slide of a cross section of mammalian spinal cord passing through the dorsal root ganglia. You will notice that it is not a perfect circle but has notches in its circumference (Fig. 16-2). The deeper notch is the **ventral fissure**. Rotate your slide so that this fissure is toward you.

The brain and spinal cord are surrounded by membranous coverings called **meninges**. There are three such membranes in mammals: an outer **dura mater** next to the skull and vertebral bones, an inner **pia mater** fitting tightly to the brain and spinal cord, and a median **arachnoid membrane**.

In the center of the cord locate a small clear space, the **central canal**. This canal runs the length of the cord and, at its anterior end, is continuous with the fourth ventricle of the medulla oblongata. It is filled with **cerebrospinal fluid**, as are all the cavities within and around the central nervous system. Looking at the cross section of the cord, you will notice an H-shaped area of **gray matter** occupying much of the central region. The gray matter contains many cell bodies and looks gray in fresh tissue. In the outer area of the cord is the more uniformly colored **white matter** containing fewer nuclei. The white matter is made up mostly of fibers that appear white in fresh tissue, owing to the glistening white sheaths around them. In your slide the white matter has a granular appearance because almost all of its fibers run lengthwise and thus are cut transversely in the slide, each fiber end appearing as a "dot."

The gray matter shows some differentiation. In the lower corner of the H-shaped area are extra-large cells—the cell bodies of **efferent (motor) neurons**. Locate their position in Fig. 16-2. Since their processes (cytoplasmic extensions) are normally very irregular, they are seldom visible in their entirety in a thin section. If you could trace them, the axons of these large neurons would be found to run out through the ventral root into the spinal nerve.

Besides the large motor neurons, numerous other cells will be found. The smallest belong to a special type of supporting cell characteristic of the central nervous system; these cells are not neurons. The intermediate-sized cells are called **association (adjustor) neurons** because they take impulses from the incoming sensory fibers and switch them to other adjustors, or up to the brain by long fibers in the cord, or directly to the motor neurons. Since the axons of adjustor neurons often branch, they can

BIOLOGICAL COORDINATION: ANIMALS

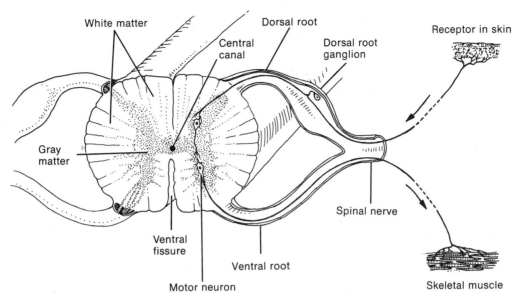

FIGURE 16-2
Diagrammatic representation of the relationship of spinal cord, spinal nerves, and the simple reflex arc.

transmit an impulse to a number of other cells and these in turn to several more. Thus, in the end, hundreds of cells may be stimulated by one entering impulse.

B. SPINAL NERVE ATTACHMENT TO SPINAL CORD

Each spinal nerve is attached to the spinal cord by two connections, a **dorsal** and a **ventral root** (Fig. 16-2). The dorsal root has an enlargement called the **dorsal root ganglion** containing numerous cell bodies. The dorsal root contains only those fibers that carry impulses toward the spinal cord; the impulses originate in sense organs from all parts (internal and external) of the body. The ventral root contains no cell bodies.

The process transmitting the nerve impulse from the sense organ to the nerve-cell body is the **dendrite**. Another process, the **axon**, transmits the impulse from the cell body and in this case to the dorsal region of the gray matter of the spinal cord. The nerve cell (neuron) that carries impulses into the spinal cord from the sense organs—for example, skin and stomach smooth muscle—is called an **afferent (sensory) neuron**.

The axon of the afferent neuron usually synapses with an **association (adjustor) neuron** found within the gray matter of the spinal cord. The parts of the association neuron are similar to those of the afferent neuron, but it has several short dendrites and a short axon that ends close to the ventral region of the gray matter.

If the reaction produced is of a simple type, the axon of the association neuron synapses with the dendrites of an **efferent (motor) neuron**. The cell body, as well as the short dendrites of motor neurons, is found in the ventral part of the gray matter. A long axon proceeds from the cell body of the motor neuron and passes ventrally out of the spinal cord into the ventral root; it then enters the spinal nerve. It passes through the spinal nerve and terminates in an effector organ, which in this case is always a striated, skeletal muscle. The pathway that the impulse takes from the sense organ through the afferent, association, and efferent neurons to the effector is known as a **reflex arc**. Label the various components of the reflex arc shown in Fig. 16-2.

1. Reflex Action

A nervous reflex may be defined as a coordinated response by effectors to a stimulus applied to a receptor, mediated through two or more neurons. Much of the behavior of an animal may be interpreted in terms of the activity of reflex circuits in the spinal cord, but these are more or less modified by influences from higher centers, to the extent that the brain has achieved physiological dominance.

a. Reflexes in the Living Frog

(1) *Blinking*. Hold a live frog securely in one hand by grasping the hind legs. Bring a blunt probe or a glass rod near one eye and observe that the frog blinks. In fact, the response at times is so strong that the eye may be pulled into the throat region. Observe, also, the "lids" covering the eyes.

(2) *"Scratch" reflex*. Grasp the head and forelimbs securely in one hand. Wash off any mucous present on the back of the frog. With a forceps place a small ($\frac{1}{4}$ inch square) piece of filter paper moistened with dilute acetic acid on the lower surface of the back. Notice how the hind leg attempts to brush off the irritant. This is the "scratch" reflex. With a stronger acid a more violent response occurs. In fact, the animal may try to use both legs to brush off the irritating substance. Wash off the acid with water.

(3) *Other reflexes*. Lay the frog on its back and stroke the throat and belly regions. When these regions are first stroked in males (males have a large swollen "thumb"), watch the distinct clasping reflex in the forelimbs. Notice the quieting, almost hypnotic effect on the frog when it is stroked longer.

b. Spinal Reflexes in the Decerebrated Frog

The frog is an excellent animal for the study of **spinal reflexes** because of the brevity of the period of spinal shock following destruction of the brain, the small blood loss in decerebration, and the adequacy of the cutaneous respiration for supporting metabolism. (See Appendix D for frog pithing and decerebration techniques.)

Holding the frog firmly in the left hand, quickly sever the spinal cord at the neck (on a line just posterior to the ear drums) with a probe. Then, with scissors, cut off the head but leave the lower jaw. Wipe off the blood and suspend the frog from a ring stand with a hook through the lower jaw, with the frog's back toward you. Wash off the mucous covering, but remember to keep the skin moist. If spinal shock has occurred, the animal gives no response; wait 5 or 10 minutes for recovery before proceeding. Such a preparation is known as a **decerebrate frog**. Since it lacks a brain it is completely without consciousness. Does the scratching reflex remain in this "decerebrate frog"?

_____

Pinch the toe of one foot with a forceps. What happens?

_____
_____

What does the other foot do?

_____

Stimulate the toe of the other foot in a like manner. Be sure to allow the frog to recover for a minute or so before applying the next stimulus. Now pinch the toe of one foot, holding it tightly with forceps so that it cannot be withdrawn. Describe what happens.

_____
_____

How did the nerve impulse reach the unstimulated leg?

_____
_____

2. Stimulation of Muscle Through Nerve

Cells are, in general, subject to stimulation from a variety of impulses. In making a study of the different agencies that may cause a muscle to react, it should be remembered that the muscle cell is different only in degree, and not in kind, in the way in which it responds to a stimulus. Cells, specialized in a different manner to perform other functions, may also respond to the same forms of stimulation, but in a different way. A nerve will transmit an impulse in response to an adequate stimulus. If the nerve is in physiological connection with an excitable muscle, the muscle will contract upon receiving an impulse from the nerve. Adequate stimulation of the nerve will have been attained when the muscle that it innervates contracts. In this part of the exercise we will examine the effect of several factors on the nerve impulse, using an

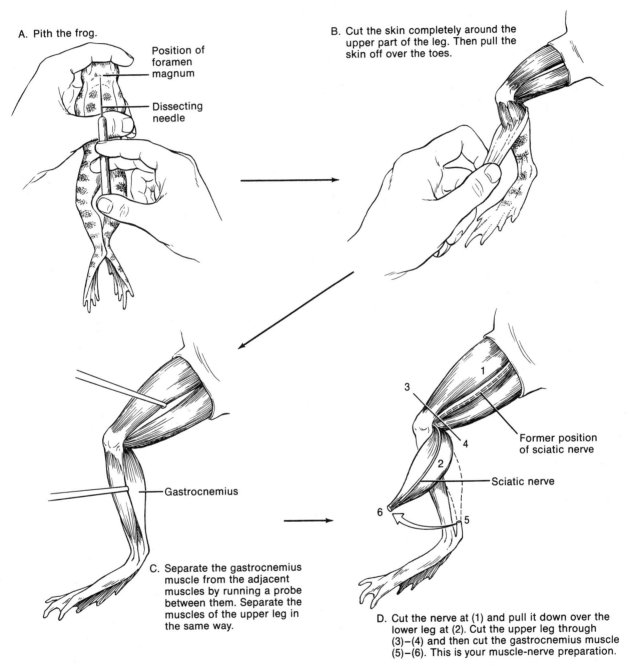

FIGURE 16-3
Muscle-nerve preparation.

isolated **muscle-nerve preparation**. Usually the isolated gastrocnemius muscle and its nerve are used.

a. To obtain your muscle-nerve preparation, kill a frog by pithing both the brain and spinal cord (see Appendix D). Remove the skin from the legs. This can be easily done if the skin is cut completely around the "thigh" of the frog leg. The entire skin of the leg may then be pulled off in the same way that you would strip a glove from your hand (Fig. 16-3). Place the animal, dorsal (back) side up, on moist filter paper. Keep the animal moist with physiological saline. Note the large "calf" muscle of the lower leg. This is the gastrocnemius muscle and is the one to be used. Run a blunt probe (the end of a forceps will do) between this muscle and the long bone of the leg, noting how easily it is freed in the region between the two ends.

Next, notice the fairly well marked groove running along the upper leg. This groove marks the position of the sciatic nerve, which innervates the gastrocnemius muscle and must be freed from the rest of the upper leg. With two pairs of forceps pull the muscles apart on either side of the groove. If this is carefully done you will expose the sciatic nerve along with blood vessels. The nerve is the faintly yellowish strand; the blood vessels are red. Without touching or pulling the nerve, remove or pull aside the muscles until you expose the nerve from a point near the knee up to the hip. Then, with a forceps, grasp the hip end of the nerve and, cutting across the nerve on the hip side of the forceps, gradually free the nerve of connective tissue down to a point near the knee joint. This is the most delicate operation of the whole dissection. You should not stretch the nerve or touch it with metal *except where it is held at the hip end.* Lay the free nerve down on top of the lower leg. Using scissors cut through the middle of the upper leg. Cut through everthing, including bone and muscle, but be sure that the nerve is laid back so you do not cut through it. Sever the tendon at the lower end of the gastrocnemius muscle. Cut the lower leg bone near the knee. Your muscle-nerve preparation is now ready to use. Place it on filter paper moistened with physiological saline, or, if you are not ready to use it at once, place it in a beaker of the solution.

b. What is the effect of mechanical stimulation? Tap the nerve lightly near its distal (cut) end with a small glass rod? What is the result?

Repeat the procedure with a harder tap. Does the muscle respond more fully than the first time?

Pinch the nerve sharply with forceps about ½ cm from its cut end. Is the response any greater?

Pinch the nerve approximately 1 cm from the cut end. What is the result?

Account for the results you observed.

c. What is the effect of thermal stimulation? Touch the nerve with the pointed end of a small glass rod so gently that there is no response. Heat the end of the rod until it is rather warm to the touch, and again touch the nerve as gently as before. What is the result?

d. What is the effect of chemical stimulation? Cut off the part of the nerve that has been used previously. Put a drop of 1 N HCl on the end of the nerve. Does the muscle move?

Wash off the terminal portion of the nerve.

e. What is the effect of electrical stimulation? When two different metals are connected and their free ends are contacting a salt solution, a current of electricity will flow in the same way as in a battery. Scrape the ends of two 6–8 inch pieces of iron and copper wire with a knife or an emery cloth. Twist together tightly at one end to make a V. Making sure the nerve-muscle preparation is moistened with physiological saline solution, touch the free ends of the wires to the uninjured portion of the nerve. Repeat several times. Describe the response observed.

Touch one end of the V to a nerve and the other to a muscle. What is the result?

Reverse the positions of the wires. What difference in the response, if any, is observed?

Repeat the above experiments, stimulating the muscle directly. Describe the response.

Contrast this to the response obtained when stimulation is via the nerve.

Would you say that nervous tissue is a better conductor than muscle? Why?

REFERENCES

Carthy, J. D. 1965. *The Behaviour of Arthropods.* W. H. Freeman and Company, San Francisco.

Hoyle, G. 1957. *Comparative Physiology of the Nervous Control of Muscular Contraction.* Cambridge University Press, London.

Katz, B. 1966. *Nerve, Muscle, and Synapse.* McGraw-Hill, New York.

Macey, R. I. 1968. *Human Physiology.* Prentice-Hall, Englewood Cliffs, New Jersey.

Marrazzi, A. S. 1957. Messengers of the Nervous System. *Scientific American,* February.

Pace, D., B. W. McCashland, and C. Riedesel. 1958. *Laboratory Manual for Vertebrate Physiology.* Rev. ed. Burgess, Minneapolis.

Roeder, K. D. 1963. *Nerve Cells and Insect Behavior.* Harvard University Press, Cambridge.

Schmidt-Nielsen, K. 1970. *Animal Physiology.* 3d ed. Prentice-Hall, Englewood Cliffs, New Jersey.

Telfer, W., and D. Kennedy. 1965. *The Biology of Organisms.* Wiley, New York.

EXERCISE 17

Mendelian Inheritance

The specific characteristics of an individual, whether plant or animal, are established at fertilization following the union of the male and female chromosome complements, which are carried by the gametes. These chromosomes carry the genes that condition the various characteristics. In the first part of this exercise you will study the developmental stages of the formation of gametes (meiosis) and through practical experience will become familiar with some of the principles of Mendelian heredity.

A. MEIOSIS

Meiosis is a singularly important biological event that not only maintains a constant chromosome number for most species of plants or animals, but also provides a means of genetic variability because of "crossing over" and the subsequent exchange of genetic material. In meiosis, immature or primordial germ cells undergo a "reduction" in the diploid number of chromosomes characteristic for the species and become mature gametes.

1. Meiosis in the Lily

In this study, meiosis will be studied as it occurs in the development of mature pollen grains of the flowering plants. These pollen grains give rise to male gametes, which fuse with an egg to produce a zygote.

As you examine the series of slides in the meiotic sequence, refer to Fig. 17-1 to help locate the stages indicated in the following description.

a. The first meiotic division is **meiosis I**. Examine a lily flower and locate the **anthers**, or pollen sacs, which contain numerous pollen mother cells (Fig. 17-1A and B). Each of these cells may undergo meiosis and produce mature pollen grains. Next examine slides of a cross section through a young lily anther and locate the pollen mother cells (Fig. 17-1C and D). The nuclei of these cells contain

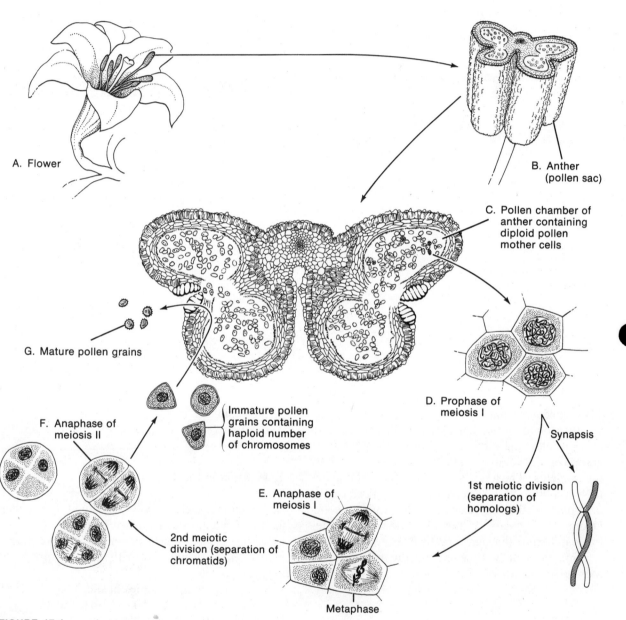

FIGURE 17-1
Meiosis in the lily.

the **diploid number** of chromosomes. Many of the pollen mother cells are in the prophase of the first meiotic division. During this phase, homologous chromosomes, each composed of two **chromatids**, lie adjacent to one another (**synapse**), form tetrads, and exchange genetic components by a process called **crossing over**. What is the significance of this process?

_____

_____

This phenomenon is diagrammatically shown in Fig. 17-2.

Subsequent to synapsis the homologous chromosomes separate during anaphase, with one member moving to each pole of the division spindle. Since this is a separation of entire chromosomes, and not chromatids, the chromosome content of the cells at the end of meiosis I has been "reduced" from the diploid to the **haploid** condition. Examine slides of a lily anther showing separation of homologous chromosomes (Fig. 17-1D and E). The diploid number in lily is 24. What is the chromosome number following the first meiotic division?

_____

b. The second meiotic division is **meiosis II**. Examine slides of anthers in which the cells resulting from the first meiotic division are in anaphase of meiosis II (Fig. 17-1F). In these cells, the chromatids that make up each chromosome now separate and migrate to the spindle poles (Fig. 17-3). As in mitosis, the chromatids when separated from each other are called **daughter chromosomes**. At the spindle poles, each group of daughter chromosomes becomes enclosed in a nuclear membrane.

Cell division follows the division of the nucleus. How many pollen grains are formed as a result of the two meiotic divisions of the pollen mother cells?

_____

What is the chromosome complement of each pollen grain?

_____

Examine lily anthers showing mature pollen grains (Fig. 17-1G).

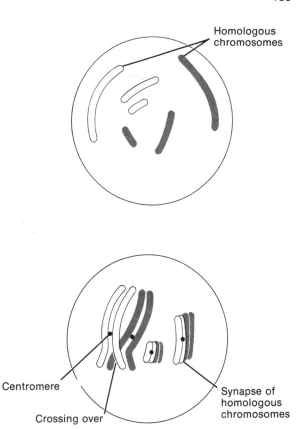

FIGURE 17-2
First meiotic division, showing crossing over.

2. Meiosis in *Ascaris*

The stages in the development of the gametes are called **gametogenesis** in animals. In males, gametogenesis occurs in the testis and, in females, in the ovaries or oviducts. In this part of the exercise the maturation of the egg (**oogenesis**) will be studied as it occurs in the parasitic roundworm *Ascaris*. Since the diploid number of chromosomes is only four, this worm is ideal for the study of this process.

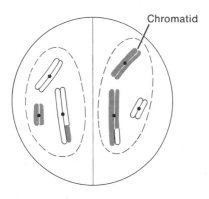

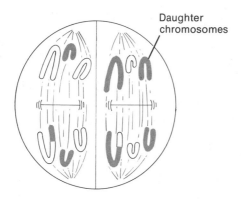

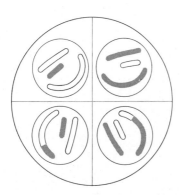

FIGURE 17-3
Second meiotic division, showing the separation of the daughter chromosomes.

The reproductive organ in *Ascaris* consists of a pair of long, highly coiled tubes that are regionally divided into the ovary, oviduct, and uterus (Fig. 17-4A). The "eggs," which are produced in the ovaries, pass into the oviducts where they are fertilized by sperm.

a. Examine composite slides of the oviduct and uterus of *Ascaris* and locate the stages of meiosis I, described below. Locate the oviduct, which characteristically has large numbers of triangular sperm interspersed among numerous "eggs" (Fig. 17-4B). The "eggs" at this stage are still diploid, since the maturation process, or oogenesis, does not occur until after the "egg" is penetrated by a sperm. Therefore, the term "egg" at this stage of development is not quite accurate. A more correct term would be **primary oocyte**, a cell that undergoes meiosis and produces the mature egg. On the slide, locate primary oocytes; some of them may have been penetrated by sperm. How many chromosomes are in the primary oocyte of *Ascaris*?

_____

After penetration by a sperm the primary oocyte secretes a thick substance that forms a shell. Suggest a function for this shell?

_____

At this time the homologous chromosomes of the primary oocyte undergo synapsis and form tetrads. What occurs at this time that makes this stage in meiosis so important?

_____

Locate primary oocytes in which synapsis has taken place (Fig. 17-4C). At anaphase of meiosis I, the homologous chromosomes, each still consisting of a pair of chromatids, move to opposite poles. At telophase an unequal division of the cytoplasm occurs (**cytokinesis**) so that the first meiotic division produces one relatively large cell (**secondary oocyte**) and a tiny one called a **first polar body**. How many chromosomes are found in the first polar body of *Ascaris*?

_____

How many are found in the secondary oocyte?

_____

Locate the first polar body on the slide (Fig. 17-4C). Describe its position with respect to the secondary oocyte.

_____

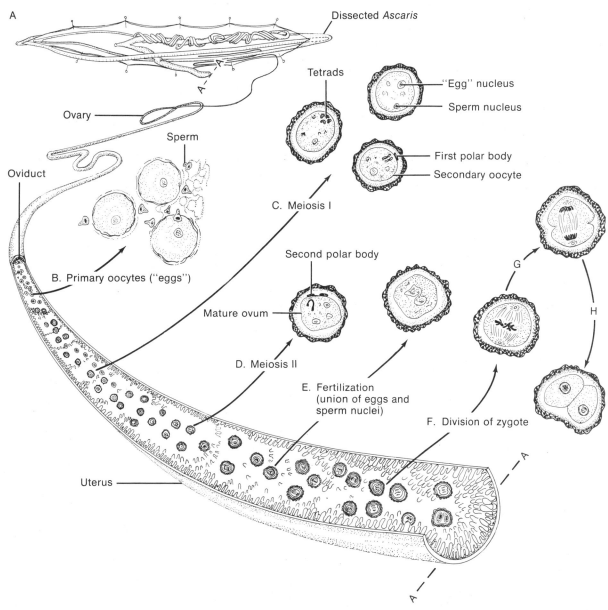

FIGURE 17-4
Meiosis in *Ascaris*.

b. After an interphase, meiosis II begins when the homologs in the secondary oocyte, which separated during the first meiotic division, line up on the equator of a division spindle. Each homolog is composed of two chromatids, which, at anaphase of meiosis II, separate and migrate to the poles. The second meiotic division of the secondary oocyte produces a tiny second polar body and a large cell, which soon differentiates into the egg cell or ovum. The first polar body produced in meiosis I may or may not go through the second meiotic division. Thus, when a diploid cell in the *Ascaris* ovary undergoes complete meiosis, only one mature ovum is produced; the polar bodies are essentially nonfunctional.

The unequal cytokinesis of oogenesis insures that an unusually large supply of cytoplasm and stored food is allotted to the nonmotile ovum for use by the embryo that will develop from it. In fact, the ovum provides almost all the cytoplasm and initial food supply for the embryo. The tiny, highly motile sperm cell contributes essentially only its genetic material.

c. During the maturation of the egg, the sperm nucleus has been lying inactive in the cytoplasm. Following the second meiotic division, the egg and sperm nuclei unite (**fertilization**) and form a single cell called a **zygote**. Locate eggs that show separate sperm and egg nuclei and eggs in which these two nuclei have fused (Fig. 17-4E). How many chromosomes does the zygote contain?

_____

If you did not know the actual number of chromosomes in the zygote of a plant or animal, how would you describe the chromosome content of this cell?

_____

_____

d. In the uterus, the zygote nucleus soon divides mitotically and forms two cells, each of which again divides until a multicellular embryo is formed (Fig. 17-4F, G, and H). Locate as many of these later stages as are evident on your slide.

B. MUTATION IN DROSOPHILA

The common fruit fly, *Drosophila melanogaster*, has proved to be a highly favored organism for use in genetic studies. In addition to ease of breeding (mature offspring are obtained in three weeks), a large number of mutation have been described in these flies. A **mutation** may be simply defined as that change in a gene or chromosome which is potentially transmissible to the offspring. Mutations may cause very obvious morphological changes—such as loss of pigmentation, which is chromosomally (gene) inherited—or they may condition very subtle changes—such as sensitivity to carbon dioxide concentration, which is cytoplasmically inherited. Hundreds of mutations of both types have been recorded in *Drosophila*. These mutant traits, many of which are obvious morphological changes, allow us to examine and study the transmission of characteristics from parent to offspring by direct observation, without using highly complicated biochemical techniques.

The fruit fly, as found in nature, is called the wild type; it characteristically has dark red eyes, a tannish, bristle-covered body, and a long, straight pair of wings reaching just beyond the tip of the abdomen (Fig. 17-5). In addition, males (symbol ♂) may be differentiated from females (symbol ♀) as follows: males have somewhat smaller bodies, a rounded, heavily pigmented abdomen, and a "sex comb" (a tuft of bristles) on the forelegs. In contrast, the abdomen of the female is somewhat pointed, is traversed by several dark stripes, and may have a terminal tuft of short bristles.

In this part of the exercise you will become familiar with some of the common mutants of *Drosophila*. Then you will be given several flies in which you are to identify the mutant traits they possess.

Obtain a vial of wild-type fruit flies. Tack a piece of cotton onto a cork that will fit the vial (Fig. 17-6). Add 2–3 drops of ether to the cotton. Gently tap the vial to force the flies to the bottom. Replace the cotton plug with the cork and ether-soaked cotton. When all of the flies have stopped moving, turn them onto a white card and carefully examine them with a dissecting microscope or hand lens. Become familiar with male and female characteristics by recording your observations in Table 17-1.

Obtain from your instructor a numbered vial containing a mixture of the mutant flies listed below:

- **Vestigial.** A wing mutant characterized by highly reduced withered wings.
- **White.** A mutant whose eyes appear white.
- **Bar.** The chief effect shown in this eye mutant

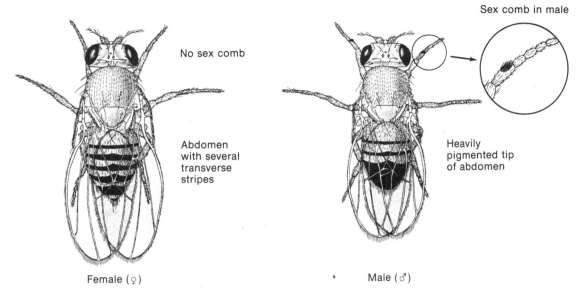

FIGURE 17-5
The adult fruit fly, *Drosophila melanogaster*.

TABLE 17-1
Comparison of male and female fruit flies.

Characteristic observed	Comparison	
	Male (♂)	Female (♀)
Which is relatively larger in overall size?		
What is the difference in banding on the abdomen?		
What is the shape of the tip of the abdomen?		
Are sex combs present or absent?		

TABLE 17-2
Unknown mutant flies.

Mutation	Sex

is the reduction in the number of facets of the eye, resulting in the appearance of a "bar" down the middle of the eye.
- **Black.** The body of this fly is black.

Etherize the flies as shown in Fig. 17-6 and determine the nature of the mutant flies in your vial. Record your observations in Table 17-2.

C. MUTATION IN CORN

1. Segregation of an Albino Mutant

Many plants reproduce by means of seeds formed following the fertilization of an egg cell by a sperm. Under proper conditions of light, temperature, and moisture each seed is capable of growing into a new plant.

Corn is known to have genes that regulate the production of chlorophyll. Therefore, if, during the development of the parent plant, a mutation occurs to the genes regulating chlorophyll synthesis, this mutation may be transmitted to the offspring resulting in plants that lack chlorophyll. Plants lacking this pigment are called **albinos**.

Obtain from your instructor the special corn seeds needed for this experiment. Plant these as directed in flats containing sand or vermiculite (Fig. 17-7). Periodically water the flats so that they are kept moist but not saturated. Examine the flats in about a week. When the seedlings have emerged, count the number of green and albino

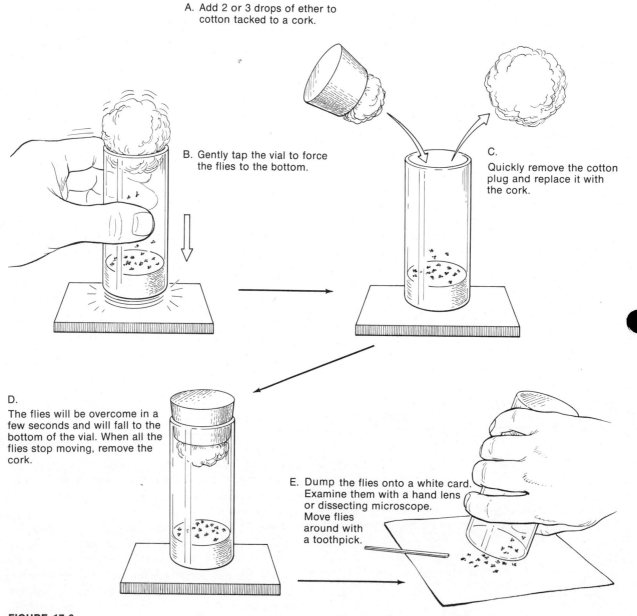

FIGURE 17-6
Procedure for etherizing fruit flies to observe body features.

MENDELIAN INHERITANCE

shoots. Tabulate your results in Table 17-3.

In the last column determine the ratio of green to albino shoots. To obtain a larger sample, tabulate the results obtained by other biology sections.

How does the green:albino ratio of your class compare with the green:albino ratio of all the classes?

_____

_____

Explain any large differences.

_____

_____

TABLE 17-3
Albino corn data.

Section number	Green shoots	Albino shoots	Green : albino ratio
Totals			

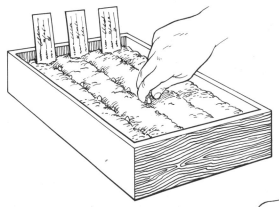

A. Plant corn seeds (previously soaked in water for 2–3 hours) about ½″ deep in sand or vermiculite.

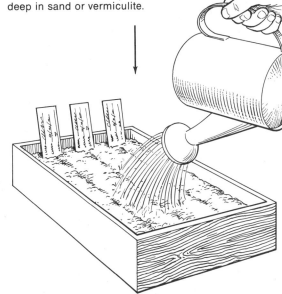

B. Water plants periodically, but not so much that water runs off the surface.

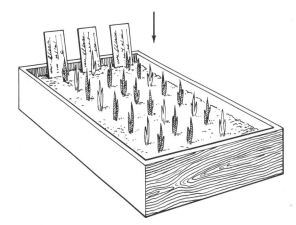

C. When plants emerge count number of green and albino seedlings and determine green:albino ratio.

FIGURE 17-7
Procedure for studying chlorophyll mutants in corn.

On the basis of your results, is the potentiality for albinism in corn transmitted as a dominant or recessive gene? Explain.

_____

_____

Under what mating circumstances would you expect to obtain a 1:1 ratio?

_____

_____

_____

Would this be as easy to obtain as the ratio you found? Explain.

_____

_____

_____

2. Temperature-Sensitive Mutants

It was suggested in the preceding part of this exercise that a mutation is an all-or-none phenomenon; that is, the corn is either green or albino. However, the expression of mutation in some cases may be modified by various treatments so that a gradation is seen in the mutation being expressed. In this study you will examine the effect that temperature exerts on the expression of pigmentation in corn.

a. Plant 10 previously soaked corn seeds provided by your instructor in each of three different trays. Label with your name, date, and section number.

b. Place one tray at each of the following temperatures: 50°F; 60°F; 75°F. Water daily.

c. Examine the plants in 5–7 days. Describe the pigmentation of the shoots grown at different temperatures.

_____

_____

_____

d. Suggest a mechanism for the effect of temperature on pigmentation in corn.

_____

_____

_____

_____

D. INHERITANCE OF HUMAN BLOOD GROUPS

The discovery of human blood groups was reported by Dr. Karl Landsteiner in 1900. This important discovery, for which Landsteiner received the Nobel Prize, led to the establishment of four blood groups: O, A, B, and AB. The basis for these groups is the presence of naturally occurring molecules (**antigens**) on the surface of red blood cells. These antigens stimulate certain white blood cells to produce a special class of globular protein molecules, called **antibodies**. The antibodies combine with the antigens to form a complex that is then eliminated from the body by the phagocytes, the ameboid white blood cells.

The two antigens discovered by Landsteiner are polysaccharides, called A and B. If you have blood type A, then the surface of your red blood cells has a specific polysaccharide, "A." Persons with type B have polysaccharide "B" on their blood cells; persons with type AB have both types of polysaccharides; and persons with type O have neither A nor B polysaccharides on their red blood cells. Individuals with type A have antibodies against type B red cells in their blood. Similarly, type B persons have antibodies against A, type O have antibodies against both A and B, and type AB have neither antibody in their blood.

If a person having blood type A is given a transfusion of blood type B or AB, the naturally occurring anti-B antibodies in his blood will **agglutinate** (clump) the red blood cells of the donor (B- or AB-type red blood cells). This reaction can be so violent that it is sometimes fatal. However, such A-type individuals can be safely transfused with O-type cells since A carries no antibodies against O. In a similar manner O can be transfused to B and AB. For this reason, individuals with O-type blood are known as **universal donors**. If you have

MENDELIAN INHERITANCE

TABLE 17-4
Human blood groups.

Phenotype	Genotype	Agglutinogen (Antigen found on surface of red blood cells)	Agglutinin (Antibody found in serum)
O	OO	none	anti-A and anti-B
A	AA or AO	A	anti-B
B	BB or BO	B	anti-A
AB	AB	A and B	none

TABLE 17-5
Determination of human blood groups.

Unknown red blood cells from individual	Serum used and reaction		Phenotype
	Anti-A	Anti-B	
1	+	−	A
2	+	+	AB
3	−	−	O
4	−	+	B

AB-type red blood cells, you are a **universal recipient** since you have no naturally occurring antibodies to A- or B-type red blood cells and therefore neither A nor B polysaccharides will be foreign to your blood.

The antigens that determine the four blood groups are the result of the expression of three genes, O, B, and A. A and B are apparently dominant over O. The genotype AA cannot be distinguished from AO and the genotype BB cannot be distinguished from BO. However, genotypes AA and AO can be distinguished from BB and BO, which leads to **phenotypes** A (for genotypes AA and AO) and B (for genotypes BB and BO). Thus while only four phenotypes can be recognized (A, B, AB and O), six genotypes occur (OO, AO, AA, BO, BB, and AB). The distribution of the antigens and antibodies in the human blood groups are shown in Table 17-4.

Many additional red blood cell types have been identified besides the O, A, B, and AB series. For example, we have identified the M, N series, the Lewis blood groups, and the Rh series, which is so important in certain pregnancies. If a woman is Rh negative (i.e., she lacks the Rh factor on her red blood cells) and her husband carries the Rh factor on his cells, then any child born to these parents will be Rh positive. Under these circumstances if any of the child's blood should cross the placenta the mother may form antibodies against the Rh factor on her child's red blood cells. If these antibodies enter the infant's circulation, agglutination of the infant's red blood cells may occur, which would endanger its life.

Cells containing antigen A are agglutinated when mixed with anti-A antibodies. A similar response occurs when cells containing antigen B are mixed with anti-B antibodies. The degree of agglutination depends upon such factors as the concentration of antibodies in the serum. Thus when blood is transfused, the antigen of the donor and the antibody of the recipient are all that is important because the antibodies in the blood of the donor are extensively diluted by the blood of the recipient and thus are essentially ineffective in agglutinating the recipient's blood.

To determine your own blood type it is necessary to have only two reagents: one containing anti-A antibodies and the other containing anti-B antibodies. Each of the two antibodies should

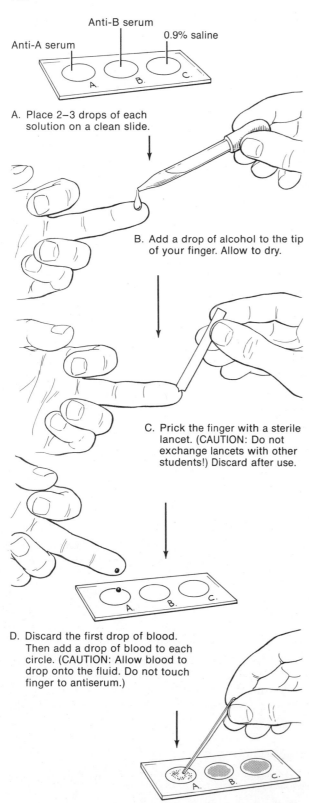

FIGURE 17-8
Procedure for determining human blood group phenotypes.

agglutinate the red blood cells containing the corresponding antigen. Table 17-5 shows the results that will be obtained if cells of the four different blood types are used. A plus (+) sign indicates agglutination of cells and a minus (−) sign indicates no agglutination.

Draw three circles with a wax pencil on a clean microscope slide. Label one circle "A" (for anti-A), one "B" (for anti-B), and the third "C." Add a drop of anti-A serum (colored blue for identification purposes) to circle "A," and the anti-B (yellow) to circle "B."

Add a drop of 0.9% saline to circle "C" (Fig. 17-8). Why is this control needed?

_____

_____

Apply alcohol to the area of the finger to be pricked. Prick the tip of the finger with a sterile, disposable lancet provided by the instructor. Squeeze the finger and wash off the first drop of blood. Continue to squeeze the finger and apply a large drop of blood to each circle. Quickly stir each drop with a separate toothpick to get a uniform mixture, and after a few minutes note if any reactions have occurred. After obtaining the blood, clean the finger with alcohol.

If agglutination has occurred only in "A," your blood type is A; if in "B," your blood type is B. If agglutination has occurred in both "A" and "B," your blood type is AB; and if there is no agglutination in either "A" or "B," your blood type is O. What is your blood group phenotype?

_____

What is your blood genotype?

_____

If you are typing your blood as part of a class project, tabulate the blood types of other students using Table 17-6. Compare the incidence (%) of blood types in your group with the incidence of human blood groups in the United States as listed in Table 17-7.

A knowledge of the blood groups is of considerable importance in medicine, biology, and forensic (legal) medicine dealing with paternity. What blood types are not possible for children of two AB or two O parents?

_____

MENDELIAN INHERITANCE

TABLE 17-6
Summary of class blood types.

Blood type	Number of students with this blood type	Incidence (%)
A		
B		
AB		
O		

TABLE 17-7
Incidence of human blood groups in the United States.

Group	Incidence (%)			
	Whites	Negroes	Chinese	American Indians
O	45	48	36	23
A	41	27	28	76
B	10	21	23	1
AB	4	4	13	1

In a case of disputed paternity, the child is group O, the mother group A, and the putative father is AB. What inference can be drawn from this knowledge?

_____

_____

What if the putative father were blood group B?

_____

_____

Suppose children's identities become mixed up in a hospital nursery. How might blood typing help straighten out the matter?

_____

_____

REFERENCES

Demerec, M., and B. P. Kaufmann. 1967. *Drosophila Guide.* 8th ed. Carnegie Institution, 1530 P Street, N.W., Washington, D.C.

Hadorn, E. 1962. Fractionating the Fruit Fly. *Scientific American,* April (Offprint No. 1166). *Scientific American* Offprints are available from W. H. Freeman and Company, 660 Market Street, San Francisco 94104, and 58 Kings Road, Reading RG1 3AA, England. Please order by number.

Herskowitz, I. H. 1962. *Genetics.* Little Brown, Boston.

Hyland Laboratories. 1965. *Hyland Reference Manual of Immunohematology.* 3d ed. Los Angeles.

Michael Reese Research Foundation. 1963. *Blood Grouping, Anti-Rh, and Anti-Human Serums.* Chicago.

Moody, P. A. 1967. *Genetics of Man.* Norton, New York.

Moore, J. A. 1963. *Heredity and Development.* Oxford University Press, New York.

Ortho Diagnostics. 1969. *Blood Group Antigens and Antibodies as Applied to the ABO and Rh Systems.* Raritan, New Jersey.

Race, R. R., and R. Sanger. 1968. *Blood Groups in Man.* 5th ed. Davis, Philadelphia.

Srb, A. M., R. D. Owen, and R. S. Edgar. 1965. *General Genetics.* 2d ed. W. H. Freeman and Company, San Francisco.

Stern, C. 1960. *Principles of Human Genetics.* 2d ed. W. H. Freeman and Company, San Francisco.

Watson, J. D. 1970. *Molecular Biology of the Gene.* 2d ed. Benjamin, New York.

EXERCISE 18

Biochemical Genetics

The science of genetics has passed through several phases during the past seventy years. Each period was initiated by some important event.

The first period began with the rediscovery of Mendel's papers in 1900. The results of this period demonstrated the universal application of the laws of heredity; that is, the transmission of heritable characteristics is brought about by the same mechanism that regulates Mendelian segregation.

The second period, beginning around 1910, saw the introduction of a new "tool" to be used in genetic research—the fruit fly, *Drosophila melanogaster*. From this period came the experimental evidence that genes are located in linear order on the chromosomes.

Further studies on the structure of chromosomes and the discovery that radiation (such as X-rays) caused mutations (a potentially transmissible change in a gene or chromosome) marked the beginning of the third period of activity. This information was developed further during the succeeding years.

The fourth period introduced microorganisms in genetic studies and provided evidence that the genes regulated the activities of cells by controlling the production of enzymes.

The formulation of the Watson-Crick model of the molecular structure of deoxyribonucleic acid (DNA) opened the present era of molecular genetics. DNA is unique in three respects. First, it is a very large molecule, having a certain uniformity of size, rigidity, and shape. Despite this uniformity, however, it has almost infinite internal variety that gives it the complexity required for information-carrying purposes. The second characteristic of DNA is its capacity to make copies of itself with remarkable exactness. The biologist or chemist would say that such a molecule can **replicate** itself. The third characteristic is its ability to transmit information to other parts of the cell. The

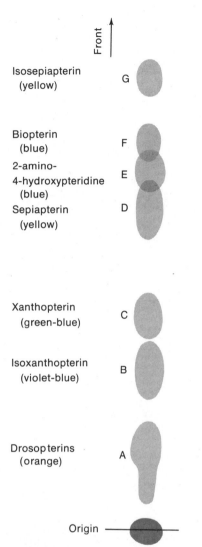

FIGURE 18-1
Pteridines of the wild-type fruit fly. The chemical structures were worked out by Max Viscontini, Institute of Organic Chemistry, University of Zurich. (From "Fractionating the fruit fly" by Ernst Hadorn. Copyright © 1962 by Scientific American, Inc. All rights reserved.)

behavior of the cell reflects the information transmitted.

The past several years have shown that this genetic information (carried in the chemical structure of DNA) consists of a four-letter alphabet of which the "words" and "sentences" of the genetic material and the role it plays in determining the structure of proteins will be regarded as one of the greatest discoveries of all time.

A. CHROMATOGRAPHIC SEPARATION OF DROSOPHILA EYE PIGMENTS

In 1941 George W. Beadle and Edward L. Tatum presented experimental evidence supporting the concept that genes control the chemical activities of the cell by controlling production of those biochemical catalysts called enzymes. Enzymes, in turn, catalyze the numerous chemical reactions occurring in cells that ultimately are expressed morphologically, physiologically, biochemically, or in the behavior of an adult organism.

Eye pigmentation in the fruit fly *Drosophila melanogaster* is under genetic control. The "normal" eye color of this insect is correlated with the presence of a characteristic series of substances called **pteridines** (derived from the Greek pteron, meaning wing). This term was chosen because the first substances in this class of compounds were extracted from butterfly wings. These compounds, when viewed under ultraviolet light, produce distinctive fluorescent patterns (Fig. 18-1). Pteridines occur in many invertebrates, in certain pigment cells of amphibians and fishes, and in plants where they may participate in photosynthesis.

In this study you will analyze chromatographically the pteridines in the normal (wild-type) fruit fly and compare it with several eye-color mutants to establish that eye-color mutations are accompanied by differences in pteridine patterns.

1. On a square sheet of thin-layer chromatography paper (about 8" × 8") draw a pencil line parallel to, and about 1 inch from, one edge (Fig. 18-2). Lightly pencil in several evenly spaced dots along the "origin" line. The number of dots will depend on the number of flies provided by your instructor. *Do not* handle the paper too much, as fingerprints may interfere with separation.

2. Obtain three wild-type flies and three of each of the mutants that you have killed by overetherization. Choose mature adult males or females. Do not mix the sexes. The wild-type fruit fly has dark red eyes, a tannish, bristle-covered body, and a long, straight pair of wings reaching just beyond the tip of the abdomen (Fig. 18-3). The adult males (symbol ♂) are differentiated from females (symbol ♀) by their somewhat smaller bodies, a rounded, heavily pigmented abdomen, and a "sex comb" (a tuft of bristles) on the forelegs. The abdomen of the female is somewhat pointed, is traversed by several dark stripes, and may have a terminal tuft of short bristles.

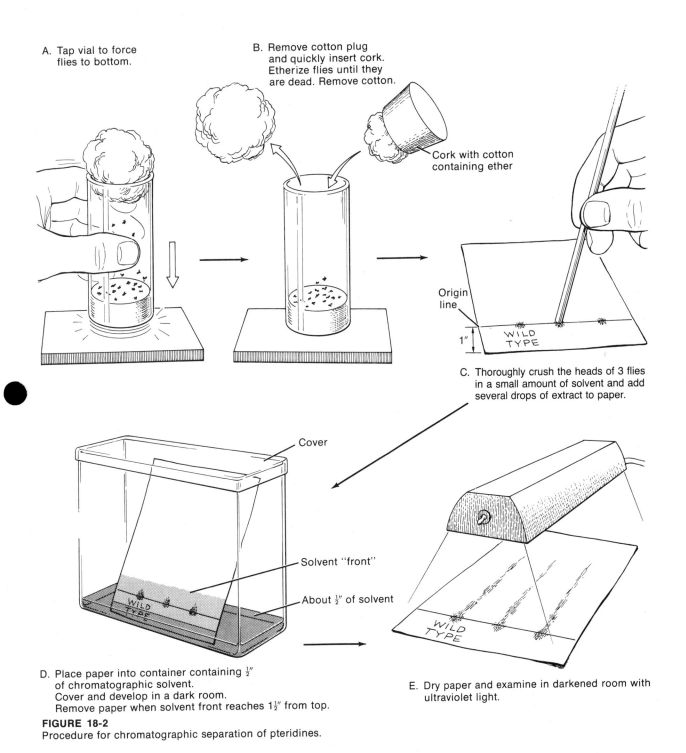

FIGURE 18-2
Procedure for chromatographic separation of pteridines.

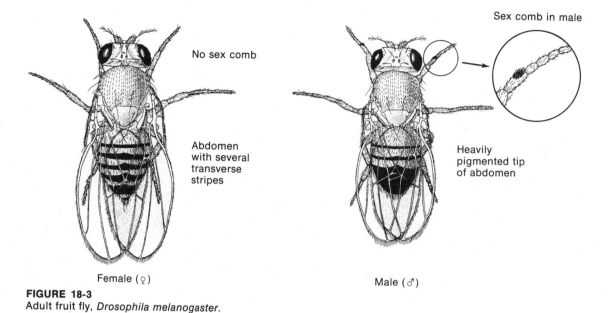

FIGURE 18-3
Adult fruit fly, *Drosophila melanogaster.*

3. Remove the heads of the three wild-type flies and place them in a vial containing about 0.25 ml of the chromatography solvent. Crush the material with a glass rod to dissolve the pigments. Using a capillary tube, add several drops of the pigment to the silica gel. Allow the spot to dry between applications. Be careful not to disturb the silica-gel coating any more than necessary. Label for identification. Repeat this procedure for each of the eye-color mutants. What is one control that should be used?

_____

Add this control to the filter paper.

4. Allow the spots to dry for several minutes. Then place the paper into the solvent in the chromatography jar (Fig. 18-2). Since the pteridines are light-sensitive, the chromatograms should be developed in a dark room (or cover the jar with foil).

5. Allow the chromatogram to develop until the solvent front reaches to within 1½ inches of the top of the paper.

6. Remove the paper from the tank and air-dry for several minutes. Examine the paper, using either ultraviolet light of "short" wavelengths (250 nm) or "long" wavelengths (360 nm). (CAUTION: Do not look directly into the UV lamp.) Note the fluorescent colors of the various pteridines (Fig. 18-1). In Table 18-1 check those pteridines present in the flies you examined.

Discuss the eye-color mutations in terms of the genetic control of enzyme synthesis and activity.

_____
_____
_____
_____
_____

If time permits you might consider using this technique to answer the following questions.

1. Do adult ♂ and ♀ flies have the same pteridine patterns?
2. Do the patterns of pteridine synthesis change during the development of the fly from the egg to the adult?
3. Will body-color mutants show pigment differences similar to those seen in eye-color mutants?

B. INDUCTION OF MUTATION BY ULTRAVIOLET LIGHT

The capacity for mutation is inherent in the genetic material of all living organisms, as well as in viruses. Although mutations may occur spontaneously, the frequency with which mutations occur can be increased by a variety of agents called **mutagens**.

BIOCHEMICAL GENETICS

TABLE 18-1
Distribution of pteridines in the wild type and eye-color mutants of *Drosophila*.

Pteridine	Fruit fly eye type					
	Wild-type	Mutants				
Isosepiapterin (yellow)						
Pterins (sky-blue)						
Sepiapterin (yellow)						
Xanthopterins (green-blue)						
Drosopterins (orange)						

One of the more commonly used mutagenic agents is ultraviolet light. Ultraviolet light produces its effects (at least in part) by causing breaks in chromosomes that result in the loss or rearrangement of parts.

In this experiment, ultraviolet radiation will be used to study the mutation rate of the antibiotic-producing mold *Penicillium*. This study will be limited to such obvious mutations as shape and pigmentation of the colony and the effects on growth rate.

1. Working in pairs obtain two Petri dishes containing nutrient medium. Label one "Control" and the other "UV," and add your name, date, and laboratory section number.

2. Add 1 ml of a *Penicillium* spore suspension to each of the plates as follows (Fig. 18-4).
   a. Gently shake the flask containing the spores to make the contents uniformly suspended.
   b. Fill the pipet in the flask to the 1-ml mark. Add the spore suspension to the surface of the agar in the "Control" dish. Then, holding the agar plate at eye level, tilt the plate, allowing the suspension to run to the far edge. (CAUTION: Do not tilt it so far that the spore suspension runs over the lip of the plate.) Then hold the plate horizontally and gently jerk it toward you. Repeat this several times, rotating the plate part way between "jerks". In this way the spore suspension will cover the entire surface of the agar.
   c. Repeat steps (a) and (b) for the second plate.

3. Set the plate aside for ¾ to 1 hour, so that the spores settle onto the agar. Do not move the plates during this period.

TABLE 18-2
Effects of ultraviolet radiation on *Penicillium*.

Penicillium	Class data		Team data	
	Control	UV	Control	UV
Total number of colonies				
Number of mutant colonies				
Percentage of mutant colonies				
Percentage of spores surviving irradiation				

4. Irradiate the "UV" plate as follows (Fig. 18-4).
   a. Turn the UV lamp on several minutes before irradiation.
   b. Place the plate, cover on, under the UV lamp.
   c. Remove the cover and expose the plate to the UV light for 35–40 seconds. Replace the cover and remove the plate.

5. Incubate the plates for one week at 28°C. Then count the number of *Penicillium* colonies that have grown on the "Control" and irradiated plates. (Each colony arises from one spore.) Record your data in Table 18-2.

6. Carefully study the appearance of the colonies on the "Control" plate. These are the normal or wild-type colonies. They should be bluish-black in color, with a narrow white fringe. If available, examine color transparencies of wild-type colonies

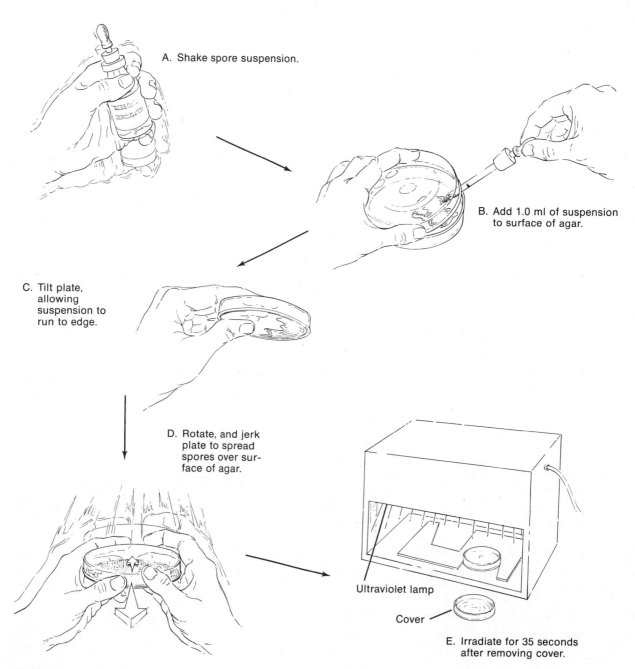

FIGURE 18-4
Ultraviolet irradiation of *Penicillium* spores. (Adapted from *Plants in Perspective* by Eldon H. Newcomb, Gerald C. Gerloff, and William F. Whittingham. W. H. Freeman and Company. Copyright © 1964.)

BIOCHEMICAL GENETICS

to aid you in your identification of the normal condition. If any of the colonies on the control plate deviate from the wild-type (for example, in colony size, margin width, color, or surface appearance), consider them to be mutants. Would you expect to find mutant colonies on the "Control" plate? Explain.

_____

_____

What evidence is there that UV radiation is indiscriminate in its ability to induce mutations?

_____

_____

_____

The mutations studied in this exercise were expressed as gross visible changes. Is it possible that some of the wild-type colonies are carrying UV-induced mutations? Explain.

_____

_____

_____

_____

_____

C. MOLECULAR BASIS OF HEREDITY

At the turn of the century the rediscovered classical work of Mendel laid the foundations of modern genetics. The nature of the **gene**—the functional unit of inheritance—was defined. Subsequently, as a result of numerous ingenious experiments, the Mendelian laws of inheritance and the chromosome theory of inheritance were developed.

About twenty years ago, studies showed that, through the mediation of a certain chemical substance, the parental cell passes its own characteristics to its daughter cells; that is, the daughter cells inherit the capacity to become nerve cells or muscle cells or to secrete hormones from "information" supplied by the parent cell. This "genetic information" was found to lie in the structure of the chemical compound known as DNA (deoxyribonucleic acid). The first evidence that DNA was the "genetic material" came from experiments using bacteria.

The chromosomes contain a **nucleoprotein complex**, which consists of DNA, RNA (ribonucleic acid), and several types of protein. It is possible to extract the nucleoproteins from bacterial cells and to separate them into DNA and protein fractions. If these fractions are added to a medium in which other bacteria are growing, the foreign DNA and protein will be absorbed by the bacteria. Avery, MacLeod, and McCarty provided evidence for the link between DNA and the genetic material, using two different strains of bacteria—one having a rough coat and the other having a smooth coat. The nucleoprotein of the rough-coated bacteria was extracted and separated into DNA and protein fractions. Each fraction was added to separate cultures of smooth-coated bacteria. The results showed that the smooth-coated bacteria, to which DNA had been added, acquired the rough-coated characteristic (Fig. 18-5). Moreover, the offspring of these bacteria gave rise to more rough-coated bacteria. This **bacterial transformation** was brought about specifically by DNA.

In this part of the exercise you will first establish that DNA is a constituent part of the chromosomes and examine changes in chromosome morphology that appear to be correlated with changes in development. Second, you will become familiar with the molecular structure of DNA and the mechanism by which "genetic information" is transferred from DNA to the cytoplasm of the cell.

1. Chromosome Morphology

In your studies of mitosis and meiosis, the chromosomes appeared as heavily-stained bodies that apparently lacked distinct morphology. However, if you examine the giant chromosomes found in the salivary glands of *Drosophila* larvae (and of certain other flies), you will become aware of a great deal of structural detail that is not apparent in "ordinary" chromosomes.

a. Examine Fig. 18-6. This is a photograph of the giant salivary gland chromosomes of the "midge" fly, *Chironomus*. (If they are available, supplement your studies with specially prepared slides of these chromosomes.) Note the darkly stained bands that are separated from each other by lighter areas.

These bands are areas of DNA and show up when these chromosomes are stained by Feulgen's reagent, a dye that stains DNA.

b. In the center of the photograph (arrow) note the appearance of a chromosome "puff." It has been shown that the pattern of "puffing" changes during the development of this organism. Recent experimental evidence suggests that these puff regions are sites of special gene activity, where a special "message-carrying" chemical is formed.

It has further been shown that the hormone involved in the insect molting process can induce puffing. This hormone, called **ecdysone**, appears to act on specific genes, which in turn generate a "message" that is transferred to the cell proper where it initiates the synthesis of specific protein. It has also been demonstrated that one of the plant growth regulators—**gibberellin**—will accelerate molting and that ecdysone, when used in one of the standard gibberellin bioassays, brings about a growth response similar to that of gibberellin.

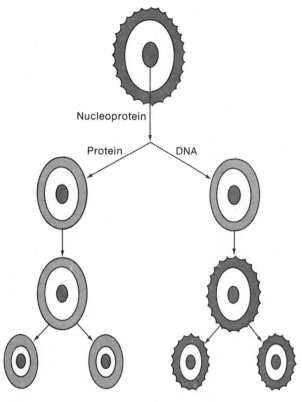

FIGURE 18-5
Bacterial transformation by DNA.

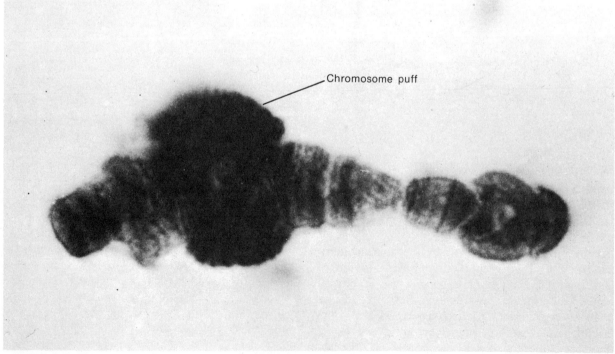

FIGURE 18-6
Giant salivary gland chromosomes of *Chironomus*. (Photomicrograph by Claus Pelling, Max Planck Institute for Biology.)

BIOCHEMICAL GENETICS

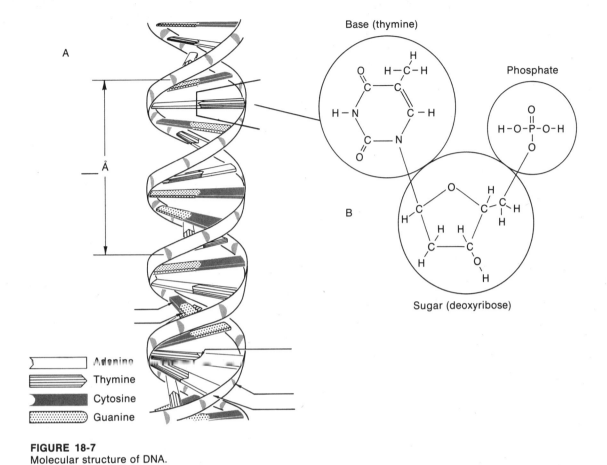

FIGURE 18-7
Molecular structure of DNA.

2. The Molecular Structure of DNA

In the early 1950's Watson and Crick formulated their classic model of the molecular structure of deoxyribonucleic acid—DNA. A three-dimensional model based upon the Watson-Crick formulation is available in the laboratory.

On the DNA model (and Fig. 18-7A) note that the molecule consists of nitrogen compounds, phosphates, and sugars, which are joined in two coiling strands that form a double helix. Note that each strand consists of alternating sugar and phosphate molecules linked together. Attached to each sugar, toward the inside, is a nitrogen base. The bases of the two strands are paired and are connected to one another by hydrogen bonds. There are four different bases: adenine (A), thymine (T), cytosine (C), and guanine (G). What pattern of base pairing is evident?

_____

The base-sugar-phosphate combination constitutes a nucleotide (Fig. 18-7B). The distance between successive base pairs is 3.4 Å units. How many base pairs are there in one complete turn of the helix?

_____

How many Å units would one turn be?

_____

Label the model of the DNA molecule diagrammed in Fig. 18-7. This model represents only a very small segment of the DNA molecule. Indeed, if it were possible to remove the DNA from a cell and stretch it out as a single, fine thread, it would cover a length of about 3 feet.

Each plant or animal begins life as a single cell that undergoes mitosis to form the multicellular adult. This suggests that DNA must also be able to replicate itself. In Fig. 18-8, suggest (by drawing) a method of DNA replication.

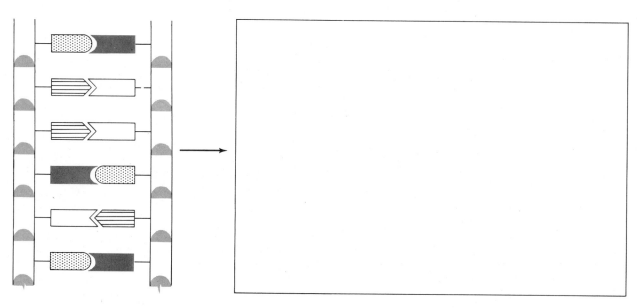

FIGURE 18-8
Suggested method of DNA replication.

In terms of your knowledge of the structure of DNA, suggest a possible mechanism of mutation at the molecular level.

_____

_____

_____

REFERENCES

Avery, O. T., C. M. MacLeod, and M. McCarty. 1944. Studies on the Chemical Nature of the Substance Inducing Transformation of Pneumococcal Types. Induction of Transformation by a Desoxyribosenucleic Acid Fraction Isolated from Pneumococcus, Type III. *J. Exp. Med.* **79**:137–157.

Barish, N. 1965. *The Gene Concept.* Reinhold, New York.

Callan, H. G. 1963. The Structure of Lampbrush Chromosomes. In *Genetics Today. Proc. Intern. Congr. Genet., 11th* **2**:323–325. Pergamon, New York.

Demerec, M., and B. P. Kaufmann. 1967. *Drosophila Guide.* 8th ed. Carnegie Institution, 1530 P Street, N.W., Washington, D.C.

Hadorn, E. 1959. Contributions to the Physiological and Biochemical Genetics of Pteridines and Pigments in Insects. *Proc. Intern. Congr. Genet., 10th* **1**:337–354. University of Toronto Press, Toronto.

Hadorn, E. 1961. *Developmental Genetics and Lethal Factors.* Wiley, New York.

Hadorn, E., and H. K. Mitchell. 1951. Properties of Mutants of *Drosophila melanogaster* and Changes During Development as Revealed by Paper Chromatography. *Proc. Nat. Acad. Sci.* **37**:650–665.

Heftmann, E., Ed. 1967. *Chromatography.* 2d ed. Reinhold, New York.

Hills, A., and A. Rosenfeld. 1963. DNA's Code: Key to All Life. *Life.* October 4, pp. 70–90. (Reprints of this article are available for $.35 each from *Life* magazine, Box D, Rockefeller Center, New York, 10020).

Leitenberg, M., and E. L. Stokes. 1964. *Drosophila melanogaster* Chromatography, I. *Turtox News* **42**:226–229.

Leitenberg, M., and E. L. Stokes. 1964. *Drosophila melanogaster* Chromatography, II. *Turtox News* **42**:258–260.

Tomasz, A. 1969. Cellular Factors in Genetic Transformation. *Scientific American*, January (Offprint No. 1130). *Scientific American* Offprints are available from W. H. Freeman and Company, 660 Market Street, San Francisco 94104, and 58 Kings Road, Reading RG1 3AA, England. Please order by number.

Van Norman, R. W. 1971. Chromatography. *Experimental Biology* 2d ed. Ch. 11. Prentice-Hall, Englewood Cliffs, New Jersey.

Wacker, A. 1963. Molecular Mechanisms of Radiation Effects. *Prog. Nucleic Acid Res.* 1:369–399, Academic Press, New York.

Watson, J. D., and F. H. C. Crick. 1953. Genetical Implications of the Structure of Deoxyribonucleic Acid. *Nature* (London) 171:964–967.

Watson, J. D. 1970. *Molecular Biology of the Gene.* 2d ed. Benjamin, New York.

EXERCISE 19

Fertilization and Early Development of the Sea Urchin

The development of the fertilized egg into the complex, interdependent system of tissues and organs that make up the adult animal is one of the more fascinating problems in developmental biology. In this exercise and the exercises that follow, the patterns of early development in an echinoderm (the sea urchin), an amphibian (the frog), and a bird (the chick) will be compared.

One important factor that determines the type of development an animal undergoes is the amount of yolk in the egg. The amount, position, and distribution of the yolk will markedly affect the patterns of cleavage and subsequent events in early development. In the more primitive eggs, the amount of yolk is small and the yolk is evenly distributed throughout the cytoplasm. This type of egg, called **isolecithal** (equal yolk), is found in the sea urchin. A second type of egg has a large amount of yolk that is displaced toward the vegetal pole. The cytoplasm is concentrated in the animal pole. This type of egg, called **telolecithal** (end yolk), is found in the frog and chick.

In this exercise you will observe fertilization and the early stages in the development of a marine echinoderm, the sea urchin. This animal, related to the starfish, can be induced to shed its gametes by injecting potassium chloride into the body cavity, or by stimulating it with a weak electric current. Each female will lay approximately one billion eggs and each male will eject several billion sperm.

A. FERTILIZATION

Prior to the laboratory meeting your instructor obtained living eggs and sperm from female and male sea urchins. The procedure used is shown diagrammatically in Fig. 19-1.

Obtain a depression slide and cover glass, or prepare a slide and cover glass as shown in Fig. 19-1. With a clean pipet, transfer a drop of the egg suspension to the slide and examine microscopically.

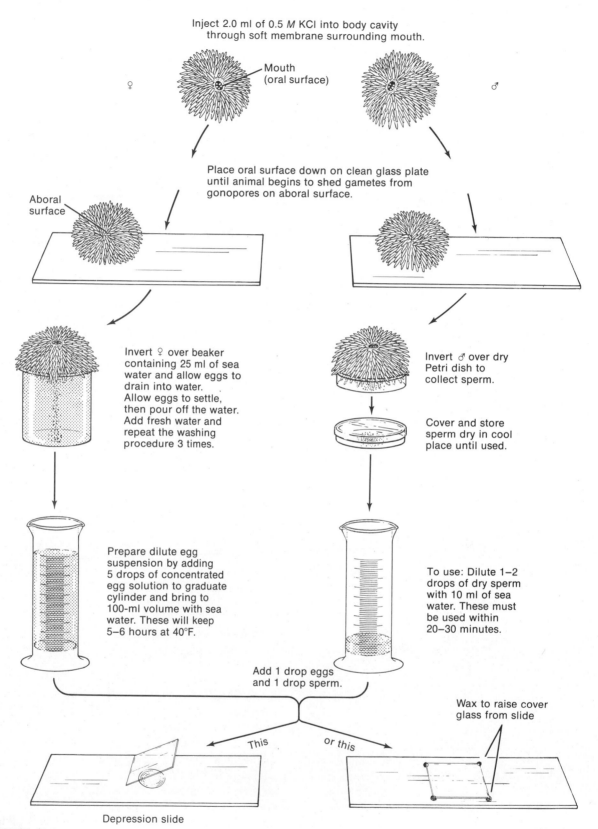

FIGURE 19-1
Preparing egg and sperm suspensions of sea urchins to observe fertilization and early development.

FERTILIZATION AND EARLY DEVELOPMENT OF THE SEA URCHIN

Add 1 drop of dilute sperm suspension to the eggs on the slide, and examine immediately under high power of the microscope. Note the time so that you will know when to expect the cleavage stages (see Table 19-1). The rate of development varies with temperature and therefore these eggs should be kept at temperatures not to exceed 70°F. What response do the sperm exhibit with respect to the eggs?

_____

_____

What might be the cause of this response?

_____

_____

Soon after a sperm penetrates the egg, a **fertilization membrane** begins to lift from the egg surface and completely surrounds the egg. The entire process—from the time the sperm touches the egg to the formation of the fertilization membrane—takes from 2 to 5 minutes.

The first cleavage does not occur until 50–70 minutes after the formation of the fertilization membrane. The changes taking place within the egg during this time are difficult to observe in living material. After the sperm enters the cell, the male nucleus migrates toward, and unites with, the female nucleus, reestablishing the diploid number of chromosomes. Subsequently, the chromatin is organized into chromosomes, and the first nuclear division, followed by cell division, occurs.

B. CLEAVAGE

The process by which the fertilized egg develops into a multicellular embryo is called **cleavage**. During this process complex events occur within the nucleus and cytoplasm. We will only be concerned with the more readily observable external changes.

While waiting for the first division, examine prepared slides of sea urchin or starfish eggs. On the slides, how can you tell if an egg is fertilized or unfertilized?

_____

_____

TABLE 19-1
Approximate time sequence for the development of fertilized sea urchin eggs.

Formation of fertilization membrane	2–5	minutes
First cleavage	50–70	minutes
Second cleavage	78–107	minutes
Third cleavage	103–145	minutes
Blastula	6	hours
Hatching of blastula	7–10	hours
Gastrula	12–20	hours
Pluteus larvae	24–48	hours

Next, prepare a slide of the living sperm. By reducing the light, you may be able to observe the flagellum by which the sperm moves. If, at this point, your eggs have not reached the first cleavage stage, continue with Part C.

From your study of meiosis you will recall that, in the development of the egg, polar bodies are formed during the first and second meiotic divisions. In the sea urchin the first cleavage begins at the region where the polar bodies are formed—the animal pole. If you carefully observe the egg, you should notice that it begins to elongate slightly before it divides. The second cleavage occurs approximately 15–20 minutes after the first cleavage. Describe the plane of this division with respect to the first cleavage.

_____

_____

Describe the plane of the third cleavage with respect to the first two cleavages.

_____

_____

If your preparation does not attain this stage of development, examine fertilized eggs that were prepared earlier by your instructor or examine your slide later in the day.

C. LATER STAGES IN DEVELOPMENT

Up to the 8-cell stage the cleavages have been uniform. Subsequent to the 8-cell stage, the 4 cells of the vegetal pole divide unequally, resulting in 4 large cells (**macromeres**) and 4 small cells (**micromeres**). Then the 4 cells of the animal pole divide into 8 equal-sized cells (**mesomeres**) (Fig. 19-2).

1. Blastula

As the cells of the embryo undergo additional cleavage, a spherical mass containing a fluid-filled central cavity is formed. With further cleavages, the cavity increases in size until the embryo, now called a **blastula**, consists of several hundred cells that become arranged in the form of a single-layered hollow ball. At this stage in the sea urchin, each cell of the embryo acquires a cilium and begins to rotate within the fertilization membrane. Approximately 8 hours after fertilization, the blastula breaks through the membrane by excreting a "hatching enzyme," which apparently digests the fertilization membrane. If you are unable to observe a living blastula, examine slides of this stage in the sea urchin or in related forms such as the starfish (Fig. 19-3A).

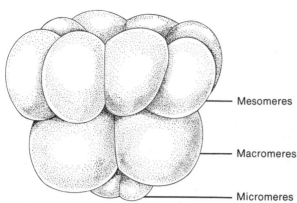

FIGURE 19-2
Unequal cell division in the sea urchin (fourth division).

2. Gastrula

For several hours after "hatching" the blastula swims actively. About 15 hours following fertilization, the single-layered blastula changes into a double-layered **gastrula** (Fig. 19-3C). In the sea urchin this process occurs by invagination of the blastula in the region of the vegetal pole. The invagination (movements of infolding or inward bending of the ectoderm) continues forward to the animal pole, where it will eventually meet the opposite wall. At this junction the mouth will form. The opening that marks the site of the original in-pocketing will become the anus in the adult.

A new cavity formed during invagination is called the **archenteron**. It is the forerunner of the digestive tract. During its formation it obliterates the blastocoel.

At the conclusion of gastrulation the embryo consists of an outer presumptive ectoderm and an inner presumptive endoderm that lines the archenteron. A third layer, the mesoderm, will develop between the ectoderm and endoderm from the proliferation of cells arising in the endodermal layer. These embryonic germ layers develop into the various tissue and organ systems of the adult animal. The skin, sense organs, and nervous system arise from the ectoderm. The muscles, bone,

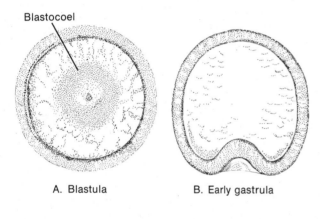
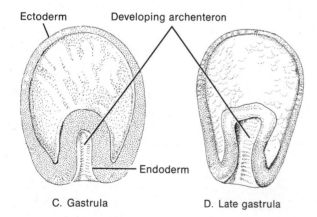

FIGURE 19-3
Blastula and gastrula of an echinoderm (starfish).

and blood originate in the mesoderm. The endoderm gives rise to the alimentary canal and its various derivatives, the pharynx, esophagus, stomach, liver, intestine and pancreas, and endocrine glands.

Examine living gastrulae or slides of this stage of development.

3. Pluteus Larva

During the later stages in the development of the gastrula, the skeleton begins to form as pairs of rodlike structures on either side of the "future anus." By 24 hours after fertilization, the larva is well formed. Examine slides or living specimens of the pluteus larva (Fig. 19-4A). When 2–2½ months old, the larva metamorphoses into the adult sea urchin (Fig. 19-4B). Examine specimens of the adult animal. Echinoderms are characterized by having numerous "tube feet." Where on the sea urchin are these "feet" located?

_____

What appears to be their function?

_____
_____
_____

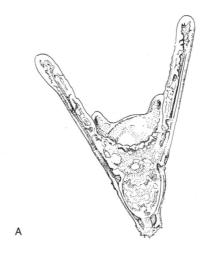

A

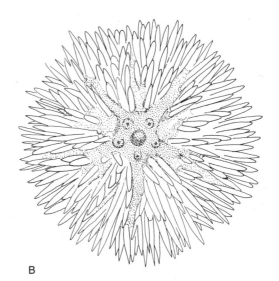

B

FIGURE 19-4
A. Pluteus larva. B. Adult sea urchin.

REFERENCES

Austin, C. R. 1965. *Fertilization*. Prentice-Hall, Englewood Cliffs, New Jersey.

Harvey, E. B. 1954. Electrical Method of Determining the Sex of Sea Urchins. *Nature* (London) **173**:86.

Harvey, E. B. 1956. *The American Arbacia and Other Sea Urchins*. Princeton University Press, Princeton, New Jersey.

Saunders, J. W., Jr. 1970. *Patterns and Principles of Animal Development*. Collier-Macmillan, London.

Tyler, A. Simple Non-Injurious Method for Inducing Repeated Spawning of Sea Urchins and Sand Dollars. *The Collecting Net*, November, 1949.

EXERCISE 20

Fertilization and Early Development of the Frog

The eggs of most frogs are formed during the summer months when the frogs are feeding heavily. By fall the eggs are fully developed but are usually retained in the ovaries until spring, at which time they are all released en masse. The release of the eggs from the ovaries is called **ovulation**. The eggs subsequently pass into the oviducts and are then passed out of the body.

Ovulation normally occurs as a result of the stimulation of the ovaries by gonadotrophic hormones produced in the pituitary gland. A mature female frog can be induced to ovulate earlier than usual by injecting whole pituitary glands into the body cavity. The number of glands needed varies with the time of year and with the sex of the frog supplying the glands. Since male pituitaries are about one-half as potent as female pituitaries, twice the number of glands must be used. Table 20-1 outlines the number of female pituitaries needed at various times of the year to induce ovulation.

In this exercise you will induce ovulation in mature female frogs and prepare a frog sperm suspension. You will use the sperm suspension to fertilize the eggs and then examine the patterns of cleavage and early development in the frog embryo. Your instructor will demonstrate how to remove and inject the pituitary glands. To have ovulating females ready at the time of the laboratory meeting, the female frogs should be injected 48 hours earlier.

A. ARTIFICIAL INDUCTION OF OVULATION

1. To anesthetize the number of frogs necessary to provide the quantity of pituitaries shown in Table 20-1, place the frogs in a sealed container

TABLE 20-1
Number of female pituitaries needed to induce ovulation in frogs.

September–December	January–February	March–April
5–6	3–5	1–3

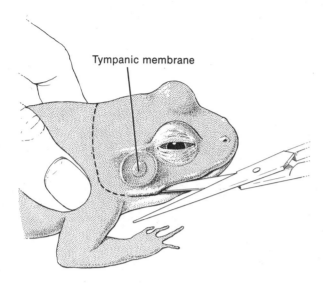
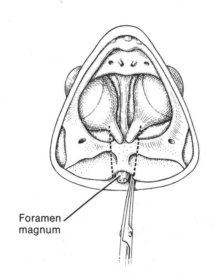
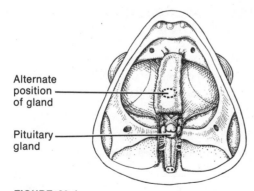

FIGURE 20-1
Procedure for obtaining pituitaries.

in which a wad of cotton saturated (but not dripping) with ether has been suspended. Use a hood or a well-ventilated room to perform this operation. When the frogs fail to move as you shake the container, remove them and dissect out the glands as illustrated in Fig. 20-1. Place the glands into a dish containing Holtfreter's solution (an optimal chemical solution for the maintenance of amphibian eggs and embryos).

2. When the required number of pituitaries has been collected, remove the needle from a 2-ml syringe and draw the glands into the barrel. Adjust the volume to 1 ml, and attach an 18-gauge needle. Hold the syringe with the needle down, so that the glands settle into the mouth of the needle. If any pituitaries stick to the sides of the barrel, tap the syringe until all of them collect on the bottom.

3. Select the female to be injected and, holding her as shown in Fig. 20-2, insert the needle into the body cavity. Keep the needle parallel to the body wall so as not to injure any internal organs. When the needle is deeply inserted (check to see if the pituitaries are at the needle end of the syringe), quickly inject the pituitary glands. Leave the needle in position for a minute and then slowly remove it, pinching the skin to prevent any loss of fluid or glands.

4. Draw more Holtfreter's solution into the syringe to determine if any pituitary material is lodged in the needle. If so, eject all but about ½ ml of solution and then inject this preparation containing the pituitary material.

5. Place the frog in a covered container with about ½ inch of water, and put in a cool room at about 68°F. Check for ovulation after 24 hours by gently squeezing the abdomen as shown in Fig. 20-3. If any eggs come out of the cloaca, ovulation has occurred. If this occurs, place the frog in a container of fresh water in a refrigerator until needed. Eggs will remain viable in the oviduct for 3 to 4 days if the animal is kept at 10°–15°C. If ovulation has not occurred, put the frog in fresh water and try again 24 hours later.

B. PREPARATION OF SPERM SUSPENSION

This should be done about ½ hour before the laboratory meeting so that active sperm will be available at the time the eggs are stripped from the female.

FERTILIZATION AND EARLY DEVELOPMENT OF THE FROG

1. Prepare a Petri dish containing 20 ml of 10% Holtfreter's solution.

2. Pith a mature male frog (see Appendix D for this procedure) and remove the paired testes (Fig. 20-4). Clean away any adhering blood and tissue, and, using the blunt end of a clean probe, thoroughly macerate the testes in the Holtfreter's solution until a milky suspension is obtained.

3. Set the sperm suspension aside for 15–20 minutes to allow the sperm to become motile.

4. Pipet a drop of the sperm suspension onto a glass slide and examine under the high power of your microscope. Describe the shape of the sperm.

_____

_____

What is the physical basis behind the motility of the sperm?

_____

_____

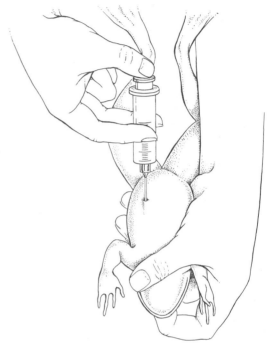

FIGURE 20-2
Method of injecting pituitary preparation.

C. FERTILIZATION

Divide the motile sperm suspension prepared earlier among two or three *clean* Petri dishes so that the bottom of each dish is just covered. Holding the frog as shown in Fig. 20-3, strip the eggs directly into the sperm suspension. Line the egg masses in rows or in a spiral, so that all of the eggs are in contact with the sperm. *Do not place them in a heaping pile in the sperm.* Shake the dishes gently. Why?

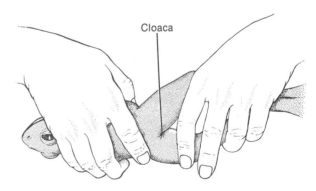

FIGURE 20-3
Method of holding frog to test if ovulation has occurred.

_____

Note the random orientation of the eggs. Some will have the pigmented animal pole uppermost; in others the light-colored vegetal pole will face up.

When the eggs are shed, the first polar body has been formed. The formation of the second polar body can be observed shortly after fertilization. To see this process, remove two or three eggs after several minutes in the sperm suspension. Remove the jelly coats (Fig. 20-5), place them in a Syracuse dish, and cover completely with Holtfreter's solution. Examine the eggs with a dissecting microscope. Adjust the light so that it strikes the egg surface at an oblique angle. With careful focusing you will see a lighter, circular area in the animal pole. By adjusting the light you may observe a pit in this light area (Fig. 20-6). This is where the second polar body will be expelled. Carefully observe the formation of this polar body. It may sometimes be seen more readily if the egg is rotated so that the pit area is at right angles to your field of vision.

After 10 minutes in the sperm suspension, flood the remaining eggs with 10% Holtfreter's solution. Rinse and completely cover the eggs with fresh Holtfreter's solution. The first cleavage will occur in about 2 hours. During this interval go on to Part D or Part E as directed by your instructor.

D. UNFERTILIZED FROG EGGS

From the previously injected female frog, gently squeeze about a dozen eggs into a Petri dish containing Holtfreter's solution. The solution should completely cover the eggs. Note that the eggs are clustered together when they first leave the cloaca but tend to separate from each other in the water. The eggs separate because the jelly that surrounds each egg swells considerably as the protein in the jelly absorbs water. Why is this "swelling" important in the development of the eggs?

_____

_____

Place one or two eggs on a glass slide, cover with water, and examine them with the low power of your microscope. How many jelly layers are present?

_____

Examine the remaining eggs using a dissecting microscope. Adjust the spot lamp so the light is striking the eggs at about a 45° angle. How does the pigmentation of the frog egg differ from the sea urchin egg?

_____

_____

_____

Remove as much jelly as possible from two or three eggs (Fig. 20-5). Place them in a Syracuse dish, animal pole up and completely covered with

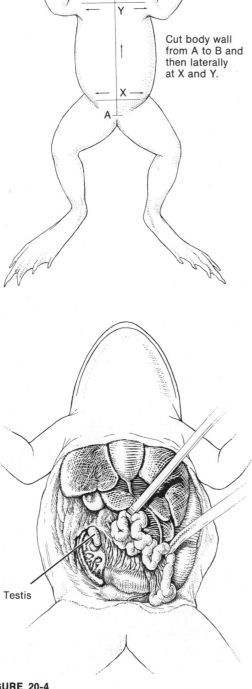

FIGURE 20-4
Procedure for removing testes.

Holtfreter's solution, and examine with the highest power available on your dissecting microscope. Adjust the spot lamp for maximum light. The dark color of the egg is due to granules of melanin pigment. While the egg may appear inactive, close observation of the cytoplasm will show intense movement of the pigment granules. What might be causing this activity?

___

___

E. CLEAVAGE STAGES

Examine the fertilized eggs in your dish. When an egg is fertilized, it secretes a proteinaceous material. As a result, a membrane, previously tightly bound to the egg surface, is forced away when this material begins to absorb water. The yolk-laden vegetal pole, being heavier, shifts downward, so that the animal pole is oriented upward. How many of the eggs in your dish have been fertilized (one-third, one-half, all)?

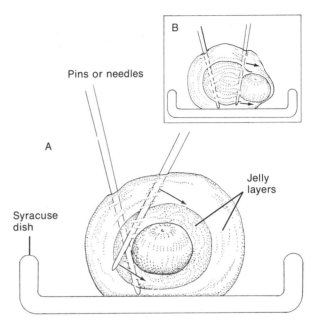

FIGURE 20-5
Procedure for stripping jelly coats from frog eggs.

In nature, why is it advantageous to the development of the egg to have the pigmented area uppermost?

___

___

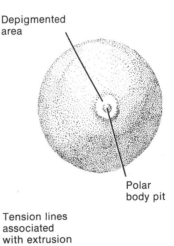

Examine the eggs with a dissecting microscope for the appearance of the first cleavage—that is, the partitioning of the egg cell into a large number of smaller cells with no increase in the size of the mass. (Be careful not to confuse degenerating eggs with normally dividing eggs; a broken, mottled surface is a sign that the egg is dead.) If your eggs have not reached this stage, obtain eggs that were fertilized before the class meeting. Continue to examine the eggs periodically. Where does the first cleavage furrow begin?

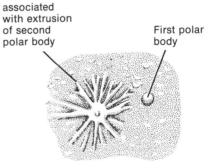

FIGURE 20-6
Location of second polar body.

___

___

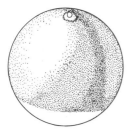
A. Unfertilized egg

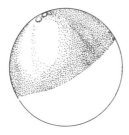
B. Fertilized egg (zygote)

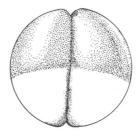
C. 1st cleavage (2-cell stage)

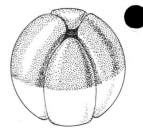
D. 2nd cleavage (4-cell stage)

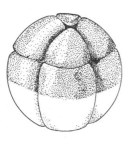
E. 3rd cleavage (8-cell stage)

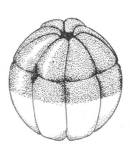
F. 4th cleavage (16-cell stage)

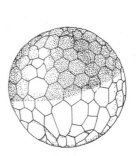
G. Blastula

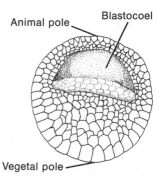
H. Cross section of blastula

FIGURE 20-7
Early stages in the frog's development.

In what plane does the first cleavage divide the egg?

_____

With patience, you will be able to observe the cleavage furrow making its way through the animal pole into the vegetal pole. Before the first cleavage furrow is completed, the second cleavage plane will begin. What is the plane of the second cleavage with respect to the first cleavage?

_____

_____

Your instructor may have examples of later stages of cleavage. If not, examine Fig. 20-7 and models of these stages.

F. LATER STAGES OF DEVELOPMENT

1. Blastula

Cleavage terminates with the formation of the blastula (Fig. 20-7). Different regions of the blastula have different prospective fates; that is, the cells of different areas will ultimately contribute to the skin, skeleton, nervous tissue, and so on. How could you follow the movement of various cells as they proceed through subsequent developmental stages in the living embryo?

_____

In the models, these presumptive areas (called ectoderm, mesoderm, and endoderm) are arbitrarily colored blue, red, and yellow, respectively. A

FERTILIZATION AND EARLY DEVELOPMENT OF THE FROG

specialized region of mesoderm, called chordamesoderm, is colored green. From your readings and the instructor's discussion, complete the following:

The presumptive mesoderm will form

_____

The presumptive endoderm will form

_____

The presumptive ectoderm will form

_____

Chordamesoderm will form

_____

At this stage of development the presumptive areas are associated with the *outside* of the embryo. However, in the mature animal the muscles, alimentary tract, nervous tissue, and others that arise from these areas are situated *inside*. This poses a unique problem for the organism—that of getting these areas of cells to the interior. The process by which this is accomplished is called **gastrulation**.

2. Gastrulation

Patterns of gastrulation vary. In the sea urchin, one side of the hollow blastula invaginates, forming a two-layered cup. Later a third tissue layer, the mesoderm, arises between the inner and outer layers.

Because the blastula of the frog has a very small cavity, gastrulation is accomplished in a different manner. This is more easily understood if models are used.

By the use of models, Fig. 20-8, and your instructor's discussion, answer the following questions.

What is epiboly?

_____

What is the first visible indication that gastrulation has started?

_____

What is the blastopore?

_____

By what morphogenetic movement is the blastopore formed?

_____

What role in development does the dorsal lip of the blastopore play?

_____

_____

What new cavity is formed during gastrulation?

_____

This cavity is eventually replaced by a third cavity that persists (with some modification) into the adult. What is this third cavity and how is it formed?

_____

_____

_____

The chordamesoderm, which will form the notochord, plays what other important role in the development of the frog?

_____

_____

3. Formation of the Neural Tube

When gastrulation is completed, the embryo is completely covered, except for the yolk plug, by ectoderm. The endoderm and mesoderm have been moved inside. Soon, two ectodermal ridges will form on the dorsal surface of the gastrula. These grow upward and toward each other, eventually fusing in the midline. When the edges fuse, a tube of ectoderm will be formed, running from the anterior to the posterior end of the embryo. This tube, covered by an outer ectoderm layer, will develop into the brain and spinal cord. Follow these events

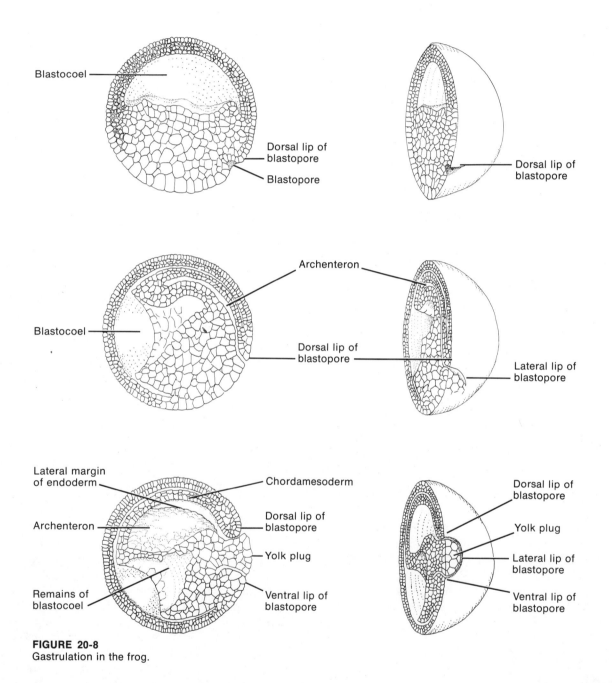

FIGURE 20-8
Gastrulation in the frog.

FERTILIZATION AND EARLY DEVELOPMENT OF THE FROG

using Fig. 20-9 and models. If available, examine living embryos at this stage of development.

Discuss the role of **embryonic induction** in the formation of the neural tube.

_____

_____

Cite other cases in the development of the frog in which induction is involved.

_____

_____

4. Development of the Tadpole

The neural tube was formed as a result of the outfolding and fusion of the ectoderm. Most other tissue-forming processes occurring in later development consist of foldings, invaginations, evaginations, or cell movements of the various germ layers. For example, the digestive glands and lungs develop as evaginations of the endoderm at various levels. The eye develops, in part, from an evagination of the brain. Examine these morphogenetic movements on the models. List any others you find.

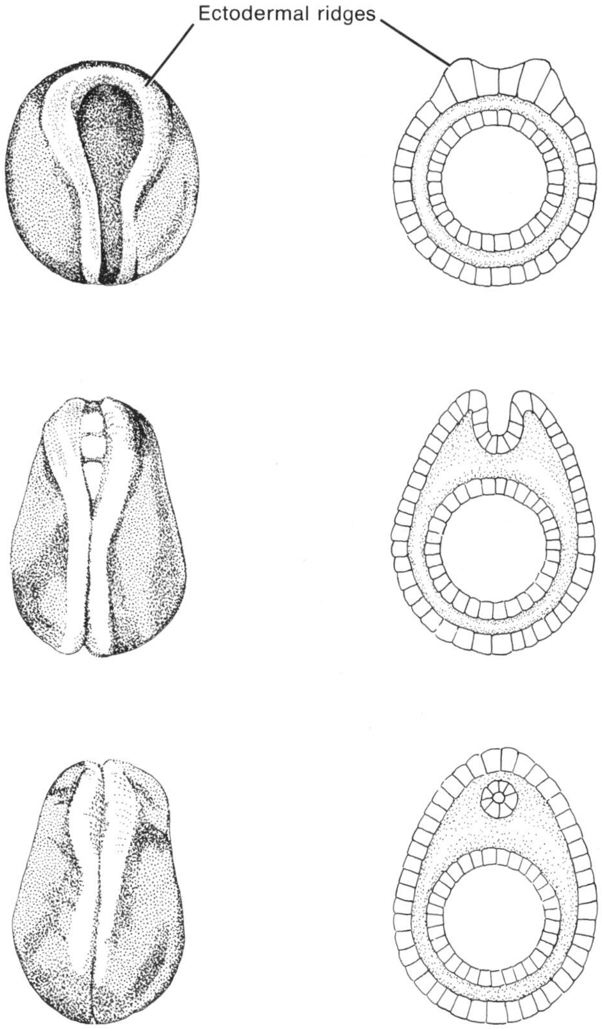

_____

_____

_____

FIGURE 20-9
Formation of the neural tube.

G. FURTHER STUDIES ON FROG DEVELOPMENT

1. Is a Sperm Necessary for an Egg's Development?

In the normal sequence of development, the mature egg is penetrated by a sperm. As a result, the egg receives a set of the genetic characteristics of the male; secondly, the egg is stimulated to develop. Since the sperm physically breaks through the surface of the egg, we might ask if mechanically breaking the surface by using a glass needle would also initiate development of the egg. This process, called **parthenogenesis** (Gr. *parthenos*, virgin, and *genes*, born), occurs naturally in some organisms. To test this question in the laboratory, the following procedures should be followed.

a. If this experiment is to be done at a time other than the normal breeding season, induce ovulation by pituitary injection as described in Part A.

b. To be sure that all instruments are clean and free of sperm, wash them in 70% alcohol and then rinse with distilled water and dry.

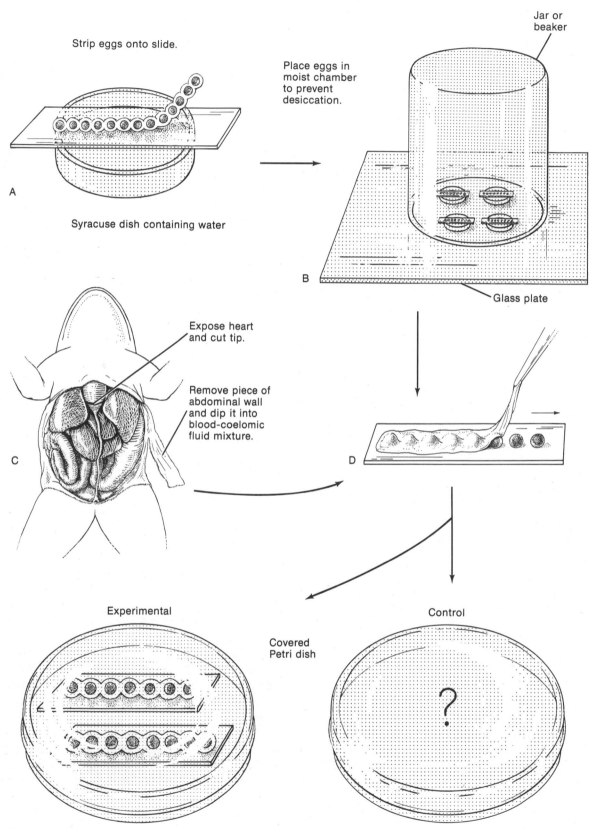

FIGURE 20-10
Procedure for inducing parthenogenetic development of frog eggs.

c. Strip the eggs from the ovulating frog in a single file along the length of a glass slide (Fig. 20-10A). Prepare at least ten slides in this way. Set each slide over a Syracuse dish containing water, and place them under a jar as shown in Fig. 20-10B.

d. Obtain a second female frog that has been segregated from male frogs for several days. Why is this necessary?

_____

_____

_____

Pith the frog (see Appendix D for this procedure) and dissect it to expose the heart (Fig. 20-10C). Cut off the tip of the heart and let the blood flow into the body cavity so that it mixes with the coelomic fluid.

e. Dissect out a piece of muscle from the abdominal wall and dip it into the mixture of blood and coelomic fluid. Remove five slides from the moist chamber and streak the eggs with this fluid (Fig. 20-10D). Using a dissecting microscope, prick each egg, slightly off center in the animal pole, with a fine glass needle.

f. Place the slides of eggs into Petri dishes containing 10% Holtfreter's solution, so that the eggs are covered by the fluid. Cover the dishes, label them "Experimental Groups," and keep in a cool place at about 68°F.

g. Describe the control group that should be tested.

_____

_____

_____

Use the remaining eggs for your controls.

h. Examine the eggs after 2 hours, and periodically thereafter. If cleavage occurs, is the pattern of cleavage similar to that found in normally fertilized eggs?

_____

If no cleavage occurs at all, does this prove that sperm are necessary for development? Explain.

_____

_____

_____

_____

2. Effects of Pressure and Temperature on Development of the Frog Egg

To conclude this study on amphibian development, devise and run adequately controlled experiments to determine (*a*) the effects of pressure and (*b*) the effects of temperature on the development of the frog egg. Prepare a report as described by your instructor.

REFERENCES

Austin, C. R. 1965. *Fertilization*. Prentice-Hall, Englewood Cliffs, New Jersey.

Balinsky, B. I. 1970. *An Introduction to Embryology*. 3d ed. Saunders, Philadelphia.

Barth, L. G. 1953. *Embryology*. Dryden, New York.

Barth, L. J. 1964. *Development: Selected Topics*. Addison-Wesley, Reading, Massachusetts.

Ebert, J. D., and I. M. Sussex. 1970. *Interacting Systems in Development*. 2d ed. Holt, Rinehart & Winston, New York.

Hamburger, V. 1960. *A Manual of Experimental Embryology*. Rev. ed. University of Chicago Press, Chicago.

King, T. J., and R. Briggs. 1956. Serial Transplantation of Embryonic Nuclei. *Cold Spring Harbor Symp. Quant. Biol.* **21**:271–290.

Moore, J. A. 1963. *Heredity and Development*. Oxford University Press, New York.

Noble, G. K. 1954. *The Biology of the Amphibia*. Dover, New York.

Rugh, R. 1951. *The Frog: Its Reproduction and Development*. McGraw-Hill, New York.

Rugh, R. 1962. *Experimental Embryology: Techniques and Procedures*. 3d ed. Burgess, Minneapolis.

Saunders, J. W., Jr. 1968. *Animal Morphogenesis*. Macmillan, New York.

Saunders, J. W., Jr. 1970. *Patterns and Principles of Animal Development*. Macmillan, New York.

Shaver, J. R. 1953. Studies on the Initiation of Cleavage in the Frog Egg. *J. Exp. Zool.* **122**:169–192.

Sussman, M. 1970. *Growth and Development.* 3d ed. Prentice-Hall, Englewood Cliffs, New Jersey.

Waddington, C. H. 1966. *Principles of Development and Differentiation.* Macmillan, New York.

Willier, B. H., and J. M. Oppenheimer. 1964. *Foundations of Experimental Embryology.* Prentice-Hall, Englewood Cliffs, New Jersey.

EXERCISE 21

Development of the Chick

The chick egg provides excellent material for the analysis of embryonic development. It has been used as an experimental "tool" for over 300 years.

In this exercise you will examine the early developmental stages of living chick embryos, supplementing your observations with specially prepared and stained whole-mounts of various stages of development.

A. UNFERTILIZED EGG

For observation place an egg in a finger bowl containing water by simply cracking and carefully removing the shell, so the unbroken yolk floats in the solution (see Fig. 21-4A and B).

Identify the albumen. This proteinaceous material is deposited around the egg as it travels down the oviduct. Of what use might the albumen be to a developing chick?

_____

Locate the spirally wound **chalazae**. Gently grasp one of these cords with a forceps and pull. Suggest a function for these cords.

_____

_____

_____

On the surface of the yolk locate a small, whitish spot, the **blastodisc**. This is the living material of the egg. The remainder of the egg is composed of inert food material and water. Normally the blastodisc faces up. If you cannot see it, gently turn the egg over.

B. FERTILIZED EGG

The blastodisc of a fertile egg undergoes a highly irregular series of cleavages (Fig. 21-1), and by the time the egg is laid the process of gastrulation

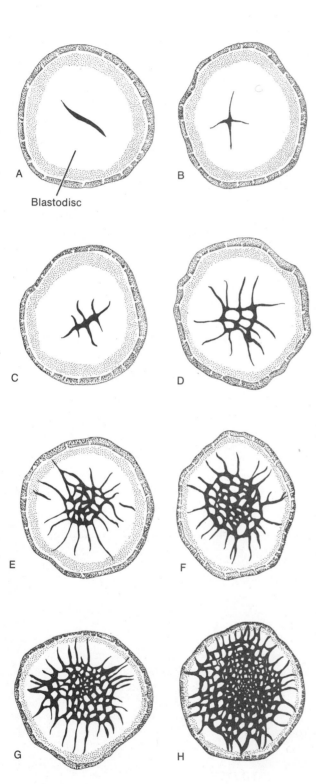

FIGURE 21-1
Successive cleavages of the blastodisc.

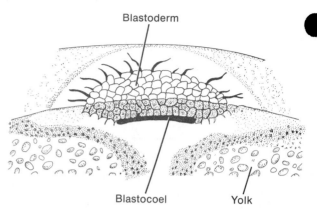

FIGURE 21-2
Cross section of the blastodisc.

has begun. Cleavage is restricted to the blastodisc; no division of the yolk occurs.

Continued division of the blastodisc results in a mass of cells that become slightly elevated above the underlying yolk. The space formed between the yolk and the blastodisc is the **blastocoel** (Fig. 21-2). The blastodisc, now called the **blastoderm**, is usually two cell layers thick at this stage of development. Subsequently, horizontal cleavages divide the blastoderm into three or four cell layers in thickness.

C. GASTRULATION

Soon after blastulation, the blastoderm becomes a two-layered structure. Despite numerous descriptive studies, it is still not clear how the two layers are formed. Some embryologists claim that the lower layer is formed by simply splitting off from the upper layer (delamination). Others suggest that the cells at the edges of the blastoderm turn under and join at the center to form the lower layer. In either case, the lower layer becomes the endoderm and the upper layer forms the ectoderm.

Following the formation of these two layers, a thickening of the ectoderm results in a narrow groove—the **primitive streak**. This groove traverses the surface of the pear-shaped blastoderm from approximately the center to the tapered edge (Fig. 21-3A). Cells of the ectodermal layer then begin to migrate toward the primitive streak, turn under, and spread out as a mesodermal layer between the ectoderm and endoderm (Fig. 21-3B and C). This description, however, is controversial. The original experimental work was based upon the placement

DEVELOPMENT OF THE CHICK

of carbon particles on the surface of the ectoderm. The movement of these particles was followed and they appeared to become positioned in the layer between the ectoderm and endoderm. The events described above were the accepted explanations for gastrulation and mesoderm formation in the chick embryo. Recently, it has been suggested that the carbon particles localized between the ectoderm and mesoderm not as a result of cell movements through the primitive streak, but because the investigator may have unknowingly pushed them through the ectoderm. In any event, when the mesoderm is fully established, gastrulation is complete. Eventually, the surface ectoderm will be folded under the embryo. When this is completed (Fig. 21-3D), the ectoderm will no longer be a single sheet of cells but a complete outer covering.

D. LATER DEVELOPMENTAL STAGES

1. The 33-Hour Embryo

Approximately 33 hours prior to the laboratory meeting, fertilized eggs were placed in an incubator.

For observation of the living embryo, place an egg in a finger bowl with warm Ringer's solution by cracking the egg and carefully removing the shell, so the unbroken yolk with its developing embryo floats in the saline solution. If only a white spot is found on the surface of the yolk, the egg has failed to develop. If this is the case, use another egg. Remove the blastoderm from the yolk mass as shown in Fig. 21-4. Examine the embryo using a dissecting microscope. Supplement your observations of the living embryo with prepared slides and models available in the laboratory. As you find the structures discussed below, label them in Fig. 21-5.

At this stage of development the blastoderm has become a thin disc about ¾ inch across. Surrounding the embryo is a transparent region—the **area pellucida**. Outside of this is the **area opaca**, where the blood vessels are beginning to develop. One of the most striking features of this stage is the heart, which can be seen lying on the right side of the embryo. The heart originates as separate masses of cells on either side of the embryonic axis. These "heart-forming regions" move to the midline and eventually fuse (Fig. 21-6). If extra fertilized eggs are available, you might incubate them for 26 hours

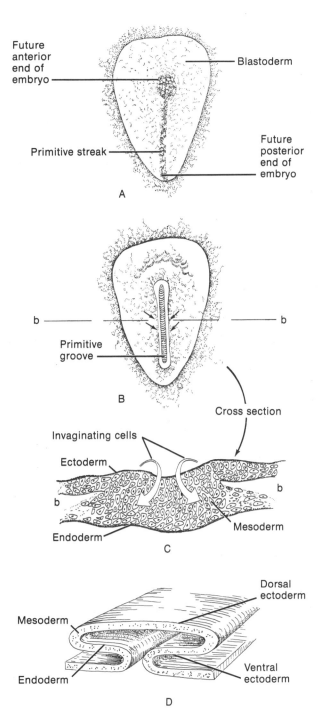

FIGURE 21-3
Gastrulation in the chick.

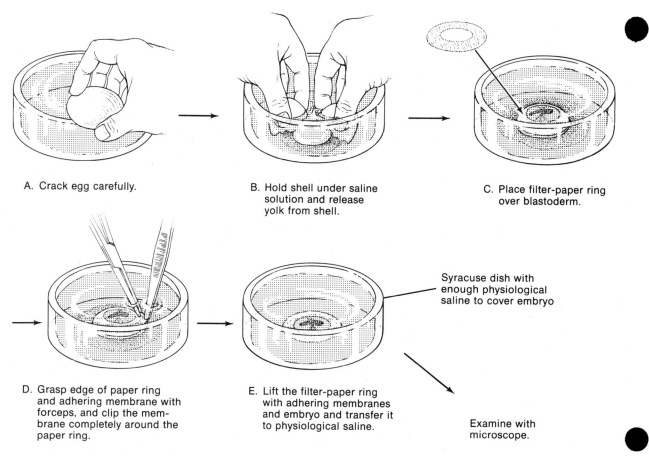

FIGURE 21-4
Procedure for removing young embryo from yolk.

and then remove the embryo as shown in Fig. 21-4. After about 26 hours the "heart mass" can be seen to beat, even though it has not yet formed a heart.

A pair of **vitelline veins** can be seen entering the heart at its posterior end. What is the function of these veins?

_____

_____

A single artery, the **ventral aorta**, leaves the heart anteriorly.

The head region is divided into three segments; the anterior **forebrain**, the **midbrain**, and the **hindbrain**, which continues for the length of the embryo as the **neural tube**. Note the lateral evaginations of the forebrain as the **optic vesicles**.

The **notochord** can be seen in the midline of the embryo, extending anteriorly to the forebrain as a solid rod. Lying on either side of the neural tube is a series of 10 to 12 **somites**. These somites will form the vertebral column and the musculature on the dorsal side, and they may contribute to the ribs and musculature of the body wall.

2. The 72-Hour Embryo

Remove the embryo from the yolk as shown in Fig. 21-4, and locate the various regions described below. At this stage the anterior part of the embryo has turned to lie upon its left side, while the posterior half remains dorsal side up (Fig. 21-7). Note the three pairs of **gill slits** and the blood vessels—the **aortic arches**—that pass through the **gill arches**. What might be the significance of the appearance of these structures in the embryos of all higher vertebrates?

_____

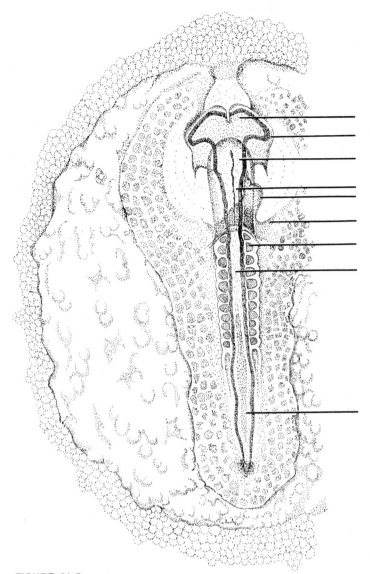

FIGURE 21-5
Dorsal view of a 33-hour chick embryo.

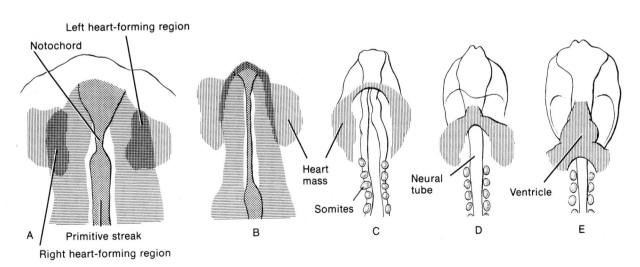

FIGURE 21-6
Formation of the heart in the developing chick (ventral view).

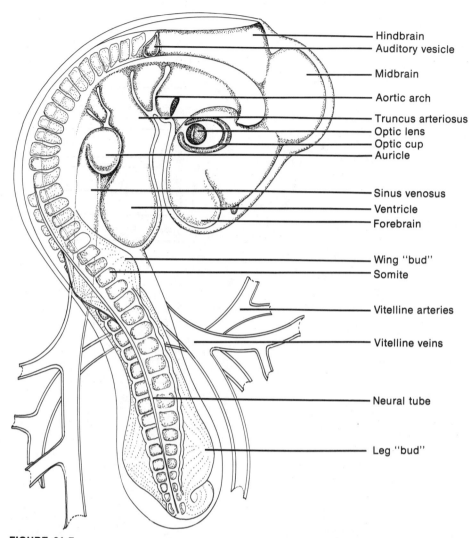

FIGURE 21-7
Dorsal view of a 72-hour chick embryo.

The central nervous system and the sense organs have advanced considerably beyond their development in the 33-hour stage. The neural tube has developed into a five-part brain anteriorly and the spinal cord posteriorly. The optic vesicles have greatly enlarged and by invaginating have given rise to the double-walled **optic cups**. Concurrently, the surface ectoderm of the head immediately over the optic cups invaginates, thickens, and detaches to form the **optic lens**. Dorsal to the gill slits on each side is an invaginated thickening of the ectoderm, forming the **auditory vesicle** ("ear").

The heart, which was tubular and single-chambered in the earlier stage, has become twisted and transformed into a two-chambered structure in this stage. That part of the heart next to the body is the **auricle** (atrium), and the part that dips ventrally is the **ventricle**. Note the **truncus arteriosus**; it arises from the ventricle and branches out into the aortic arches, which unite to form the **dorsal aortae**. Locate the **vitelline arteries**, a pair of large transverse blood vessels that leave the embryo about one-third of the distance from the posterior end and branch out into the **yolk sac**, and the **vitelline veins** that enter the auricle.

REFERENCES

Ebert, J. D., and I. M. Sussex. 1970. *Interacting Systems in Development.* 2d ed. Holt, Rinehart & Winston, New York.

Hamburger, V. 1960. *A Manual of Experimental Embryology.* Rev. ed. University of Chicago Press, Chicago.

Holtfreter, J. 1943. A Study of the Mechanics of Gastrulation. I. *J. Exp. Zool.* **94**:261–318.

Patten, B. M. 1951. *Early Embryology of the Chick.* McGraw-Hill, New York.

Patten, B. M. 1964. *Foundations of Embryology.* 2d ed. McGraw-Hill, New York.

Rugh, R. 1962. *Experimental Embryology: Techniques and Procedures.* 3d ed. Burgess, Minneapolis.

Saunders, J. W., Jr. 1968. *Animal Morphogenesis.* Macmillan, New York.

Saunders, J. W., Jr. 1970. *Patterns and Principles of Animal Development.* Macmillan, New York.

Spratt, N. T. 1964. *Introduction to Cell Differentiation.* Reinhold, New York.

Sussman, M. 1970. *Growth and Development.* 3d ed. Prentice-Hall, Englewood Cliffs, New Jersey.

Waddington, C. H. 1956. *Principles of Embryology.* Allen & Unwin, London.

Waddington, C. H. 1966. *Principles of Development and Differentiation.* Macmillan, New York.

EXERCISE 22

Plant Growth and Development

A higher plant proceeds through a continuous series of changes during its life cycle. The outward expression of this activity begins with the embryo in the seed and proceeds, sequentially, through germination, the appearance and subsequent enlargement of stems, leaves, and roots, to the production of flowers, fruits and seeds. The location of these organs and their size and shape are the visible manifestations of the correlated activities of those cells produced by meristems. This complex and highly ordered series of events is called development.

Development may be divided into separate phases called growth, cell differentiation, and organogenesis. **Growth** may be defined as an irreversible increase in volume, which is usually—though not always—paralleled by an increase in weight. **Cell differentiation** includes all those events involved in the specialization of structure and function. **Organogenesis** is the development of the various plant organs.

A. COMPARATIVE STUDY OF SEEDS

The process of development in the higher plants begins with the seed. You will begin the study of plant growth and development by investigating the structure and function of seeds.

Seeds develop from ovules and are formed following the sexual union of the sperm and egg, an event that occurs in the ovary of the flower. If you are not familiar with floral anatomy, refer to Fig. 22-1 and a flower model or living flowers. The ovary and, in some cases, other parts of the flower become the fruit. Within the seed is the embryo, which consists of three parts: the **cotyledons** (sometimes called seed leaves), the **epicotyl**, that part of the embryo which is located above the point of attachment of the cotyledons, and the **hypocotyl**, found below the cotyledons. The mature seed also may have an **endosperm**, which functions as a nutritive tissue for the developing embryo.

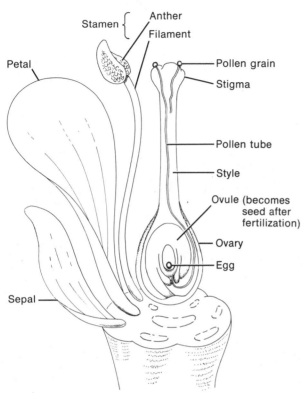

FIGURE 22-1
Diagram of a flower. (Adapted from *Biology, Its Principles and Implications* by Garrett Hardin. W. H. Freeman and Company. Copyright © 1961.)

Obtain previously soaked bean, pea, and corn seeds. (The so-called "seed" of corn is really a fruit. The seed coat is intimately fused with a hard outer tissue called the **pericarp**, which is part of the fruit.) Carefully remove the outer seed coats from each seed. Then separate the other parts of each seed, comparing them with each other and with Fig. 22-2.

1. Cotyledons

Scrape the surface of the cotyledons of the bean and pea, and add a drop of iodine. On the basis of your observations, what appears to be the role of the cotyledons in the development of these seeds?

_____

2. Epicotyl

This part of the embryo gives rise to the shoot system of the mature plant. What indication is there, if any, that this region will produce stem and leaf tissues?

_____

3. Hypocotyl

This region will primarily give rise to all root tissues and, in some cases, to the lower part of the stem.

4. Endosperm

The endosperm (although found in many seeds) is present only in the corn "seed" of the group you are examining. Depending on whether the endosperm is composed of starches or sugars, corn is "starchy" or "sweet." Remove the endosperm from several corn "seeds," and cut it up into fine pieces. Then add it to 5 ml of Benedict's solution in a test tube, and heat in a hot water bath for several minutes. If a reducing sugar is present, the solution will change in color (from blue to green to orange to red to brown), depending on the amount of sugar present. Cut through a kernel of corn, as shown in Fig. 22-2. Scrape the surface of the endosperm with a razor blade and add a drop of iodine. The appearance of a blue color indicates the presence of starch. What commercial value does this corn have?

_____

_____

B. GERMINATION

The seed is a resting stage in the development of a plant, serving to carry the plant over periods of unfavorable environmental conditions. When provided with optimal growing conditions, the seed, if living, will germinate and produce another plant. The production of viable seeds is required for maintaining the existence of any plant species in nature. It is also commercially important to seed growers that they be able to judge the viability of the seed

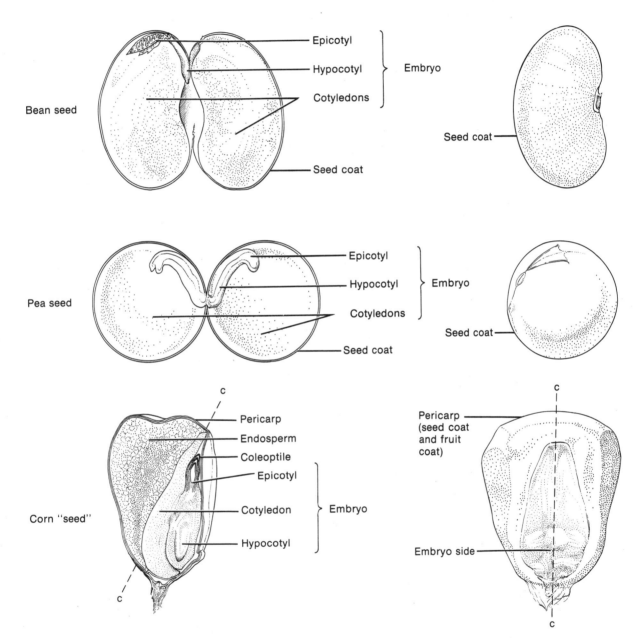

FIGURE 22-2
Comparative anatomy of various seeds.

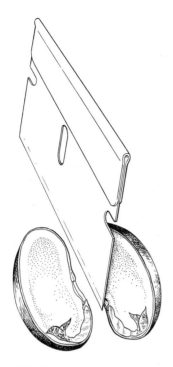

FIGURE 22-3
Procedure for cutting bean seeds for tetrazolium test.

from any given crop. For a biologist who grows research plants from seed, it is important that he be assured that the percentage of germination will be high. In this part of the exercise you will determine the degree of viability of a batch of seeds. One way of doing this is to germinate sample batches of seed under standardized conditions. This **germination test** has one major drawback—it takes from 7 to 10 days for many seeds to germinate. A quicker and easier method is to use the fact that embryos in viable seeds respire and are able to change colorless dyes, such as tetrazolium, into highly colored forms.

1. Germination Test

Obtain 100 bean seeds that have been soaked in water. Boil 50 of these seeds for 10 minutes to kill them. Plant 25 living seeds and 25 killed seeds about ½ inch deep in separate rows in the planting tray. The remaining boiled and unboiled seeds will be used for the tetrazolium test. Label each row as to section number, team number, date, number of seeds, and treatment (boiled or unboiled). After a week, determine the total number of seeds in each treatment that have germinated and are growing and record this data in Table 22-1. Record your team results and (in parentheses) those of the combined class.

Examine the tray again after 2 weeks and determine the percentage of germination. Record your data in Table 22-1. How do you account for the difference, if any, between the results obtained after 1 week and that after 2 weeks?

_____

2. Tetrazolium Test

Take the boiled and unboiled seeds not used in the germination test and cut them longitudinally through the embryo as shown in Fig. 22-3. Place each unboiled seed in a Petri dish with the cut surface up. Pour in enough tetrazolium chloride solution to cover the seeds. Repeat this procedure with seeds that have been boiled. *Make sure each dish is labeled as to contents.* Set the seeds aside in a dark place for 30–45 minutes and then examine them. If the seed is viable, a red or pink color will be evident on the cut surface. Determine the percentage of viability and record your data in Table 22-1. Record your results and (in parentheses) those of the combined class.

How does the percentage of viability obtained by the germination test compare with that obtained by the tetrazolium test?

_____
_____
_____
_____

If results obtained by your team were different from those of the class, account for the difference.

_____
_____
_____

PLANT GROWTH AND DEVELOPMENT

TABLE 22-1
Germination and tetrazolium data.

Data	Germination test				Tetrazolium test	
	Unboiled seeds		Boiled seeds		Unboiled seeds	Boiled seeds
	Week 1	Week 2	Week 1	Week 2		
Total number of seeds	( )	( )	( )	( )	( )	( )
Number of seeds germinated	( )	( )	( )	( )	( )	( )
Germination (%)	( )	( )	( )	( )	( )	( )

TABLE 22-2
Factors affecting the germination of seeds.

Data	Seed treatment		
	Untreated	Seed coat cut	Treated with H_2SO_4
Total number of seeds	( )	( )	( )
Number of seeds germinated	( )	( )	( )
Germination (%)	( )	( )	( )

3. Factors Affecting Germination

The seeds used in the first part of this exercise germinated in a short period of time when placed under conditions suitable for germination. Viable seeds of many plants, however, fail to germinate even when provided with ample water and oxygen and a suitable temperature. Such seeds are in an arrested state of development called **dormancy**. In this part of the exercise you will study one factor involved in the dormancy of seeds.

Place moistened filter paper in the bottoms of three Petri dishes. Obtain 40 dry honey locust seeds (other possible seeds are sweet clover, okra, and alfalfa) and 20 honey locust seeds that have been soaking for 2–3 hours in a 70% sulfuric acid (H_2SO_4) solution. Carefully pour off the acid into a sink containing running water. With a rubber band, fasten cheesecloth over the beaker and wash the seeds in running water for 5 minutes.

Place the 20 acid-treated seeds in one of the Petri dishes. Put 20 of the untreated seeds into a second Petri dish. With a sharp razor blade remove a small chip from the 20 untreated seeds, exposing the inner tissues. Place these seeds with their cut surface down in the third dish. Cover all dishes, tape them shut (to retard evaporation), and label each as to contents. Put the dishes in the dark at room temperature (or they may be wrapped in aluminum foil and taken home for observation). Periodically examine the seeds over the next 3–5 days and determine the percentage of germination for each group of seeds. Record your team data and (in parentheses) those of the class in Table 22-2. What is the "factor" that controls dormancy in these seeds?

_____

_____

Of what advantage is dormancy to seeds?

_____

_____

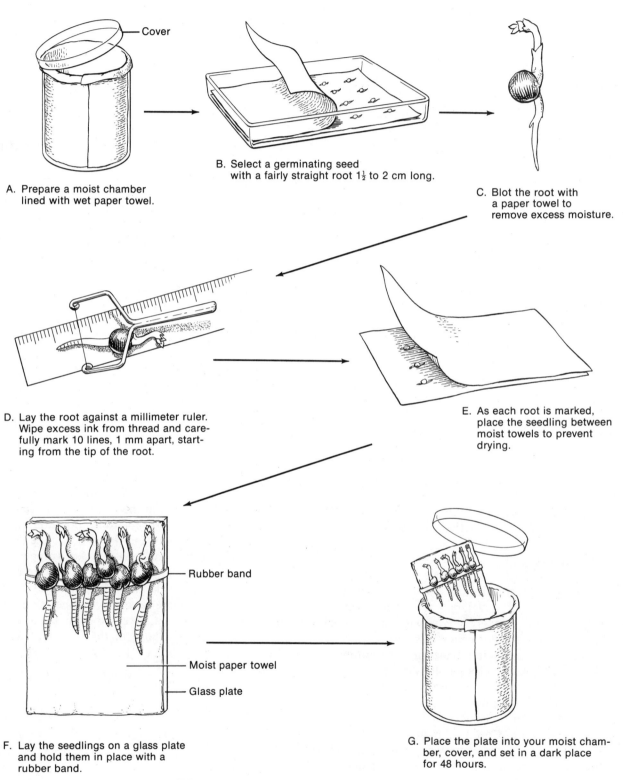

FIGURE 22-4
Procedure for determining locus of growth in roots.

PLANT GROWTH AND DEVELOPMENT

C. MEASUREMENT OF PLANT GROWTH

Growth may be defined as an irreversible increase in volume generally accompanied by an increase in weight. Growth, therefore, is quantitative and can be measured. In these experiments you will determine the locus of growth, and plot a growth curve for a specific plant organ.

1. Localization of Plant Growth

Prepare a moist chamber as shown in Fig. 22-4A. Cover a glass plate ($3\frac{1}{2} \times 4$ inches) with moist toweling, and place it in the chamber. Cover the container to prevent drying.

From the germinating seeds provided by the instructor, select one having a fairly straight root about $1\frac{1}{2}$–2 cm long. Blot the root to remove any excess moisture and then lay it against a millimeter ruler. Carefully mark ten lines, each 1 mm apart, *starting from the tip of the root* (Fig. 22-4D). Make sure you wipe the excess ink from the thread, or the mark will become smudged, making accurate observations difficult. Repeat the marking operation until five roots have been marked. Avoid drying of the roots during marking. Lay the seedlings on the paper-covered glass plate and hold them lightly in place with a rubber band so that the whole length of each root is touching the moist paper (Fig. 22-4F). Place the plate and seedlings in the moist chamber and then cover and keep in a dark place.

After 48 hours measure the distance between the ink marks on each root. Average the lengths for each interval. Record your data in Table 22-3, and plot your data in Fig. 22-5.

On the basis of your results where does most of the growth occur?

_____

If the ink lines were initially sharp and clear, how do you account for the smudging of the first, and possibly the second, line?

_____

If you were to cut a longitudinal section through a young root and examine it microscopically, what would you expect to see in the first millimeter or so that would account for the results you have observed?

_____

_____

Examine a prepared slide of a longitudinal section through an onion root tip (compare with Fig. 22-6). Locate a cone-shaped mass of loosely arranged cells covering the tip of the root. This is a root cap, which serves to protect the meristematic region of the root. Closely examine the cells of the meristematic region. What occurred in this region when the tissue was living to account for the results obtained in the growth measurement experiment?

_____

_____

_____

_____

Locate the region of the root in which cell differentiation is occurring. Approximately how far back from the tip does this region occur? (The diameter of your low-power field, 10× objective, is approximately 2 mm.)

_____

TABLE 22-3
Data for locus of root growth.

Root tip No.	Interval									
	1	2	3	4	5	6	7	8	9	10
1										
2										
3										
4										
5										
Total										
Average										
Control										

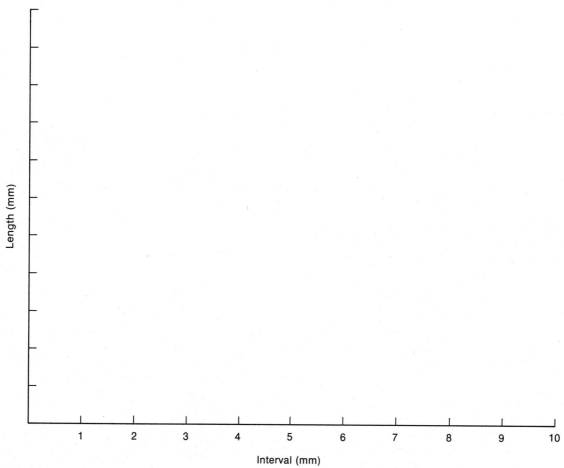

FIGURE 22-5
Localization of root growth.

What is the consequence of cell division and the subsequent enlargement of cells at the root tip.

_____

_____

_____

_____

_____

_____

2. Growth Curve of Leaves

Select three bean seeds that have been soaking in water for several hours. Split them open and, with a millimeter ruler, measure the length of the embryonic foliage leaves (Fig. 22-7A). Determine the average length and record this figure under Date "0" in Table 22-4. This is the first in a series of measurements to be made of the growth of the first two foliage leaves.

Plant 25 of the soaked bean seeds about ½ inch deep in the container provided. Water thoroughly and place them in the greenhouse. In two or three days dig up three of the seeds and measure the length of the leaves, including the petiole (Fig. 22-7B). Enter the average length in Table 22-4 and discard the young plants.

In succeeding laboratory periods, select three plants and measure the leaves as above (Fig.

PLANT GROWTH AND DEVELOPMENT

TABLE 22-4
Growth curve data.

	Date of measurement					
	0					
Average length (mm)						

22-7C). *Do not remove these plants from the container. Use the same three plants for all measurements beginning with the third measurement.*

Plot your data in Fig. 22-8. List several other organisms (or organs) that would show a similar growth curve.

_____

_____

_____

_____

REFERENCES

Black, M., and P. F. Wareing. 1954. Photoperiodic Control of Germination in Seed of Birch (*Betula pubescens* Ehrh.). *Nature (London)* **174**:705.

Borthwick, H. A., S. B. Hendricks, E. H. Toole, and V. K. Toole. 1954. Action of Light on Lettuce Seed Germination. *Botan. Gaz.* **115**:205–225.

Cutter, E. G. 1969. *Plant Anatomy: Experiment and Interpretation.* Addison-Wesley, Reading, Massachusetts.

Esau, K. 1960. *Anatomy of Seed Plants.* Wiley, New York.

Galston, A. W. 1960. *The Life of the Green Plant.* 3d ed. Prentice-Hall, Englewood Cliffs, New Jersey.

Kahn, A. 1960. Promotion of Lettuce Seed Germination by Gibberellin. *Plant Physiol.* **35**:333–339.

Leopold, A. C. 1964. *Plant Growth and Development.* McGraw-Hill, New York.

Levitt, J. 1969. *Introduction to Plant Physiology.* Mosby, St. Louis.

Steward, F. C. 1968. *Growth and Organization in Plants.* Addison-Wesley, Reading, Massachusetts.

Toole, E. H., H. A. Borthwick, S. B. Hendricks, and V. K. Toole. 1956. Physiology of Seed Germination. *Ann. Rev. Plant Physiol.* **7**:299–324.

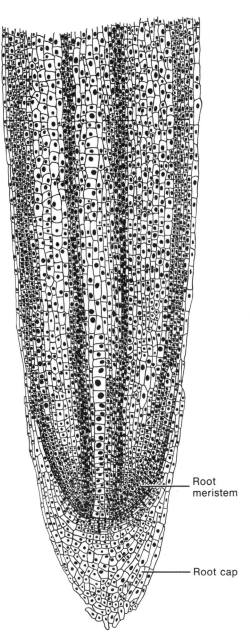

FIGURE 22-6
Longitudinal section of onion root tip.

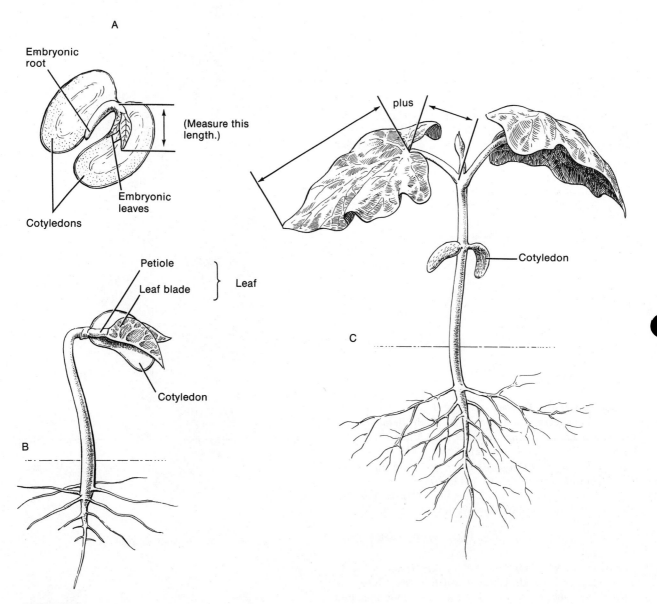

FIGURE 22-7
Measuring the growth of a bean seedling. (Adapted from *Plants in Action* by Leonard Machlis and John G. Torrey. W. H. Freeman and Company. Copyright © 1956.)

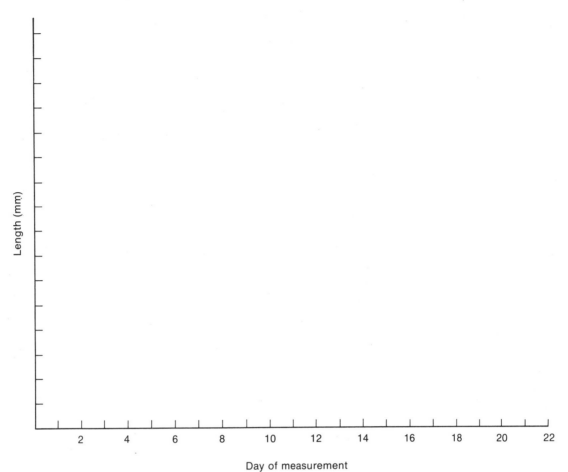

FIGURE 22-8
Growth curve of bean plant leaves.

EXERCISE 23

Plant Development: Hormonal Regulation

One of the challenging problems facing biologists is that of understanding the basic mechanisms that regulate the development of plants and animals. The systematic differentiation of millions of cells into various tissues and organs, and the development of specific form and shape, requires precise coordination between the individual parts of the organism during each stage of growth.

The development of a seed into a mature plant is one aspect of this remarkable process. It consists of growth by cell division and cell elongation, differentiation of new organs such as roots, stems, leaves, and flowers, and a complex series of chemical changes. The final form of the plant is a blend of the plant's genetic "blueprint" and modifying effects of the environment. When the seed begins to germinate, it absorbs large amounts of water, and cells at the meristems begin to divide. For reasons not yet understood, the root almost always begins to develop before the shoot. At both root and shoot ends of the seed new cells are formed by the meristematic (dividing) areas of the growing points, followed by elongation and differentiation of these cells.

Ultimately, the control of growth and differentiation is in the DNA of the nucleus, which controls the production of hormones and other regulatory chemicals. Our present knowledge of the mechanisms of regulation of the various patterns of growth and differentiation is quite limited.

Some of the growth regulating substances in plants are auxins, gibberellins, cytokinins, and various growth inhibitors.

- **Auxins** are produced in meristematic tissues. They influence cell elongation, inhibit the growth of lateral buds, promote the initiation of roots, and, in a few plants, regulate the differentiation of flower buds.
- **Gibberellins,** also apparently synthesized in meristems, are highly complex substances that affect cell elongation, cell division, and flowering responses in some plants.
- **Cytokinins,** which have been suggested to be

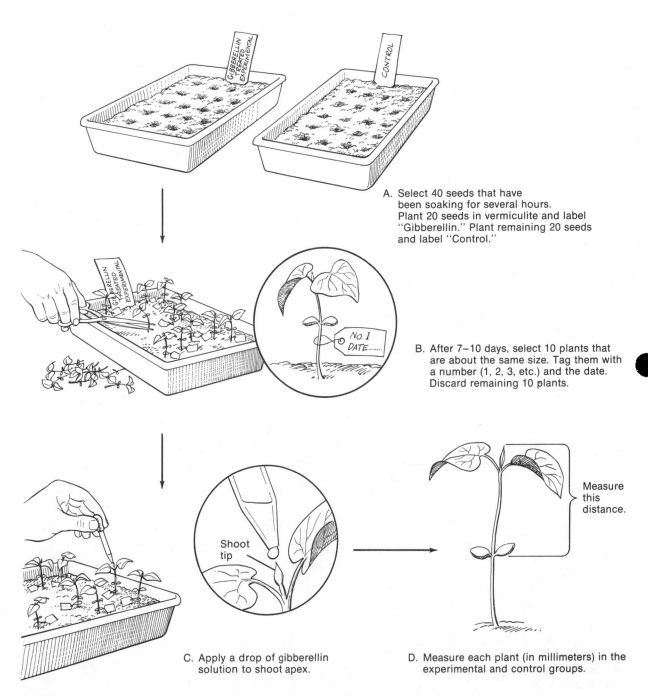

FIGURE 23-1
Procedure for determining effect of gibberellin on plant growth.

PLANT DEVELOPMENT: HORMONAL REGULATION

related to a component of RNA, promote cell division, and, in the presence of auxin, induce the differentiation of roots and shoots.
- **Inhibitors** of different types regulate such responses in plants as flowering, dormancy of buds and seeds, and rates of growth.

In the following parts of the exercise, the effects of certain environmental and chemical factors on the patterns of growth and differentiation in plants will be studied.

A. GIBBERELLIN

Gibberellins, first discovered by Japanese scientists in the 1920's went largely unnoticed until the early 1950's, when English and American biologists became interested in these compounds. They were first isolated from a fungus, *Gibberella*. To date, more than sixteen gibberellins have been isolated from higher plant tissues and appear to participate in normal plant development.

In this part of the exercise you will determine some of the more obvious effects of gibberellic acid on plant development.

1. Working in teams of three, obtain 40 bean seeds that have been soaking in water for several hours.

2. Plant 20 seeds (about ½ inch deep) in moist vermiculite in a tray. Label the tray "Gibberellin" (Fig. 23-1A). Plant the remaining 20 seeds in a second tray labeled "Control."

3. When the plants are several centimeters tall (about 3 inches), select 10 plants in each tray that are about the same size. Label each individual plant with a number, along with the date. Cut the remaining plants at the ground level and discard the parts you have cut off (Fig. 23-1B).

4. Measure the height of each plant (in millimeters) from the cotyledons to the tip of the shoot apex. Can you think of other measurements you might take? If so, use these in place of the one suggested.

5. Record the date, height, and appearance of the plants in Table 23-1 under Day "0."

6. Apply a drop of gibberellin to the shoot apex of each plant in the tray labeled "Gibberellin." What will you apply to the "Control" plants?

7. Apply gibberellin to the plants weekly and record the height of each plant and the general appearance of all plants in Table 23-1. As the plants continue to grow, it may be advisable to "stake" the plants using thin bamboo poles and twine. Be careful not to crush the stems when tying them to the stakes.

8. At the conclusion of the experiment (as determined by your instructor), plot the data in Fig. 23-2. Analyze the data and determine whether the growth response obtained with gibberellic acid is significantly different from that of the controls. (See Appendix F for a method of statistical analysis of the data.)

B. PLANT GROWTH INHIBITORS

The idea that inhibitors were important in the regulation of plant growth first gained importance when it was discovered that dormant buds of ash trees contained large amounts of inhibitors. When dormancy ended the inhibitor concentration declined. More recently, a substance called **dormin** (now called abscisic acid) has been isolated. It is believed to cause plants to stop growing and enter into their dormant state. Dormin is identical with another substance, **abscissin,** that is responsible for causing leaves to fall off (absciss) plants late in the growing season. Thus a single substance isolated from different plants not only controls leaf fall but also regulates dormancy and many other growth processes.

You will be provided with several bean or sunflower seedlings. Select three plants that are in about the same stage of development (only the first two leaves should have expanded). Tag them (1) "phosfon," (2) "gibberellin," and (3) "control," along with your name and the date (Fig. 23-3B). Record the average beginning height of the plants in Table 23-2. In Fig. 23-5 draw the plants as they appear before treatment.

Using a wooden match or similar applicator, apply a ring of phosfon-lanolin paste around the stem of the plant labeled "phosfon" about ½ inch below the first pair of leaves (Fig. 23-3C). Then place the plant in bright sunlight.

Using a different applicator, apply gibberellin-lanolin paste to the pot labeled "gibberellin." If phosfon and gibberellin are prepared in a lanolin paste, what should be applied to the control plants?

Examine your plants and those of other students every 2 to 3 days for the next 3 weeks. Record the

TABLE 23-1
Effect of gibberellic acid on plant growth.

Day	Date	Treated with gibberellic acid		Controls	
		Height of plants	Appearance of plants	Height of Plants	Appearance of plants
0		1.____ 6.____ 2.____ 7.____ 3.____ 8.____ 4.____ 9.____ 5.____ 10.____ Avg.____		1.____ 6.____ 2.____ 7.____ 3.____ 8.____ 4.____ 9.____ 5.____ 10.____ Avg.____	
		1.____ 6.____ 2.____ 7.____ 3.____ 8.____ 4.____ 9.____ 5.____ 10.____ Avg.____		1.____ 6.____ 2.____ 7.____ 3.____ 8.____ 4.____ 9.____ 5.____ 10.____ Avg.____	
		1.____ 6.____ 2.____ 7.____ 3.____ 8.____ 4.____ 9.____ 5.____ 10.____ Avg.____		1.____ 6.____ 2.____ 7.____ 3.____ 8.____ 4.____ 9.____ 5.____ 10.____ Avg.____	
		1.____ 6.____ 2.____ 7.____ 3.____ 8.____ 4.____ 9.____ 5.____ 10.____ Avg.____		1.____ 6.____ 2.____ 7.____ 3.____ 8.____ 4.____ 9.____ 5.____ 10.____ Avg.____	
		1.____ 6.____ 2.____ 7.____ 3.____ 8.____ 4.____ 9.____ 5.____ 10.____ Avg.____		1.____ 6.____ 2.____ 7.____ 3.____ 8.____ 4.____ 9.____ 5.____ 10.____ Avg.____	
		1.____ 6.____ 2.____ 7.____ 3.____ 8.____ 4.____ 9.____ 5.____ 10.____ Avg.____		1.____ 6.____ 2.____ 7.____ 3.____ 8.____ 4.____ 9.____ 5.____ 10.____ Avg.____	
		1.____ 6.____ 2.____ 7.____ 3.____ 8.____ 4.____ 9.____ 5.____ 10.____ Avg.____		1.____ 6.____ 2.____ 7.____ 3.____ 8.____ 4.____ 9.____ 5.____ 10.____ Avg.____	

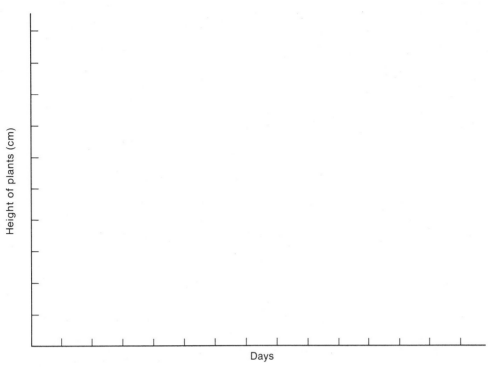

FIGURE 23-2
Effect of gibberellic acid on plant growth.

TABLE 23-2
Effect of phosfon and gibberellin on growth.

Date	Phosfon			Gibberellin			Control		
	Average height (mm)	Color of leaves	Other	Average height (mm)	Color of leaves	Other	Average height (mm)	Color of leaves	Other
0									

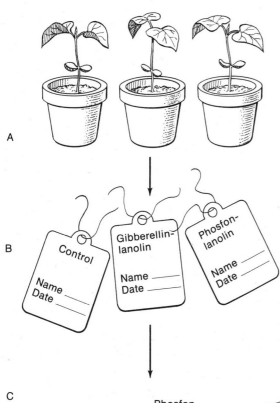

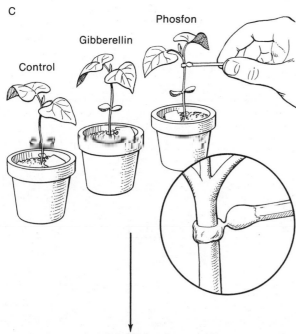

D. Examine the plants every 2 to 3 days for the next 3 weeks.

FIGURE 23-3
Effect of growth inhibitors on plant development.

average height of the plants, and other information, in Table 23-2.

At the conclusion of the study plot your data in Fig. 23-4. In Fig. 23-5 draw the plants as they appear at the end of the study.

C. DETECTION OF PLANT GROWTH

Hormonal regulation of growth in animals has been known for a number of years. By 1930 it was commonly accepted that plant growth was also under the regulation of special hormonal substances, now called **auxins**. The principal naturally occurring auxin is called **indoleacetic acid**, or **I.A.A.** (Fig. 23-6).

Before indoleacetic acid was discovered, "auxin" was used when referring to the natural growth hormone. The term auxin now is used to describe any of a large number of compounds having physiological activity similar to indoleacetic acid. I.A.A. does have, or is suggested to have, several effects on growth; a critical effect, however, is the ability to induce the elongation of cells.

The isolation and identification of auxins was simplified by the use of a biological testing procedure called a **bioassay**. There are numerous biologically active compounds present in living organisms in amounts so minute that they cannot be detected by the usual chemical procedures. In the bioassay the whole organism, or some part of it, is used to detect the presence of, and to measure the amounts of, biologically active substances.

The most sensitive bioassay for auxins is the *Avena* curvature test. This test makes use of the coleoptile of the *Avena* (oat) seedling (Fig. 23-7). The **coleoptile** is a cellular sheath that surrounds the embryonic leaves of the developing seedling. If oat seeds are germinated in the dark, cells of the coleoptile continue to divide until the coleoptile is approximately 1 cm long. For the next 3 or 4 days, when the coleoptile reaches its maximum length of 5–6 cm, extension of the coleoptile is by cell elongation only. This elongation is brought about by auxin produced in the tip of the coleoptile. When grown in the dark, the auxin is directed downward, causing the embryonic cells of the coleoptile to elongate. If the tip is removed, no cellular enlargement occurs. If the decapitated tip is replaced by an agar block containing auxin, growth of the coleoptile resumes. If the block of agar is displaced to one side, the cells on that side

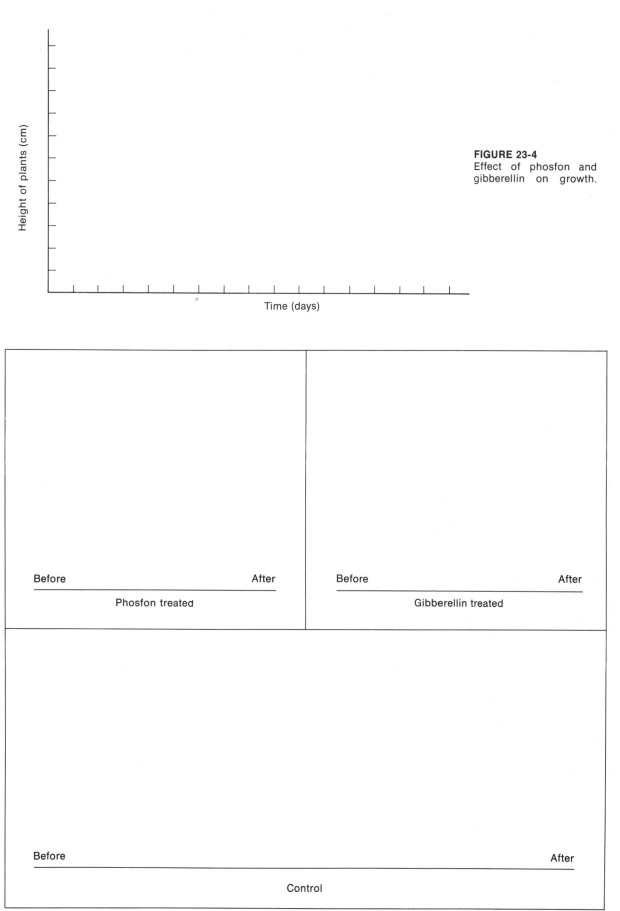

FIGURE 23-4
Effect of phosfon and gibberellin on growth.

FIGURE 23-5
Effect of phosfon and gibberellin on plant growth.

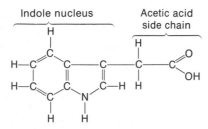

FIGURE 23-6
Structure of indoleacetic acid.

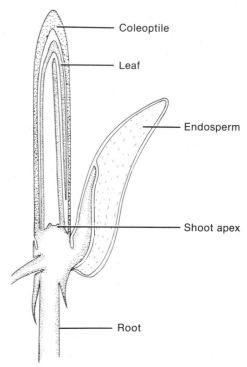

FIGURE 23-7
Avena seedling. (From *Plants in Action* by Leonard Machlis and John G. Torrey. W. H. Freeman and Company. Copyright © 1956.)

elongate more rapidly than the cells on the opposite side with the result that the coleoptile curves. This is the basis for the curvature test. It has been shown that the degree of curvature is, within limits, proportional to the concentration of auxin in the agar block. This test is a highly reliable and sensitive quantitative test for the detection of substances having auxin activity.

Although the *Avena* curvature test is very sensitive, it also requires rigidly controlled conditions of temperature and humidity and accurate measurements of curvature. For these reasons it is difficult to conduct such a test in most introductory laboratory courses. There is a simpler bioassay using oat coleoptiles, which is less sensitive in terms of the concentration of auxin measurable. It is called the *Avena* **straight-growth test.** The test measures straight growth as reflected in an increase in length of coleoptile sections. To save time the preliminary steps of soaking and planting the oat seeds will be carried out prior to the laboratory meeting.

1. The *Avena* Coleoptile Straight-Growth Test

a. For each laboratory section, soak 600–700 Brighton hull-less oat seeds for 2 hours in a liter of distilled water. After soaking, rinse the seeds two or three times with distilled water. Plant the oats on moist germinating paper (or Kleenex) in a tray; cover and pack tightly with about ½ inch of moist vermiculite; cover and place in the dark at room temperature. Germinate for 70–72 hours, exposing the seeds to 1 hour of red light out of each 24.

b. Label six test tubes and fill them as indicated in Table 23-3. The I.A.A. dilutions and incubating medium (a buffered nutrient solution containing sucrose) will be provided.

Using the information given in the note at the bottom of Table 23-3, describe how you would prepare the various dilutions used in this experi-

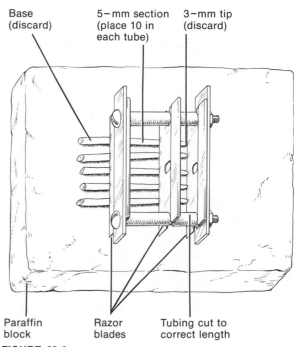

FIGURE 23-8
Method of cutting coleoptiles for *Avena* straight-growth test.

PLANT DEVELOPMENT: HORMONAL REGULATION

TABLE 23-3
Protocol for the *Avena* straight-growth test.

Tube contents	Tube No.					
	1	2	3	4	5	6
Incubation solution (ml)	1	1	1	1	1	1
Distilled H$_2$O (ml)	1	–	–	–	–	–
10^{-4} M I.A.A. solution (ml)	–	1	–	–	–	–
10^{-5} M I.A.A. solution (ml)	–	–	1	–	–	–
10^{-6} M I.A.A. solution (ml)	–	–	–	1	–	–
10^{-7} M I.A.A. solution (ml)	–	–	–	–	1	–
Unknown I.A.A. solution (ml)	–	–	–	–	–	1

Note: The dilution series uses solutions of varying molar (*M*) concentration. A molar solution is made by dissolving the molecular weight of a compound, in grams, in a liter of solvent. Thus, the molecular weight of I.A.A. is 175.2. A 1.0 *M* solution of I.A.A. contains 175.2 g/1000 ml of water. In a similar manner:

```
1/10 M         = 10⁻¹ M = 17.52000000 g/liter = 17520.00000 mg/liter
1/100 M        = 10⁻² M =  1.75200000 g/liter =  1752.00000 mg/liter
1/1000 M       = 10⁻³ M =  0.17520000 g/liter =   175.20000 mg/liter
1/10,000 M     = 10⁻⁴ M =  0.01752000 g/liter =    17.52000 mg/liter
1/100,000 M    = 10⁻⁵ M =  0.00175200 g/liter =     1.75200 mg/liter
1/1,000,000 M  = 10⁻⁶ M =  0.00017520 g/liter =     0.17520 mg/liter
1/10,000,000 M = 10⁻⁷ M =  0.00001752 g/liter =     0.01752 mg/liter
```

TABLE 23-4
Data for *Avena* straight-growth test.

	Tube number and contents					
	1 Incubating medium	2 10^{-4} M	3 10^{-5} M	4 10^{-6} M	5 10^{-7} M	6 Unknown
Average coleoptile length (mm)						

ment. (*Hint*: Notice that 1 ml of each dilution series is further diluted with 1 ml of incubating solution.)

c. Approximately 70 hours after planting, the coleoptiles will be 20–30 mm in length. Working in a room illuminated with red light (or use a room darkened as much as possible—but work quickly) select 60 coleoptiles measuring 25–30 mm in length. Cut off the root and seed and place the coleoptiles on moistened filter paper in a Petri dish.

d. Place four or five coleoptiles at a time on a paraffin block and cut them with the special cutter, as shown in Fig. 23-8. This cutter will divide each coleoptile into three parts: a 3-mm tip portion, a 5-mm section, and a base of about 20 mm. Discard the 3-mm tips and the bases. Place 10 of the 5-mm sections in each of the six test tubes.

e. Stopper the test tubes with cotton or styrofoam plugs and place in the dark for 24 hours at 25°C. After 24 hours remove the sections from the tubes and measure them with a millimeter ruler to the nearest 0.5 mm. Record the average length in Table 23-4.

Plot your data in Fig. 23-9 to obtain the standard curve for these concentrations. Does the curve show a proportional increase in growth with increasing I.A.A. concentration? Explain.

_____

_____

_____

_____

_____

_____

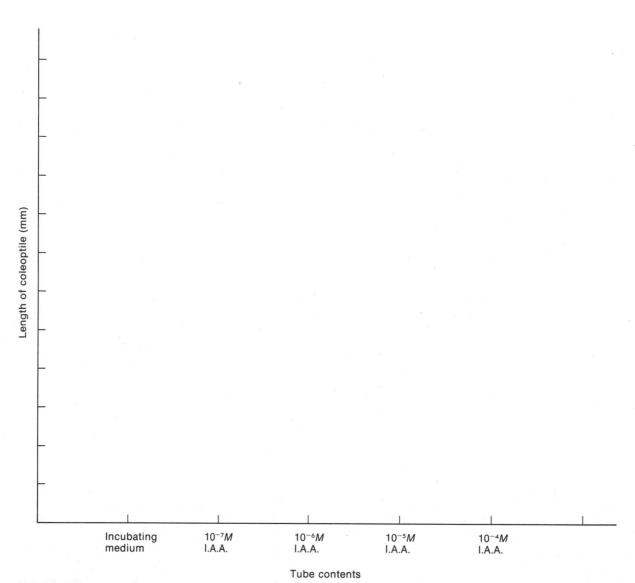

FIGURE 23-9
Standard curve for *Avena* straight-growth test.

TABLE 23-5
Data for colorimetric determination of I.A.A.

	Tube contents					
	0 mg/liter Control	1 mg/liter	10 mg/liter	20 mg/liter	40 mg/liter	Unknown
Transmittance (%)						

PLANT DEVELOPMENT: HORMONAL REGULATION

● Plot the "Unknown I.A.A. Concentration" on the graph. On the basis of the position of the unknown, estimate the concentration of the I.A.A. in the unknown solution.

What is the reason for having tube 1?

_____

_____

_____

2. Colormetric Test for I.A.A.

A simpler chemical test for detecting one of the natural auxins (such as I.A.A.) is much less sensitive than either of the bioassays, but is useful for in vitro (test tube) studies of auxin concentrations well above those normally found in plant tissue. For example, there is an enzyme found in plant tissues called indoleacetic acid oxidase, which breaks down I.A.A. This test would be useful for studying the action of such an enzyme.

The color test you will perform makes use of the fact that I.A.A., in the presence of iron chloride in Salkowski's reagent, forms a red color complex that may be quantitatively measured with a colorimeter (see Appendix B for a discussion of colorimetry).

You will be provided with a stock solution of I.A.A. at a concentration of 100 mg/liter. Make 10-ml samples of the following concentrations: 40, 20, 10, 1, and 0 mg/liter. This is done as follows. Collect 10 ml of the stock solution in a test tube. Pipet 4 ml of this solution into a test tube and dilute with distilled water to 10 ml. Label this tube 40 mg/1. Repeat this procedure using 2 ml, 1 ml, and 0.1 ml of the stock solution, diluting each to 10 ml. Mark these tubes 20 mg/1, 10 mg/1, and 1 mg/1, respectively. For the 0 mg/liter concentration, merely pipet 10 ml of distilled water into a tube. This dilution series will be used to establish a standard curve. In addition, you will be given a solution containing an unknown concentration of I.A.A.

● To conduct the color test, add 2 ml of the solution being tested to a test tube containing 8 ml of the Salkowski reagent. (Note: Since full color development is a function of time, stagger your tests to allow time for each colorimetric measurement.) Shake the mixture thoroughly (*and carefully*, as the reagent contains sulfuric acid!) and set aside for exactly 30 minutes to allow the color to develop. While waiting for the color reaction, warm up the colorimeter and adjust the wavelength to read 510 nm (see Appendix A for instructions on the use of the Bausch & Lomb Spectronic 20 Colorimeter).

Standardize the colorimeter, using a solution consisting of 2 ml of distilled water and 8 ml of Salkowski's reagent. Immediately transfer about 3 ml of this mixture to a colorimeter tube and place in the Spectronic 20. Adjust the instrument to read 100% transmittance. Why is this step required?

_____

_____

_____

At the end of 30 minutes, record the percent transmittance for each of the I.A.A. concentrations and the Unknown, and record the data in Table 23-5. Plot your data in Fig. 23-10. What is the concentration of the unknown I.A.A. solution?

_____

Of the three methods discussed in this exercise, which would you use if you had to assay accurately for the total amount of auxin in 500 mg of plant tissue? Explain.

_____

_____

_____

_____

_____

_____

_____

_____

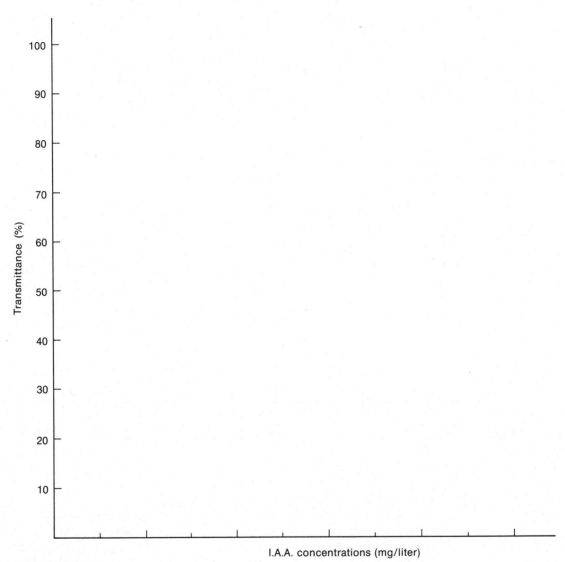

FIGURE 23-10
Colorimetric determination of I.A.A.

REFERENCES

Ebert, J. D., and I. M. Sussex. 1970. *Interacting Systems in Development*. 2d ed. Holt, Rinehart & Winston, New York.

Kefford, N. P. 1963. Natural Plant Growth Regulators. *Science* **142**:1495–1505.

Marth, P. C., W. V. Audra, and J. W. Mitchell. 1956. Effect of Gibberellic Acid on Growth and Development of Various Species in Plants. *Plant Physiol. Suppl.* **31**(43).

Mitchell, J. W. 1961. Fundamental Developments in the Field of Plant Growth Regulators. *Bull. Torrey Botan. Club,* **88**:299–312.

Paleg, L. G. 1965. Physiological Effects of Gibberellins. *Ann. Rev. Plant Physiol.* **16**:291–322.

Salisbury, F. B., and C. Ross. 1969. *Plant Physiology*. Wadsworth, Belmont, California.

Stowe, B. B., and T. Yamaki. 1959. Gibberellins: Stimulants of Plant Growth, *Science* **129**:807–816.

Thimann, K. V. 1957. Growth and Growth Hormones in Plants. *Am. J. Botany* **44**:49–55.

Torrey, J. C. 1967. *Development in Flowering Plants*. Macmillan, New York.

Wain, R. L., and F. Wightman. 1956. *The Chemistry and Mode of Action of Plant Growth Substances*. Academic Press, New York.

EXERCISE 24

Plant Development: Effect of Light

The morphogenetic effects of light on plants have been known since the latter part of 19th century. As early as 1880, experiments conducted with carbon arc lamps showed that artificially prolonged days promoted plant growth. Subsequent to the invention of the electric light bulb a large amount of data began to accumulate on the morphogenetic effects of light on plants. These observations ranged from such anatomical effects as variations in fiber and vascular tissue development to establishing the presence of a pigment called **phytochrome**, which may regulate many of the responses of plants or plant parts to light.

In this exercise you will study the effects of light on development and the effect of day length on flowering and dormancy of plants.

A. PLANT GROWTH IN THE ABSENCE OF LIGHT

Obtain two pots or trays that contain vermiculite. Label one "dark" and the other "light." Include your name, the date, and section number on each label. Plant 10 bean seeds about $\frac{1}{2}$ inch deep in each container. Water thoroughly and then place one of the pots in the sunlight and the other in the dark (Fig. 24-1). (Note: *Water plants daily throughout the period of the study.*)

After about 4 or 5 days examine the plants that have been growing in the dark. Compare them with those growing in the light. Record your observations in Table 24-1.

Reverse the conditions under which the plants have been growing (that is, place the "dark" plants in the light and the "light" plants in the dark). Examine the plants after 4 or 5 days. Describe the appearance of the plants.

_____

_____

Record your observations in Table 24-2.

The growth response observed in the plants grown in the dark is called **etiolation**. What advantage does such a response have for plant survival?

_____

_____

A. Plant 10 bean seeds about ½" deep in each of 2 containers labeled "light" and "dark" (include your name and the date).

B. Place the "dark" plants in a cabinet and the "light" plants where they will get sunlight.

FIGURE 24-1
Effect of darkness on plant growth.

B. EFFECT OF DAY LENGTH ON PLANT GROWTH

Day length varies with the seasons. In the Northern Hemisphere the day length increases during spring to a maximum on June 21, and thereafter it decreases to a minimum on December 21. This day length is called the **photoperiod**, and the response of plants to photoperiods of varying lengths is called **photoperiodism** or a **photoperiodic response**. In this part of the exercise you will examine the effects of the photoperiod on two aspects of plant development.

1. Regulation of Flowering

There are basically two types of photoperiodic flowering responses. These are called **long-day** and **short-day responses**. Plants that exhibit these responses are called long-day (L-D) and short-day (S-D) plants. L-D plants flower when the photoperiod *exceeds* some critical value, which is generally in excess of 14 hours (Table 24-3). At day lengths less than the critical value the plant usually grows only vegetatively. An S-D plant (really a long-night plant) will flower when the photoperiod is *less* than some critical value, usually about 10 hours, and grows vegetatively when this critical day length is exceeded. Other plants, not responsive to day length, are called **day-neutral** plants.

The photoperiodic control of flowering involves a complex mechanism about which few facts are known. These responses appear to be under the

TABLE 24-1
Effect of darkness on plant growth.

Observations and measurement	Treatment	
	Plants grown in dark	Plants grown in light
Color of shoot		
Average number of leaves		
Average length of first pair of leaves		
Average length of shoot between cotyledons and first leaves		

PLANT DEVELOPMENT: EFFECT OF LIGHT

control of the light-sensitive proteinaceous pigment called phytochrome. This bluish green pigment has been extracted and isolated with a high degree of purity.

The effect of phytochrome on flowering appears to be related to its having two forms in the plant. One form, called P_{660}, absorbs red light (with maximum absorption at 660 nm) and is converted into a second form, called P_{730}. The P_{730} form of the pigment strongly absorbs far-red radiation (maximally at 730 nm) and may then be reconverted to the P_{660} form. The photoperiod responses in various plants may be a result of differences in the rate of conversion of one form of phytochrome to another. There is also some evidence for a second photomorphogenic receptor involving a system of pigments absorbing blue and far-red radiations. These appear to operate independently of the red, far-red phytochrome system. For more information on phytochrome, refer to Wilkins (see References).

In this study you will be given two different species of plants and will attempt to determine whether they are L-D or S-D plants.

The L-D and S-D conditions are shown in Fig. 24-2.

a. Obtain two trays of plants provided by your instructor. Label one of the trays "L-D," along with your name, section number, and the date. Label the other tray "S-D" (Fig. 24-2A).

b. Place the trays under light conditions so that the L-D plants receive 16 hours of light and 8 hours of darkness (Fig. 24-2B). The S-D plants should receive 8 hours of light and 16 hours of darkness (Fig. 24-2C).

c. Examine the plants weekly and watch for the appearance of flower buds. The detection of flowers is difficult in the early stages of formation, so close inspection may be necessary. However, avoid excessive handling of the plants.

d. If, at the conclusion of the experiment (as determined by your instructor), no flowers are visible, remove the buds from each plant and place them in separate, labeled Petri dishes containing moistened filter paper.

e. Using a dissecting microscope, dissect the buds and determine if flower primordia are present or if the buds are still producing leaves. Your instructor will have "floral buds" and "vegetative buds" on demonstration to help you distinguish between the two.

f. Determine the photoperiod response of the plants and confirm your results with your instructor.

What economic use can be made of the photoperiodic response?

There are many other aspects of plant growth that are under photoperiodic control. Based on your reading, list several of these here.

2. Dormancy

In the broadest sense, and without reference to the factors involved, dormancy can be defined as a temporary suspension of visible growth and development. In this study you will determine the effects of the photoperiod in regulating the initiation of the dormant, winter bud in a woody plant. If you are unfamiliar with the appearance of a winter bud, examine the demonstration of dormant buds set up by the instructor.

a. Working in teams, obtain two potted seedlings that are approximately the same height. Label each with your name, section number, the date, and treatment (L-D or S-D).

b. Place the plants on the L-D and S-D benches in the greenhouse. Measure the length of the main shoot on each plant and record this and any other observations that seem appropriate under Day 0 in Table 24-4.

c. Examine the plants during succeeding laboratory periods and record the date and stem length. What evidence is there that the plant (under one treatment or the other) is forming a dormant bud? Examine the plants of other students, but do not handle them excessively. Make similar measurements and observations over the next several weeks, or until your instructor concludes the experiment.

A. Label "L-D" (long-day) or "S-D" (short-day), along with your name, date, and section number.

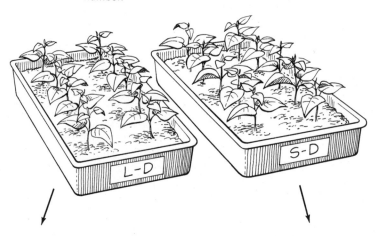

B. Place one group of plants under L-D conditions.

C. Place the second group under S-D conditions.

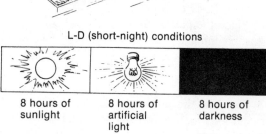

L-D (short-night) conditions

8 hours of sunlight | 8 hours of artificial light | 8 hours of darkness

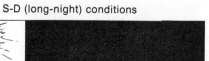

S-D (long-night) conditions

8 hours of sunlight | 16 hours of darkness

D. Examine the plants weekly for the appearance of flowers. After 3 to 4 weeks (if no flowers are visible) dissect the buds to determine if they have flowers or leaves in them.

FIGURE 24-2
Effect of photoperiod on flowering.

PLANT DEVELOPMENT: EFFECT OF LIGHT

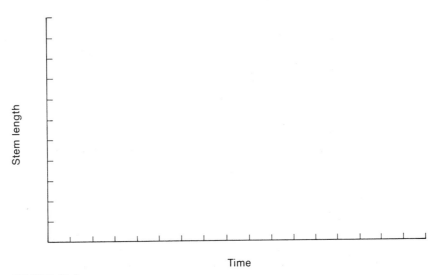

FIGURE 24-3
Stem length data.

TABLE 24-2
Effect of placing "light grown" plants in the dark and "dark grown" plants in the light.

Observations	Plants grown in dark ⟶ light	Plants grown in light ⟶ dark
Color of shoot		
Other observations		

TABLE 24-3
Flowering responses of long-day, short-day, and day-neutral plants.

Plant	24-hour period		Flowers
	Light	Dark	
Long-day	→ (Critical day length)		no
	———→	→	yes
Short-day	→		yes
	———→	→	no
Day-neutral	→		yes
	———→	→	yes

TABLE 24-4
Observations of plant response to photoperiod.

Day	Date	L-D treatment		S-D treatment	
		Length of shoot (cm)	Other observations	Length of shoot (cm)	Other observations
0					

d. At the conclusion of the experiment, plot the "stem length" data in Fig. 24-3. Suggest reasons for the differences in stem length, if any, between the plants given the L-D and S-D treatments.

Under which photoperiod has a dormant bud been formed?

In view of your knowledge of growth regulators, suggest a mechanism by which the photoperiod is inducing a state of dormancy in these plants.

e. Remove all the leaves of the dormant plant and return the plant to the same photoperiod that induced dormancy. Examine the plant weekly for 2–3 weeks. List your observations below and discuss how the plant may receive the light stimulus.

REFERENCES

Downs, R. J., and H. A. Borthwick. 1956. Effect of Photoperiod upon the Vegetative Growth of *Weigela florida*, var. *variegata*. *Proc. Am. Soc. Hort. Sci.,* **68**:519–521.

Galston, A. W., and P. F. Davies. 1970. *Control Mechanisms in Plant Development.* Prentice-Hall, Englewood Cliffs, New Jersey.

Nitsch, J. P. 1957. Photoperiodism in Woody Plants. *Proc. Am. Soc. Hort. Sci.* **70**:526–544.

Phillips, I. D. J., and P. F. Wareing, 1959. Studies in Dormancy of Sycamore. II. The Effect of Day Length on the Natural Growth-Inhibitor Content of the Shoot. *J. Exp. Botany* **10**:504–514.

Van der Veen, R., and G. Meyer. 1959. *Plant Growth and Light.* Macmillan, New York.

Wareing, P. F. 1956. Photoperiodism in Woody Plants. *Ann. Rev. Plant Physiol.* **7**:191–214.

Waxman, S. 1957. The Development of Woody Plants as Affected by Photoperiodic Treatments. Ph.D. dissertation, Cornell University, Ithaca, New York. (*University Microfilms*, Ann Arbor, Michigan, Pub. No. 23150.)

Wilkins, M. B., Ed. 1970. *Physiology of Plant Growth and Development.* McGraw-Hill, New York.

Winthrow, R., Ed. 1959. *Photoperiodism and Related Phenomena in Plants and Animals.* American Association for the Advancement of Science, Washington, D.C.

EXERCISE 25

Procaryotes

In this exercise you will study examples of procaryotic organisms. The nuclear material in the cells of these organisms is not bound by a nuclear membrane. This distinguishing feature, however, is not visible at ordinary magnifications of the light microscope but can be seen with the electron microscope.

A. BACTERIA

Despite their minute size and apparently simple structure, bacteria have become adapted to a wide variety of environments. In fact, they are more widely distributed in nature than any other group of organisms.

Bacteria are especially abundant in the soil. Some species play an important role in the nitrogen cycle—where nitrogen gas in the atmosphere is converted into nitrogenous salts that are used by plants for growth.

Bacteria are also responsible for the decay of organic matter, such as dead plants and animals. During the breakdown of this material carbon dioxide is released and may then be reused in photosynthesis. While most bacteria are **heterotrophs**, some are **autotrophs** and can make their own sugars, fats, amino acids, and so forth. One group of bacteria has the ability to obtain energy for their synthetic activities by converting light energy into chemical energy by a process similar to photosynthesis. Photosynthesis in bacteria, however, differs from that in eucaryotic organisms in that water (H_2O) is not used as the reductant for carbon dioxide. Rather, hydrogen sulfide (H_2S) may be used. Thus, in bacterial photosynthesis sulfur is evolved rather than oxygen. They obtain their nitrogen from ammonia or nitrate, and their carbon from carbon dioxide in the air.

A second group of autotrophic bacteria produces its own organic materials from carbon dioxide, ammonia, or nitrate by obtaining the energy for these syntheses from the oxidation of inorganic substances. For example, a bacterium present in the soil is capable of oxidizing ammonia to nitrate,

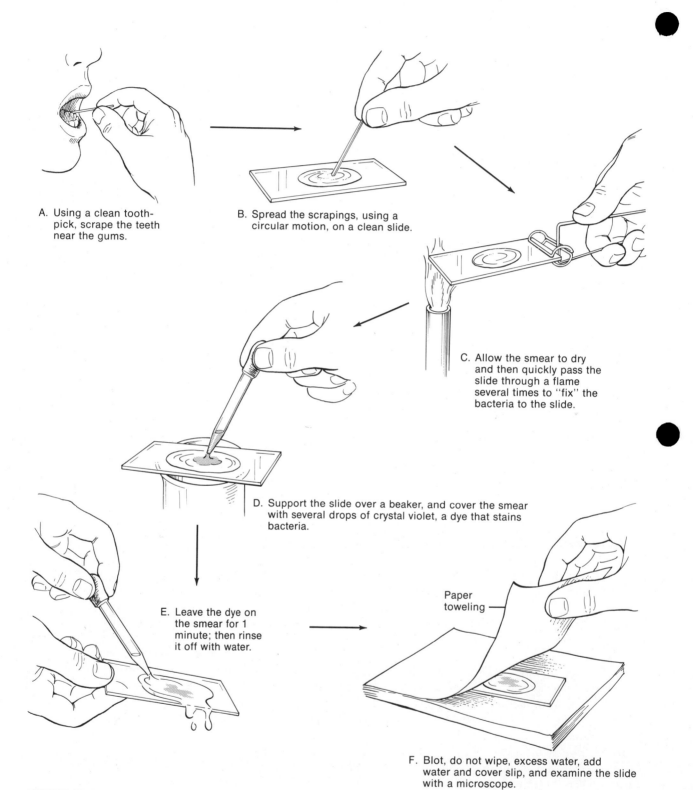

FIGURE 25-1
Procedure for staining bacteria.

PROCARYOTES

thus generating useful energy. This organism is a typical **chemosynthetic** autotroph. Since these bacteria contain carbohydrates, fats, proteins, nucleic acids, and vitamins, they represent magnificent synthetic factories for making protoplasm.

Other bacteria are widely used in the commercial production of bakery goods, alcohol, vinegars, chemicals, enzymes, antibiotics, and many other products. These microorganisms are also responsible for food spoilage, food poisoning, and for many plant and animal diseases including tuberculosis, scarlet fever, pneumonia, and diphtheria. Bacteria, therefore, directly or indirectly influence the survival of mankind.

The functional unit of these organisms is a single cell, the smallest of which is barely visible with the light microscope. For many years bacteria were thought to reproduce asexually by a process called fission, during which the cell pinches in two. It is now known that bacterial cells also exchange genetic material through sexual reproduction. Although sexual mechanisms have been found in only a few species, future research in this area may reveal that sexual reproduction is actually a widespread occurrence among bacteria.

1. Bacterial Identification

Bacterial cells generally assume three basic shapes: rods (**bacilli**), minute spheres (**cocci**), and corkscrews (**spirilla**). The cells of many species tend to adhere to each other and form relatively simple colonies. However, each cell in the colony still functions as an independent unit and maintains the ability to carry out all of the processes necessary for survival.

In this exercise you will examine some bacteria and stain them to identify the three cell types listed above.

a. Using the wide end of a clean toothpick that has been dipped in alcohol, scrape your teeth near the gums. Spread the scrapings in a thin film on a clean glass slide (Fig. 25-1A and B).

b. Allow the smear to dry, then heat it gently by quickly passing the slide back and forth several times over the flame of an alcohol lamp or Bunsen burner (Fig. 25-1C). Support the slide over a beaker or other small dish and apply several drops of crystal violet—a dye that stains bacteria (Fig. 25-1D).

c. Leave the dye on the smear for 1 minute. Wash off the dye under a gentle stream of tap water or water squirted from an eye dropper (Fig. 25-1E). Using a paper towel, wipe off any dye that may be sticking to the bottom of the slide. Blot the excess water from the upper surface, avoiding only the smear (Fig. 25-1F).

d. Place a drop of water on the stained smear, add a cover slip and examine your slide with the high power of your microscope.

In Fig. 25-2 draw the different types of bacteria you observed. Which of the bacterial types is most common?

_____

FIGURE 25-2
Types of bacteria.

2. Control of Bacterial Growth

The control of bacterial infections is a major medical problem. The use of antibiotics is an effective method of controlling infectious bacteria and other disease-causing organisms. The story of antibiotics began in 1929 when the British biologist Alexander Fleming found that a Petri dish containing a bacterial culture had become contaminated by a mold called *Penicillium*. Noticing that the growth of the bacteria around the mold colony was inhibited, he concluded that the mold produced a diffusable chemical agent capable of inhibiting bacterial growth. Fleming later isolated this antibacterial chemical, calling it penicillin. Such chemicals, which are produced by living organisms and have the ability to retard the growth of certain other organisms, are called antibiotics. Among the wide variety of such antibiotics are streptomycin, chloromycin, and terramycin. The use of such antibiotics has significantly lowered the occurrence of a number of infections.

In this part of the exercise you will examine the effect of different antibiotics on the growth of one of the more common bacteria.

a. Obtain a Petri dish containing nutrient agar—a mixture of chemicals that is optimal for the growth of the bacteria you are to study. Divide each plate into four sections labeled 1, 2, 3, and "Control" by marking the bottom of the dish with black pencil (Fig. 25-3A).

b. Lift the cover of the Petri dish slightly and add 10 drops of the bacterial suspension (provided by your instructor) to the plate (Fig. 25-3B and C). Holding the plate at eye level, tilt the plate, allowing the suspension to run to the far edge (Fig. 25-3D). Then, holding the plate level, jerk it toward you. Repeat this several more times, rotating the plate partway between each jerk (Fig. 25-3E). In this way the bacterial suspension will cover the entire surface of the agar.

c. Incubate the plate for 24 hours at 37°C (or 48 hours at room temperature) before proceeding with the next step. After incubation, partially lift the cover. Then, using flamed forceps place three antibiotic discs in the numbered sections of the plate (Fig. 25-3G). Use a different antibiotic in each section.

What should be placed in the "Control" section of the plate?

_____

Incubate the plate. Examine daily for the next 3 or 4 days. Record the effects of the antibiotics in Fig. 25-4. Why might incubation for a period of time be desirable before adding the antibiotic discs?

_____
_____
_____
_____

Some bacteria have an enzyme called penicillinase. Would the growth of such bacteria be inhibited by penicillin? Explain.

_____
_____
_____
_____
_____
_____

3. Bacteria in Milk

Milk, the most nearly "perfect food" for man, is also an excellent medium for the growth of bacteria. Therefore, great care must be used in its processing. Harmful bacteria in milk are killed by the process of **pasteurization**. In this exercise you will examine milk from different sources and determine the quality of the milk in terms of its bacterial population.

a. Bring to class three to four samples of milk of varying age or from different sources (fresh milk from a dairy farm, milk in an unopened carton or bottle, milk in a carton or bottle that has been opened and in the refrigerator for one to three or more days, powdered milk, canned milk, etc.).

b. Fill separate test tubes one-third full with each of these milk samples (Fig. 25-5A). Label each tube and then add 1 ml (20 drops) of methylene blue solution (Fig. 25-5B). Mix by shaking.

c. Plug each tube with sterile cotton and incubate or place the tubes into a water bath at 37°C (Fig. 25-5C). Record the time in Table 25-1.

When bacteria are actively growing in milk, they consume oxygen. The reduction of oxygen can be detected by using methylene blue, which becomes colorless as the oxygen content of the milk diminishes. If the number of bacteria present

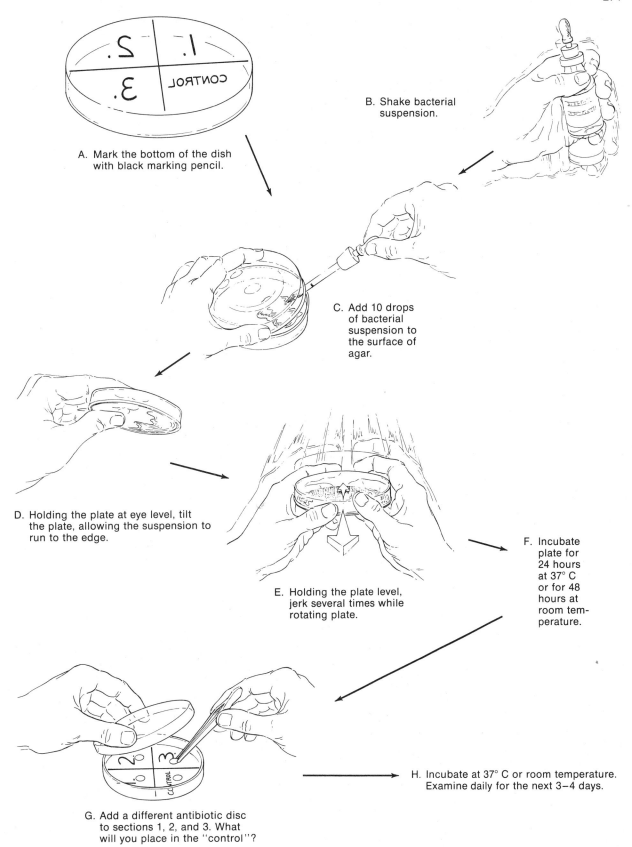

FIGURE 25-3
Effect of antibiotics on bacterial growth.

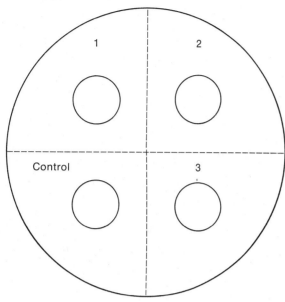

FIGURE 25-4
Effect of antibiotics on bacterial growth.

is high, the methylene blue solution will rapidly lose its color. If the bacterial population is low, more time will be required for decolorization.

For the purposes of this study, the quality of the milk with respect to the number of bacteria present can be rated by the time it takes to decolorize the methylene blue solution. Periodically examine the milk samples. In Table 25-1 record the time it takes to decolorize each sample. Rate each sample according to the description given in Table 25-2. Which sample rated the highest in bacterial contamination?

_____

TABLE 25-1
Decolorization times.

Time to decolorize methylene blue	Rating
Less than 20 minutes	Highly contaminated
20 minutes to 2 hours	Poor
2–5½ hours	Fair
5½–8 hours	Good
More than 8 hours	Excellent

Which rated the lowest?

_____

From your results what appears to lead to the contamination of milk?

_____

What appears to reduce contamination?

_____

B. BLUE-GREEN ALGAE

This group of organisms exhibits two types of plant body structure—unicellular and colonial. The unicellular condition is considered to be more primitive. In addition, blue-green algae, although considered to be procaryotic organisms on the basis of their cellular structure, utilize water in photosynthesis as the reducing agent for CO_2 and therefore evolve oxygen as a by-product of photosynthesis. In this respect they are biochemically similar to photosynthetic eucaryotic organisms. In this study

TABLE 25-2
Rating of milk samples of different ages or different sources.

Contents of tube	Time methylene blue added	Length of time to decolorize methylene blue	Rating
1			
2			
3			
4			

PROCARYOTES

you will examine blue-green algae to become familiar with their morphology, cytology, and range of complexity.

1. Unicellular Forms

Prepare a wet mount of *Chroococcus* and *Gloeocapsa*. To see the characteristic gelatinous sheath that surrounds the cells of blue-green algae, add a drop of India ink to the slide. The sheath will stand out against the dark background. In those cases where several cells are clustered together, do they share a common sheath?

_____

Although these are unicellular forms, they frequently form small clusters of cells. Would you consider these clusters as representing multicellular organisms? Explain.

_____

_____

_____

2. Colonial Forms

In colonial forms the shape of the colony is largely determined by the plane of cell division and may result in filaments, plates, or spheres. Examine the following algae and classify each as to body form (e. g., filamentous, etc.) and plane of cell division (e. g., single plane, double plane, or irregular) that resulted in the form of the colony.

    a. Merismopedia. Locate the gelatinous sheath. Frequently cells within the colony may be seen in cytokinesis.

    b. Oscillatoria. What evidence is there of cellular differentiation within this colony?

_____

_____

Describe any motility you observe.

_____

_____

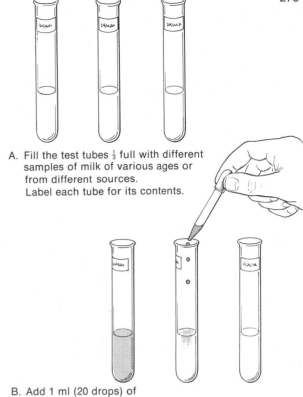

A. Fill the test tubes ⅓ full with different samples of milk of various ages or from different sources. Label each tube for its contents.

B. Add 1 ml (20 drops) of methylene blue solution to each tube. Mix by shaking.

C. Plug the tubes with cotton and incubate (or place them in a hot water bath) at 37°C.

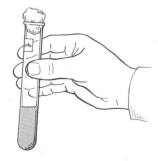

D. Examine each tube periodically. Rate the quality of the milk according to its bacterial population as described in the procedure.

FIGURE 25-5
Procedure for estimating bacterial contamination of milk.

How does this organism reproduce?

_____

_____

_____

c. Rivularia. How does this alga differ from *Oscillatoria*?

_____

_____

_____

Locate the heterocyst. What is its function?

_____

_____

_____

d. Gloeotrichia. How does this alga differ from *Rivularia*?

_____

_____

e. Anabaena. Crush the cells of the water fern *Azolla* to release this alga. What function might this alga carry out in the cells of *Azolla*?

_____

_____

_____

C. ALGAE IN WATER SUPPLIES

The increasing human population and rapid development of industry and agriculture has resulted in a phenomenal increase in the use of water. This need for water has produced many difficult problems in procuring an adequate and suitable water supply.

As population and industrial demands increase, villages and cities are turning from ground water sources to surface waters, such as lakes, streams, and reservoirs, for their water supplies. Ground waters are essentially free from contaminating organisms. Surface waters, on the other hand, contain many organisms that in one way or another contribute to the unpalatability of the water supply. Such organisms affect the odor and taste of water, clog filters, grow in pipes, cooling towers, or on reservoir walls, form mats or blooms on the surface of the water, produce toxic materials, and so forth. Furthermore, the present methods of waste disposal are intensifying the problems of such nuisance organisms in water. Materials such as sewage and organic wastes from paper mills, fish processing factories, slaughter houses, and milk plants, to name a few, greatly increase the growth of algae and other organisms. Many of these produce the problems cited above when they become abundant.

Blue-green algae are among the common nuisance inhabitants of surface waters. Collect surface water samples from different sources—i.e., lakes, streams, reservoirs (including the walls), swimming pool filters, or water treatment plants, etc. Table 25-3 characterizes some of the more common problems

TABLE 25-3
Problems caused by blue-green algae in water supplies.

Problem	Organism
Taste and odor	Anabaena circinalis Anacystis cyanea Aphanizomenon flos-aque Cylindrospermum muscicola Gomphosphaeria lacustris
Filter clogging	Anabaena flos-aque Anacystis dimidiata Gloeotrichia eschinulata Oscillatoria princeps Oscillatoria chalybea Oscillatoria splendida Rivularia duro
Reservoir wall growth	Calothrix braunii Nostoc pruniforme Phormidium uncinatum Tolypothrix tenuis
Polluted water	Anabaena constricta Anacystis montana Arthrospira jenneri Lyngbya digireti Oscillatoria chlorina Oscillatoria putrida Oscillatoria lauterbornii Phormidium autumnale

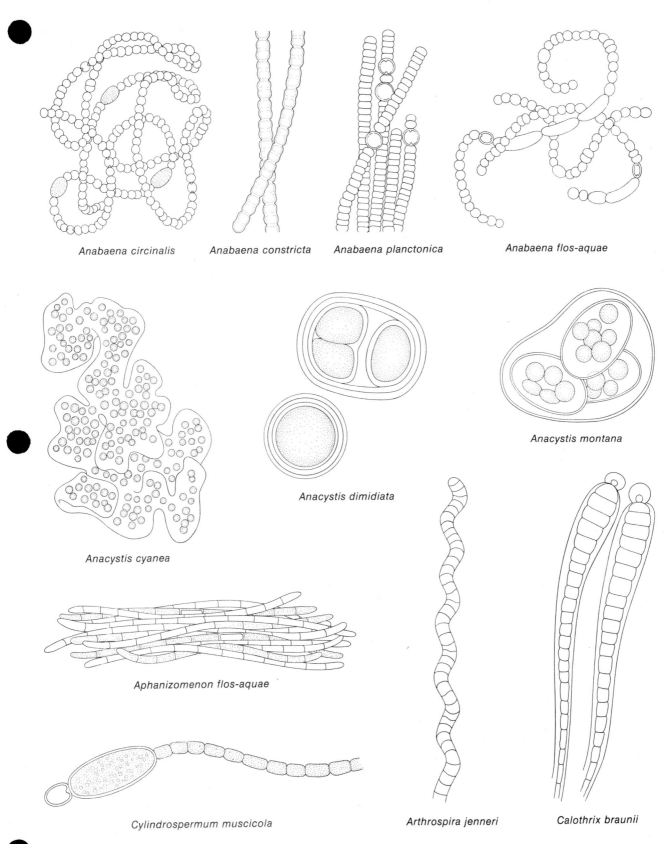

FIGURE 25-6
Blue-green algae contaminating water supplies.

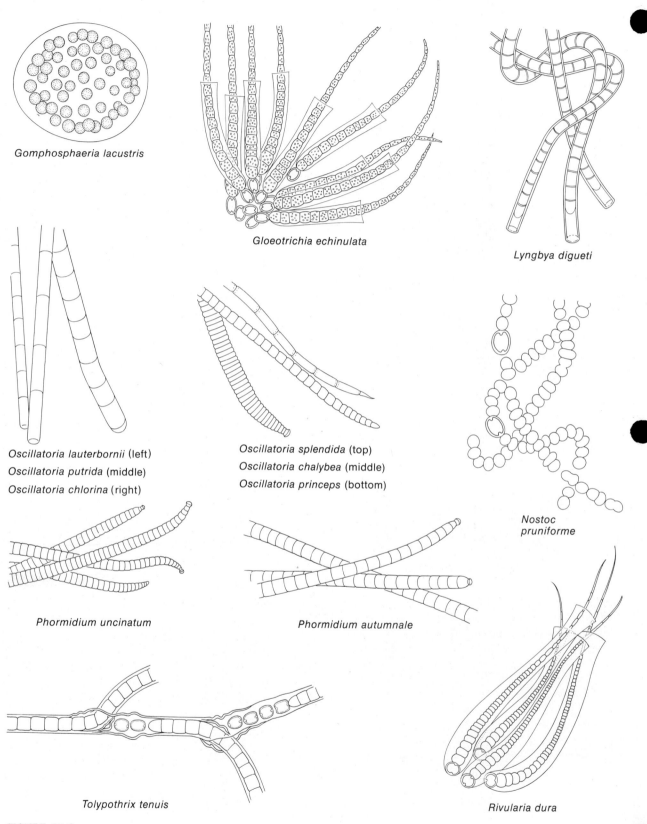

FIGURE 25-7
Blue-green algae contaminating water supplies.

associated with an overabundance of blue-green algae. Use Figs. 25-6 and 25-7 to identify some of the algae in your water samples.

REFERENCES

Brock, T. 1961. *Milestones in Microbiology.* Prentice-Hall, Englewood Cliffs, New Jersey.

Cairns, J. 1966. The Bacterial Chromosome. *Scientific American,* January (Offprint No. 1030). *Scientific American* Offprints are available from W. H. Freeman and Company, 660 Market Street, San Francisco 94104, and 58 Kings Road, Reading RG13AA, England. Please order by number.

Davis, B. D., R. Dulbecco, H. N. Eisen, H. S. Ginsburg, and W. B. Wood, Jr. 1968. *Principles of Microbiology and Immunology.* Harper & Row, New York.

DeKruif, P. 1959. *Microbe Hunters.* Pocket Books, New York.

Delevoryas, T. 1966. *Plant Classification.* Holt, Rinehart & Winston, New York.

Dubos, R. J. 1945. *The Bacterial Cell.* Harvard University Press, Cambridge, Massachusetts.

Echlin, P. 1966. The Blue-Green Algae. *Scientific American,* June (Offprint No. 1044).

Frobisher, M. 1962. *Fundamentals of Microbiology.* 7th ed. Saunders, Philadelphia.

Greulach, V. A., and J.E. Adams. 1962. *Plants: An Introduction to Modern Botany.* Wiley, New York.

Prescott, G. W. 1968. *The Algae: A Review.* Houghton Mifflin, Boston.

Pringsheim, E. G. 1949. *Pure Cultures of Algae: Their Preparation and Maintenance.* Cambridge University Press, Cambridge, England.

Ratcliff, J. D. 1945. *Yellow Magic, the Story of Penicillin.* Random House, New York.

Sharon, N. 1969. The Bacterial Cell Wall. *Scientific American,* May (Offprint No. 1142).

Smith, G. M. 1950. *The Freshwater Algae of the United States.* 2d ed. McGraw-Hill, New York.

Smith, G. M. 1955. *Cryptogamic Botany.* Vol. 2. McGraw-Hill, New York.

Thimann, K. V. 1963. *The Life of Bacteria.* 3d ed. Macmillan, New York.

Wollman, E. L., and F. Jacob. 1950. Sexuality in Bacteria. *Scientific American,* July (Offprint No. 50).

EXERCISE 26

Diversity in Plants: Fungi

The tremendous diversity in life forms of plants and animals in existence today makes it extremely difficult to attempt a random study of them all. In order to simplify such a study, a system of classification based upon different morphological, and more recently biochemical, criteria has been established.

While there are several methods of studying relationships or evolutionary trends within or between groups of organisms, the following exercises dealing with the plant kingdom will be concerned with changes in the complexity of the vegetative body and variations in the reproductive structures.

The fungi include a large variety of organisms that are without chlorophyll and lack roots, stems, and leaves. Nutrition is by heterotrophic means; that is, complex, organic compounds are obtained from the environment. Among the fungi, there are two types of **heterotrophs**: **parasites,** which obtain food from living organisms, and **saprophytes,** which obtain their food from nonliving, organic materials. In this study representatives of several classes of fungi will be examined.

A. SLIME MOLDS

Slime molds are organisms that have been classified as both plants and animals. Confusion arises because the vegetative structure (feeding phase) is an amoeboidlike mass that migrates over or through soil and decaying wood, feeding on food particles. On the other hand, its fruiting body (a type of reproductive structure) resembles the fruiting body of fungi and produces spores. Because the fruiting body in these organisms was first studied by mycologists, slime molds have traditionally been classified as fungi.

Slime molds are divided into two groups based upon the form of the feeding phase. The vegetative phase of the *cellular slime molds* consists of masses of single amoeboid cells. The acellular, or *plasmodial slime molds,* have a vegetative phase consisting of naked masses of protoplasm (plasmodia) of indefinite size and shape. Both types live predominantly on decaying plant material, primarily microorganisms (especially bacteria). In this part of the exercise you will study the plasmodial slime molds by observing their growth and production of fruiting structures.

Obtain a small piece of slime mold **sclerotia** (a dry, resting phase of the organism) and place it on the surface of the agar in the Petri dish (Fig. 26-1).

Sprinkle a few oatmeal flakes over the sclerotium. Add 2 or 3 drops of water to moisten the oatmeal.

Replace the cover on the dish and set it aside in a dark place. After 24 hours examine the plates for growth of the slime mold. Record your observation by drawing the growing organism in the spaces below.

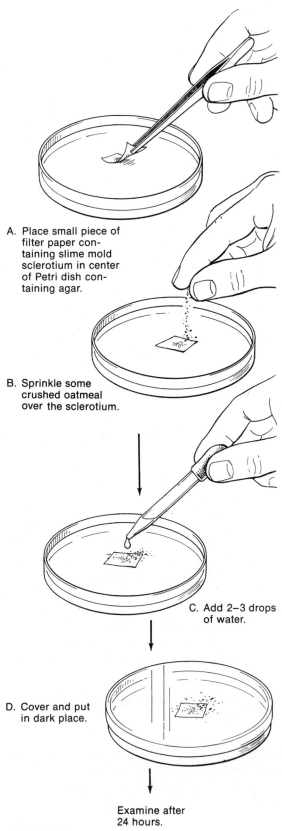

A. Place small piece of filter paper containing slime mold sclerotium in center of Petri dish containing agar.

B. Sprinkle some crushed oatmeal over the sclerotium.

C. Add 2–3 drops of water.

D. Cover and put in dark place.

Examine after 24 hours.

FIGURE 26-1
Procedure for growing slime mold.

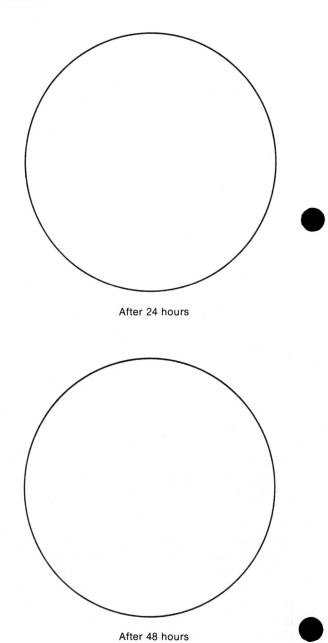

After 24 hours

After 48 hours

Once growth has begun, examine the slime mold with a hand lens, a dissecting microscope, or the low power of your compound microscope. Describe the unique pattern of cytoplasmic movement observed in the branching of the mold.

_____

_____

Puncture a branch of the slime mold with a needle and watch it for a few minutes.

After the dish becomes covered with the plasmodium of the slime mold, partially remove the cover. Under these conditions the slime mold will begin to dry out and in the process will initiate the formation of fruiting bodies. Examine the culture during the next few days and describe the shape of the fruiting body that is formed.

_____

_____

B. ALGALLIKE FUNGI (PHYCOMYCETES)

The black bread mold (*Rhizopus stolonifer*) is one of the commonest of all fungi and can be found growing on many foods that contain starches.

1. Asexual Reproduction

Examine a Petri dish containing black bread mold. The mold is not growing on bread here, but on agar containing various organic compounds needed for growth. [If you wish, bring some bread from home (but remember that many mass-produced breads contain mold inhibitors) and place a piece in a Petri dish. Moisten it with several drops of water and then place a small piece of the mold from the agar onto the bread. Cover the dish and set it aside for 1–2 days.]

Examine the mold with your dissecting microscope. *Do not remove the cover!* Note the whitish mass of filaments growing over the surface of the agar (Fig. 26-2A). Each filament is called a **hypha** (plural **hyphae**). The total mass of hyphae is a **mycelium**. Some hyphae grow upward and form small black, globelike structures called **sporangia** (Fig. 26-2C). Inside the sporangia are cells called **spores**. They are released when the sporangia open. What is the function of the spores?

_____

_____

Many hyphae penetrate the agar. Turn the dish over and focus downward through the agar to locate small rootlike hyphae called **rhizoids** growing into the agar (Fig. 26-2C). Suggest a function for the rhizoids?

_____

_____

Remove a small piece of the mold and place it in a drop of water on a slide (Fig. 26-2B). Add a cover slip and examine with your microscope. Locate sporangia, spores, hyphae, and rhizoids.

2. Sexual Reproduction

The black bread mold requires two different strains before sexual reproduction can occur. Your instructor has inoculated an agar plate with a (+) strain and a (−) strain. The growth of each strain has brought them into contact. At the points of contact, gametes are formed. Fusion of the gametes (**fertilization**) has resulted in the formation of a line of black, thick-walled **zygospores** (Fig. 26-3B and C) across the culture. Keeping the dish closed, locate the zygospores with your dissecting microscope. In what ways, if any, do the plus and minus strains look different?

_____

_____

Would you consider each strain to have a different sex? Explain.

_____

_____

_____

_____

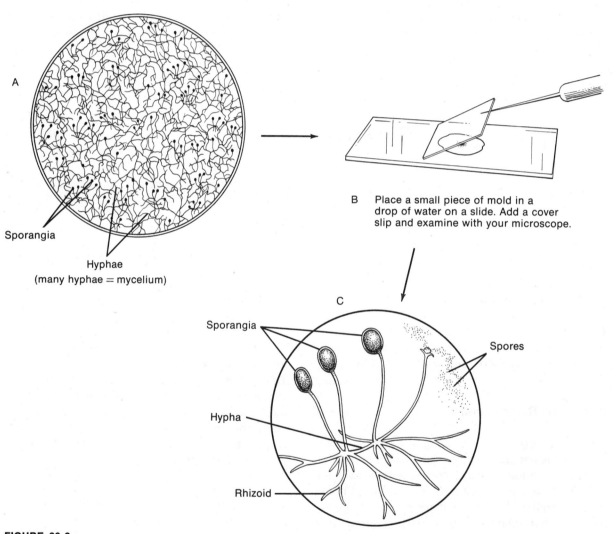

FIGURE 26-2
Structure of the black bread mold.

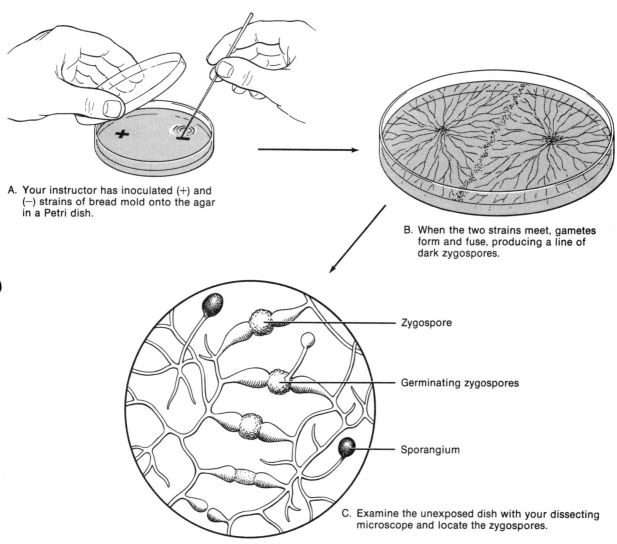

FIGURE 26-3
Sexual reproduction in the black bread mold.

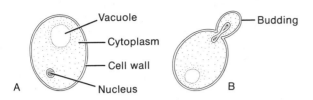

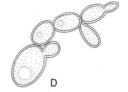

FIGURE 26-4
Budding in yeast.

C. SAC FUNGI (ASCOMYCETES)

The members of this group produce spores in a saclike container called an **ascus**. These spores, usually eight in number, are called **ascospores**.

1. Yeasts

Yeasts are among the simpler of the ascomycetes. Mount some living yeast cells on a slide and examine microscopically. Look for nuclei and small glistening food granules. Note that some of the yeast cells may exhibit small rounded projections called **buds** (Fig. 26-4). What do these buds represent?

_____

_____

If none of the yeast cells are budding, examine a demonstration slide that shows this process.

Yeasts are classified as ascomycetes because at some period in their life cycle an ascus is formed. Examine a demonstration slide that shows the ascus of yeast and its contents. List several ways in which yeasts are economically important to man.

_____

_____

_____

2. Powdery Mildew (*Microsphaera*)

Examine leaves of a lilac plant that are infected with this fungus. Why is this organism called powdery mildew?

_____

_____

_____

_____

The mycelium that grows over the surface of the leaf occasionally penetrates the epidermal cells (Fig. 26-5). These hyphae are called **haustoria** (singular haustorium). What is their function?

_____

Examine demonstration slides that show haustorial penetration of the epidermal cells of the host plant. What kind of nutrition does this fungus exhibit?

_____

During late spring large numbers of asexual spores (**conidia**) are produced and disseminated by the wind to spread the infection to the same or other lilac plants. Examine slides of powdery mildew and locate the conidia.

Toward the end of summer large numbers of spherical **fruiting bodies** are formed as a result of sexual reproduction. Examine an infected leaf with a dissecting microscope and locate some of these fruiting bodies. Scrape the surface of the leaf and mount the material in a drop of water on a slide. Examine microscopically. Locate a fruiting body and note the elaborate appendages. How might these appendages be useful to the fungus?

_____

_____

While examining the fruiting body, gently apply pressure to the cover slip. Note the saclike asci that are extruded. Locate an ascus that contains ascospores. How many spores does each ascus contain?

_____

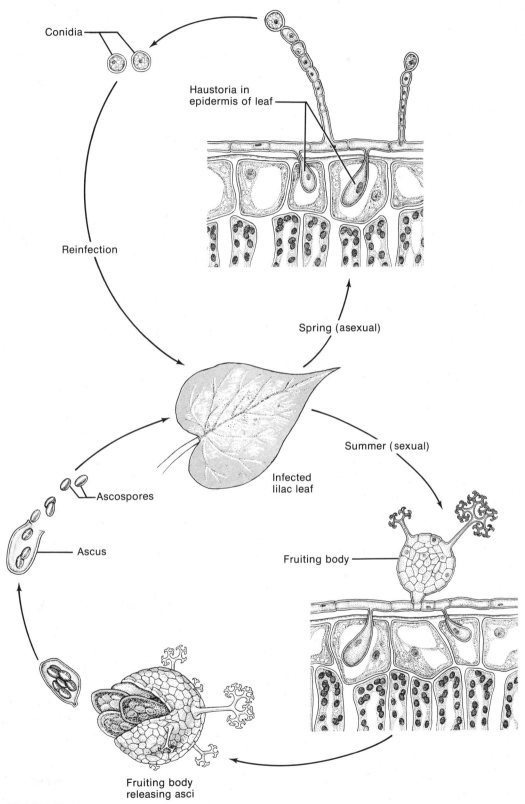

FIGURE 26-5
Life cycle of a powdery mildew.

What becomes of the ascospore after it is released?

_____

_____

_____

3. *Penicillium*

Much of the spoilage of food, leather, and cloth occurs as a result of this fungus. Examine a living culture of *Penicillium*. Why is it sometimes referred to as a blue-green mold?

_____

Examine prepared slides showing the special hyphal branches that produce asexual spores or conidia (Fig. 26-6). How does *Penicillium* differ from *Rhizopus* in the way in which spores are produced?

_____

_____

_____

How is it similar to *Microsphaera*?

_____

_____

_____

One species, *Penicillium notatum*, was found to produce a potent antibiotic, penicillin. It has proven to be one of the most effective antibiotics for combating bacterial infections.

D. CLUB FUNGI (BASIDIOMYCETES)

The basidiomycetes represent another large and varied group of fungi that have both saprophytic and parasitic members. They are characterized by having a **basidium** [a club-shaped structure upon which are usually produced four spores (**basidiospores**)] formed at some point in the life cycle.

1. Mushrooms

Mushrooms are characterized by plates or gills on the under surface of a fruiting body. The basidia extend from the free surface of the gills (Fig. 26-7).

The mushroom mycelia grow beneath the surface of their substrate. When haploid mycelia of different strains unite, another mycelium is formed that is diploid. This mycelium may produce a fruiting body, commonly called a mushroom, which consists of a stalk and cap.

Examine the fruiting body of *Agaricus campestris,* the common commercial mushroom. Note the stalk and cap. Examine the undersurface of the cap and locate the gills. If the mushroom is young, the gills may be covered by a thin membrane that extends from the stalk to the outer margin of the cap.

Mount one of the gills in a drop of water on a slide and microscopically examine the edges of the gills. Locate basidia and basidiospores. If available, examine a prepared slide of *Coprinus*, showing a cross section through the cap. Identify gills, basidia, and basidiospores.

2. Bracket Fungi

These fungi are parasitic or saprophytic on various trees. The mycelium of the fungus may grow within the trunk several years before forming the characteristic woody fruiting body on the outside of the tree. Examine several bracket fungi on display and note the several growth layers that are evident. How could these growth layers be used to estimate the age of a bracket fungus?

_____

_____

_____

Examine the undersurface of one of the fruiting bodies with a dissecting microscope. How does the undersurface of a bracket fungus differ from a mushroom?

_____

_____

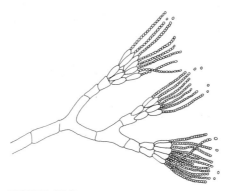

FIGURE 26-6
Conidia of *Penicillium*.

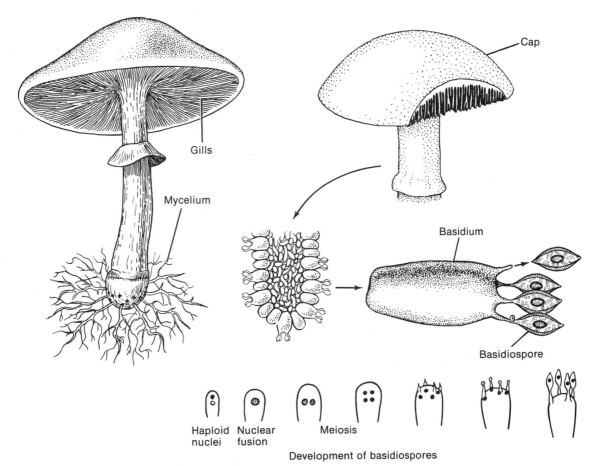

FIGURE 26-7
Details in the life of a mushroom.

Examine a prepared slide showing a cross section of the fruiting body of a bracket fungus. What do the circular openings in cross section represent?

_____

Locate basidia and basidiospores.

REFERENCES

Alexopoulos, C. J., and H. C. Bold. 1967. *Algae and Fungi*. Macmillan, New York.

Bold, H. C. 1967. *Morphology of Plants*. Harper, New York.

Delevoryas, T. 1966. *Plant Diversification*. Holt, Rinehart & Winston, New York.

Doyle, W. T. 1965. *Nonvascular Plants: Form and Function*. Wadsworth, Belmont, California.

Fenn, R. H. 1962. A Simple Method for Growing *Rhizopus nigricans*. *Turtox News* **40**:226–227.

Levine, M. 1954. *An Introduction to Laboratory Technique in Bacteriology*. 3d ed. Macmillan, New York.

Scagel, R. F., R. J. Bandoni, G. E. Rouse, W. B. Schofield, J. B. Stein, and T. M. C. Taylor. 1965. *An Evolutionary Survey of the Plant Kingdom*. Wadsworth, Belmont, California.

EXERCISE 27

Diversity in Plants: Algae

Algae are commonly found in fresh and salt water, although some may inhabit damp soil, stagnant waters, or the bark of trees.

Green algae, like higher plants, have chlorophylls *a* and *b* in plastids. In addition, carotenes and xanthophylls are present. Nutrition is autotrophic, with reserve carbohydrates stored in the form of starch.

A variety of body forms are observed in this group. Some members are composed of a single cell, while others may consist of numerous cells arranged in the form of long filaments that may or may not exhibit internal specialization. Still others form spherical colonies.

Reproduction is accomplished asexually by fragmentation of the parent body or through the production of spores that are capable of developing into other individuals. In addition, new individuals may arise as a result of the union of two gametes. The zygote that is formed as a result of this union may develop directly into another alga or may produce spores.

In this exercise the green algae will be studied from the viewpoint of evolutionary relationships within the group. One obvious evolutionary development is an increase in complexity of the vegetative body—that is, from single-celled organisms to those in which the cells are united to form specialized colonies. Another evolutionary trend is toward increasing differences between the gametes and/or reproductive organs.

A. UNICELLULAR GREEN ALGAE

1. *Chlamydomonas*

Chlamydomonas is a primitive, single-celled, motile alga commonly found in damp soil, lakes, and ditches. It is typically egg-shaped and has a large cup-shaped chloroplast containing a proteinaceous body, the **pyrenoid**, which functions in starch formation. The nucleus is difficult to observe in living material.

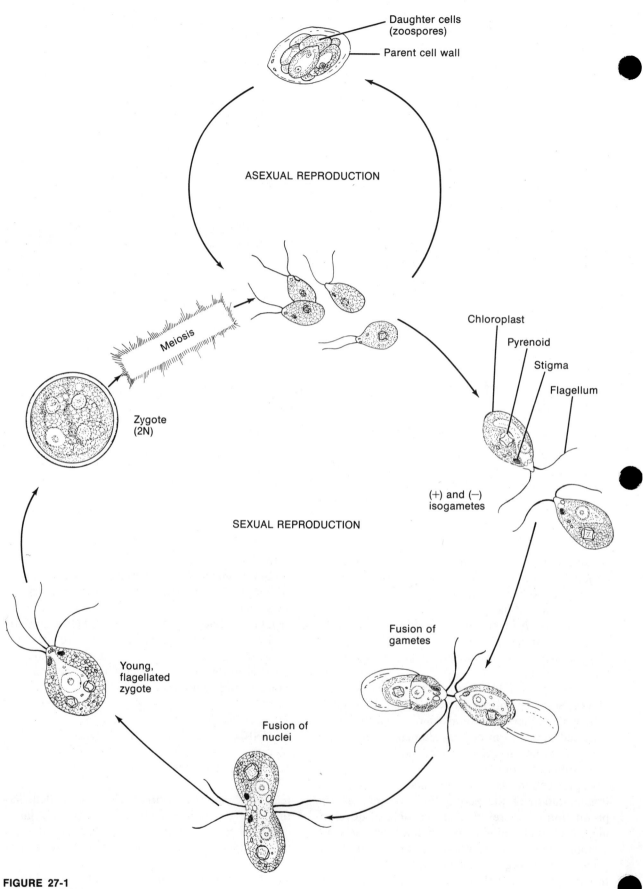

FIGURE 27-1
Life cycle of *Chlamydomonas*.

DIVERSITY IN PLANTS: ALGAE

Prepare a slide of living *Chlamydomonas* and examine microscopically (Fig. 27-1). Add a drop of methylcellulose to slow the movement of this alga. Locate the conspicuous chloroplast and the pyrenoid. The activity of threadlike structures, **flagella**, located at the anterior end of the cell, enables the organism to move through water. The flagella can be more easily observed by reducing the light.

Locate a reddish pigment spot (**stigma**) near the flagellar end of the cell. Suggest a function for this structure.

_____

At the start of asexual reproduction, the flagella are retracted and movement ceases. During this quiescent period, mitosis results in the formation of daughter protoplasts, which may divide a second and third time (Fig. 27-1). A cell wall is formed around each of the daughter protoplasts, resulting in the formation of temporary colonies that soon rupture and release individual daughter cells called **zoospores**. Locate nonmotile, vegetative colonies showing two, four, or eight daughter cells.

In sexual reproduction, haploid gametes are formed in the majority of the green algae. In *Chlamydomonas*, however, vegetative cells may function as gametes, one functioning as the male and another as the female gamete. The gametes are identical in size and appearance, though in some species the female gamete may be slightly larger. Gametes that are morphologically indistinguishable from each other are called **isogametes**.

Sexuality in *Chlamydomonas* can be demonstrated by using mating strains having separate sexes characterized as (+) and (−). Place a drop of each of the mating types next to each other on the slide. *Do not mix.* While examining them with the dissecting microscope, mix the drops together. Note the peculiar "clumping" phenomenon that precedes union of the gametes. Using a compound microscope, locate cells that have paired. At which end of the cell has union occurred?

_____

As a result of sexual union a diploid zygote is formed that secretes a thick, spiny wall and then enters a period of dormancy. Examine the preparation closely and observe zygotes. Under favorable conditions, the zygote nucleus undergoes meiotic division to form four haploid nuclei. The protoplast then divides and forms four uninucleate zoospores, which soon escape from the old zygote wall, develop flagella, and swim away.

2. Pleurococcus

Pleurococcus is commonly found growing on the north side of buildings, trees, fences, and so on. How does the habitat of this alga differ from that of *Chlamydomonas*?

_____

Obtain a small sample of *Pleurococcus* from the surface of a piece of bark and examine microscopically. Is *Pleurococcus* a unicellular or multicellular organism?

_____

Suggest a reason why this alga is usually found growing on the north side instead of on the south side of trees in the northern hemisphere?

_____

B. FILAMENTOUS GREEN ALGAE

1. Spirogyra

Spirogyra is a free-floating alga found in small freshwater pools in the spring. It is frequently referred to as "pond scum." Prepare a fresh mount and examine microscopically. Is any branching evident?

_____

Examine a single cell under high power. What is the shape of the chloroplast?

_____

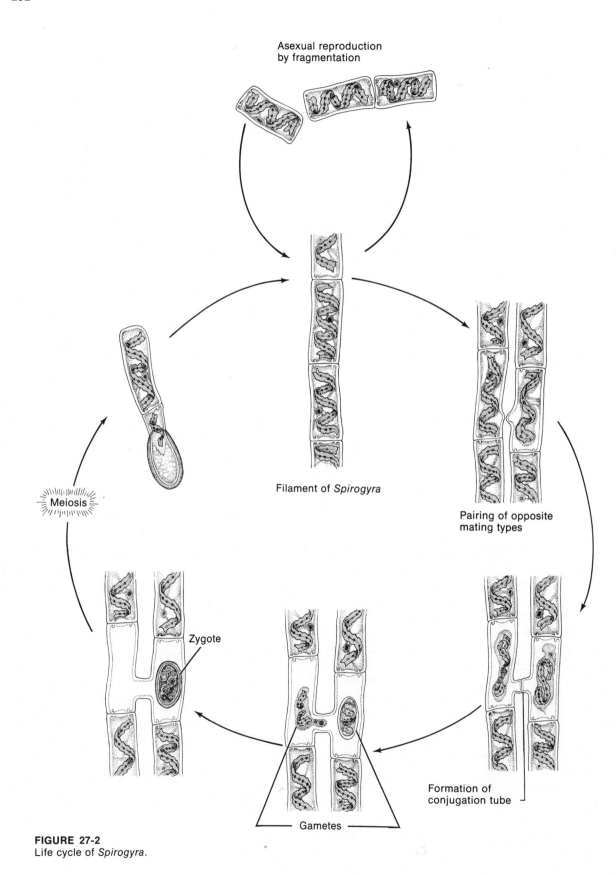

FIGURE 27-2
Life cycle of *Spirogyra*.

DIVERSITY IN PLANTS: ALGAE

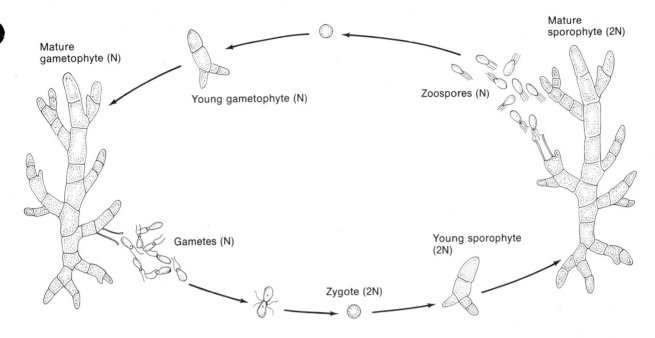

FIGURE 27-3
Life cycle of *Cladophora*.

Locate several small pyrenoids within the chloroplast. The nucleus, suspended in the center of the cell by cytoplasmic strands, is difficult to observe unless stained. Apply a drop of methylene blue to the edge of the cover glass. After a few minutes, reexamine and locate the nucleus, which will now appear as a bluish body in the central part of the cell.

In sexual reproduction filaments of opposite mating types come to lie adjacent to one another and small projections appear in opposing cells of each filament (Fig. 27-2). The projections increase in length and eventually contact each other. At the point of junction, the cell walls dissolve, forming a **conjugation tube**. The protoplasts of the conjugating cells become isogametes. One gamete functions as the male and migrates through the conjugation tube to unite with the nonmotile female gamete. Fusion of the gametes produces a zygote, which is released upon disintegration of the filament. Prior to germination, the nucleus of the zygote undergoes meiosis and three of the four haploid nuclei that are formed disintegrate. Upon germination of the zygote a short protuberance is formed, which contains the fourth nucleus.

Mitosis, followed by cell division, results in a filament of cells similar to the parental filaments. Examine your living material and look for various stages in gametic union. If your preparation does not show any stages of conjugation or zygote formation, examine prepared slides showing this process.

No means of asexual reproduction is known to occur in *Spirogyra* other than fragmentation of the individual filaments.

2. Cladophora

Cladophora may be found in fresh water where it usually grows attached to rocks or other objects by means of a specialized cell, a **holdfast**. Examine some filaments of *Cladophora*. Locate the holdfast. Note the extensive branching of the filaments.

Cladophora exhibits a phenomenon in which any one plant produces either motile zoospores or gametes but never both (Fig. 27-3). The plant that produces spores is called a **sporophyte** and is diploid. The gamete-producing plant is haploid and is called a **gametophyte**. This condition, called **alternation of generations**, has long been known to occur in land plants and, therefore, in combination with certain biochemical similarities (e.g., pigments, cell wall structure, storage products) suggests an evolutionary relationship between the green algae and the higher plant groups.

Zoospores in *Cladophora* are produced in cells that are located near the ends of branches. Following release from the parent plant, the zoospores

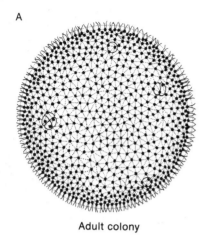

Adult colony

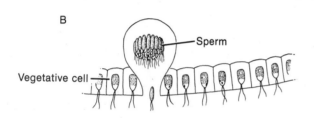

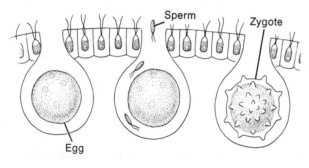

FIGURE 27-4
Sexual reproduction in *Volvox*.

settle down, retract their flagella, and develop into filaments that are morphologically similar to the filaments of the sporophyte plant. How are the filaments cytologically different?

_____

_____

Germination of the zoospore results in the formation of a gametophyte plant that produces gametes. Fusion of the gametes results in a zygote that germinates and develops into a diploid sporophyte plant.

C. COLONIAL GREEN ALGAE

Volvox represents a colonial type of plant body in which the cells are bound together in a common matrix. Individually the cells exhibit many of the features seen in *Chlamydomonas*—that is, they have stigma, flagella, and large chloroplasts. However, in *Volvox* only specialized cells are able to function during reproduction.

Asexual reproduction is accomplished by the enlargement and subsequent division of certain of the cells within the colony. In early development, these cells form a flat plate, which soon rounds up into a sphere (daughter colony) having a small pore at the posterior end. The colony continues to increase in size and then the sphere evaginates through the pore and turns itself inside out. The daughter colonies are released upon disintegration of the parent colony.

Examine a sample of living *Volvox* with a dissecting microscope. Describe the motility of *Volvox*.

_____

_____

_____

What is the shape of the colony?

_____

Examine living material and prepared slides of *Volvox* for presence of daughter colonies.

The green algae studied thus far all produced isogametes. In contrast, *Volvox* produces gametes that are morphologically differentiated into sperm and eggs (Fig. 27-4). These are developed from cells that become differentiated as the colony grows. In forming an ovum, a cell of the colony increases greatly in size, takes on a rounded form, and becomes filled with food materials, especially lipids. The male gametes are formed from other cells, which give rise to flat bundles of flagellated sperm. Both kinds of gametes are released into the hollow, fluid-filled center of the colony, where fertilization may occur. The resulting zygote soon secretes a spiny wall. Germination of the zygote occurs in the spring and results in the formation of a new colony. Examine prepared slides of *Volvox* and locate sperm, ova, and zygotes.

DIVERSITY IN PLANTS: ALGAE

D. RED, BROWN, AND GOLDEN BROWN ALGAE

The members of these algal groups differ from the green algae chiefly in the nature of the pigments and, except for the golden brown algae, are marine organisms.

1. Red Algae

Red algae are primarily found in tropical seas. The coloration of the red algae is due to the abundance of **phycoerythrin**, a red pigment that masks the photosynthetic pigments. None of the red algae attain the large sizes attained by some of the brown algae. Examine the red algae on demonstration. How are the red algae economically important?

_____

_____

_____

2. Brown Algae

Brown algae are marine organisms that range in size from microscopic forms to plants in excess of 150 feet in length. In addition to chlorophyll and carotene, a golden pigment, **fucoxanthin**, is present in sufficient amount to mask the other pigments. Examine specimens of brown algae, noting particularly the variations in size and complexity in the plant bodies. Note that many of the specimens have small, rounded "floats." How are these floats of benefit to the plant?

_____

_____

_____

3. Golden Brown Algae

Although there are three classes in this group, only the diatoms will be studied.

Diatoms are abundant in both fresh and salt waters, making up a large part of the plankton (free-floating microscopic plants and animals) of lakes and oceans. Obtain a sample of living diatoms and examine microscopically. Note the variety of forms and sizes and the intricate sculpturing of the cell walls. What kind of movement do these organisms exhibit?

_____

Closely examine the protoplast. What color are the plastids?

_____

What pigment is responsible for the characteristic coloring of the diatoms?

_____

The walls of diatoms consist of two overlapping halves (much like a Petri dish) that contain large amounts of silica and thus fail to decay after death. In some areas of California large deposits of these cell walls accumulate and form thick beds of **diatomaceous earth**. This material is economically important as filtering material, insulation, and abrasives. Mount some diatomaceous earth on a slide and examine microscopically. Note the large variety of forms. If available, examine prepared slides of members of this group.

REFERENCES

Fritsch, F. E. 1935. *The Structure and Reproduction of the Algae.* Vol 1. Cambridge University Press, Cambridge, England.

Fritsch, F. E. 1945. *The Structure and Reproduction of Algae.* Vol 2. Cambridge University Press, Cambridge, England.

Prescott, G. W. 1954. *How to Know the Freshwater Algae.* Wm. C. Brown, Dubuque, Iowa.

Prescott, G. W. 1968. *The Algae: A Review.* Houghton Mifflin, Boston.

Pringshcim, E. G. 1949. *Pure Cultures of Algae: Their Preparation and Maintenance.* Cambridge University Press, Cambridge, England.

Scagel, R. F., R. J. Bandoni, G. E. Rouse, W. B. Schofield, J. B. Stein, and T. M. C. Taylor. 1965. *An Evolutionary Survey of the Plant Kingdom.* Wadsworth, Belmont, California.

Smith, G. M. 1950. *The Freshwater Algae of the United States.* 2d ed. McGraw-Hill, New York.

Starr, R. G. 1964. Culture Collection of Algae. *Am. J. Botany.* **51**:1013–1044.

EXERCISE 28

Diversity in Plants: Liverworts and Mosses

In contrast to the algae, the liverworts and mosses have become adapted to a terrestrial existence. Characteristics of this group are as follows:

- They are green, have rhizoids (rootlike structures), and may have stemlike and leaflike parts (mosses).
- They lack true vascular tissues and as a consequence are restricted to moist habitats.
- The plant body (**thallus**) may be dorsoventrally flattened and bilaterally symmetrical (liverworts), or it may be erect and radially symmetrical (mosses).
- The reproductive organs, in contrast to the algae, are multicellular. As with the algae, the presence of water is necessary for fertilization.
- In addition to reproduction by asexual means, the members of this group also reproduce sexually and have a well-established alternation of generations, consisting of an independent gametophyte (haploid) phase and a somewhat dependent sporophyte (diploid) phase.

In this exercise several representatives of the liverworts and mosses will be studied, with special emphasis on form, habitat, and reproductive processes. In the algae and fungi, the plant body is reproduced by fragmentation, by production of asexual structures (spores), or by the union of isogamous or anisogamous gametes. In each case the reproductive structure is single-celled. In the liverworts and mosses the gametes are produced in multicellular sex organs—the **archegonia** (female) and **antheridia** (male).

A. LIVERWORTS

Examine living gametophyte plants of *Marchantia*. Note the characteristic Y-shaped (dichotomous) branching of the thallus. Meristematic tissue found in the notch of the Y produces the peculiar growth pattern seen here.

Remove a small piece of the thallus and examine the ventral surface with a dissecting microscope.

At the edges locate slender, hairlike structures, the **rhizoids**. What is the function of the rhizoids?

_____

_____

_____

Associated with the rhizoids are long, flattened structures, the **scales**. Locate these structures.

Examine the dorsal surface. Observe the minute "pores" in the center of small diamond-shaped areas. Since *Marchantia* is photosynthetic, what role might be ascribed to these openings?

_____

_____

_____

What structures in the higher plants appear to be morphologically and functionally similar to these openings?

_____

_____

Examine prepared slides of a cross section of the thallus of *Marchantia* (Fig. 28-1). Locate the upper epidermis and the "pores" that were seen in the living material. Below the epidermis, find a series of small air chambers that are partitioned by branching filaments arising from the floor of the chambers. In living material the filaments contain chloroplasts. What is the function of the air chambers?

_____

_____

_____

The tissue underlying the air chambers is several cells in thickness. Many of the cells contain leucoplasts; a few large cells contain mucilage, a gelatinous substance that tends to absorb water. What is the advantage of this material?

_____

_____

_____

On the lower epidermis locate the rhizoids and scales. How do scales differ morphologically from rhizoids?

_____

_____

_____

1. Asexual Reproduction

The gametophyte of *Marchantia* reproduces asexually by means of small "plantlets" (**gemmae**), produced from cells located at the bottom of gemma cups. These are found on the dorsal surface of the thallus. Examine a demonstration of the gemma cups using a dissecting microscope. Are there any gemmae present in the cups?

_____

_____

If not, how could you account for their absence?

_____

_____

2. Sexual Reproduction

The reproductive (sex) organs of *Marchantia* are borne on special upright branches (Fig. 28-2). Each branch is composed of a stalk and terminal disc. Many species of *Marchantia* are **dioecious**; that is, there are separate male and female gametophytes and therefore two kinds of reproductive branches. Examine a male gametophyte, which bears an antheridial branch. What is the shape of the terminal disc?

_____

_____

DIVERSITY IN PLANTS: LIVERWORTS AND MOSSES

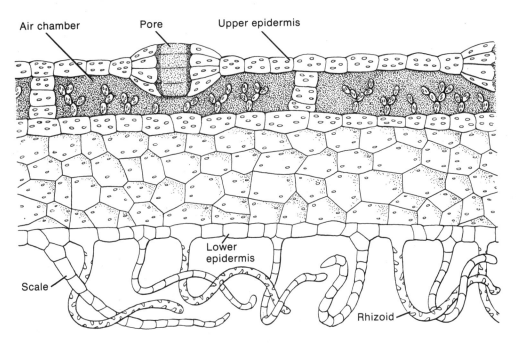

FIGURE 28-1
Diagrammatic cross section of a *Marchantia* thallus.

Compare the antheridial branch with the archegonial branch of the female gametophyte. Describe the female reproductive branch.

_____

_____

_____

_____

Examine slides showing cross sections of the antheridial branch. Locate the antheridia just below the upper surface of the disc (Fig. 28-2). Note that the antheridium is contained within a chamber that opens to the surface of the disc through a pore. Examine an antheridium closely, and note that it is attached to the base of the chamber by a stalk. The body of the antheridium consists of an outer layer of cells, the jacket. Within the jacket are found numerous small cells, which will develop into male gametes called sperm. If available, examine slides of *Marchantia* sperm.

Obtain a slide showing a cross section of an archegonial disc, on the ventral surface of which are located the archegonia. Locate an archegonium that shows the following parts: an elongated neck, an enlarged venter (containing the egg), and the stalk.

3. Fertilization

When the antheridia are mature, the sperm are released and fertilize the egg. The zygote (Fig. 28-2), which remains within the venter, undergoes cell division and develops into a multicellular embryo (young sporophyte). The embryo differentiates into three distinct parts—the foot, stalk, and capsule. Examine a prepared slide of a developing sporophyte. Locate the foot embedded in gametophytic tissue. What are some functions of the foot?

_____

_____

Note the large, terminal capsule within which are found spores. By what process were the spores formed?

_____

_____

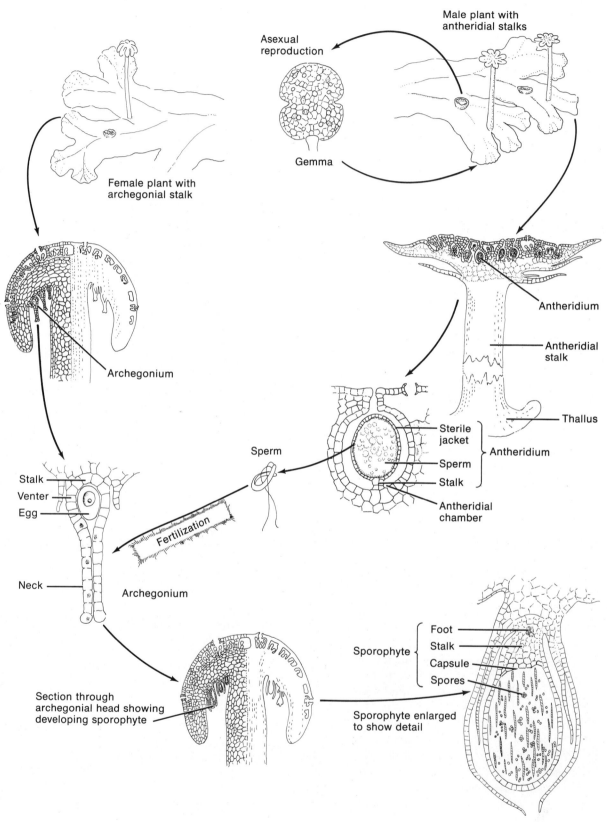

FIGURE 28-2
Life cycle of *Marchantia*.

DIVERSITY IN PLANTS: LIVERWORTS AND MOSSES

What is the chromosomal condition of these spores?

_____

_____

Into what structure will the spores develop?

_____

Among the spores, locate elongated cells with spirally thickened walls. These are called **elaters** and aid in the dispersal of the spores when the capsule breaks open. Note that the sporophyte matured within the archegonium. Prior to the release of the spores, the capsule is pushed free of the gametophyte by the elongation of the stalk.

B. MOSSES

Mosses differ from the liverworts in that the gametophyte of the moss begins as a filamentous, branching structure—a **protonema**—and is differentiated into "stems" and "leaves." In addition, the capsule of the sporophyte contains a central, sterile region of cells—the **columella**—which greatly reduces the amount of sporogenous tissue. The capsule contains a highly complex series of "teeth," which regulate the dispersal of the spores.

1. Sporophyte of *Polytrichum*

Examine living specimens of the common moss, *Polytrichum*, or, if this is not available, study preserved specimens that consist of both the gametophytic and sporophytic generations. The sporophyte is readily distinguishable since it consists of a terminal capsule (often covered by a hairy cap, the **calyptra**), a slender stalk (the **seta**), and a foot that is embedded in the top of the "leafy" gametophyte (Fig. 28-3). Carefully separate the sporophyte from the gametophyte. Remove the calyptra and save for later examination. Note that the capsule has a lid. Gently remove the lid and examine the exposed surface of the capsule with a hand lens. Describe what you see.

_____

_____

Suggest a function for these structures?

_____

_____

Crush the capsule in a drop of water on a slide. Examine microscopically. If the sporophyte is mature, numerous spores will be present. These spores represent the first cells of the gametophyte generation. What process precedes the formation of these spores?

_____

What is another name for the capsule?

_____

2. Gametophyte

Following release from the capsule, and providing that suitable environmental conditions exist, the spore soon germinates and grows into the filamentous structure of the protonema. Examine living or prepared slides of a moss protonema. Note the similarity of the protonema to the filamentous algae. What evolutionary significance is indicated here?

_____

_____

Look for small budlike structures along the length of the filament. These will later develop into mature, "leafy" gametophytes (Fig. 28-3).

Examine the leafy part of *Polytrichum* that remains after removal of the sporophyte. This gametophyte is photosynthetic. The mature gametophyte consists of "leaves," a "stem," and rhizoids. How could you determine if these are true stems and leaves?

_____

_____

Polytrichum is a dioecious moss, with the reproductive organs located at the tips of separate male and female plants. With a forceps carefully remove some of the leaves from the apex of a female plant. After removing as many leaves as possible, cut off the stem tip and mount in a drop of water on a slide. With your probe gently tease the tip into small

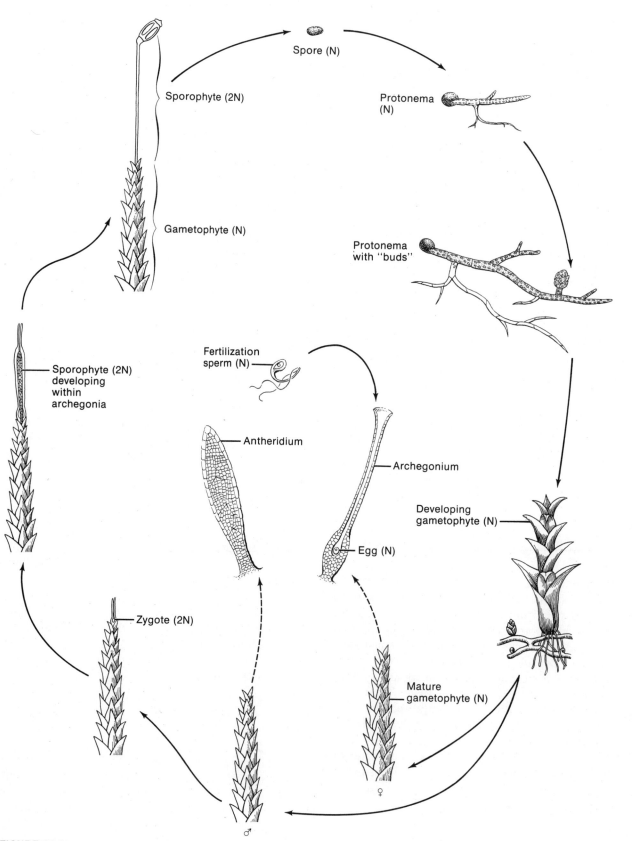

FIGURE 28-3
Life cycle of *Polytrichum*.

DIVERSITY IN PLANTS: LIVERWORTS AND MOSSES

TABLE 28-1
Advances in complexity shown by liverworts and mosses over the algae.

Character	Algae	Liverworts and mosses
Morphology		
Habitat		
Asexual reproduction		
Sexual reproduction		

fragments. Examine microscopically. Locate the archegonia (Fig. 28-3). Note the canal that leads through the neck and terminates in the venter of the archegonium. Where is the egg located in the archegonium?

In a similar manner mount a stem tip of a male plant. Locate the antheridia, which are the elongate, saclike structures consisting of an outer sterile jacket and an inner mass of cells destined to become antherozoids. If available, also examine prepared slides of moss antheridia and archegonia.

3. Fertilization

The nonmotile egg is retained within the venter of the archegonium. When the archegonium is mature, the cells lining the inside of the stalk disintegrate and a canal is opened leading to the female gamete. The male gamete swims through this canal and unites with the egg. What physical factor must be present for fertilization to occur?

The zygote undergoes a series of divisions and becomes a mass of undifferentiated cells, still held in the venter. The developing embryo differentiates into three parts—the foot, seta, and capsule. The remains of the archegonium is retained as the calyptra and may be found enclosing the capsule. Is the calyptra sporophytic or gametophytic tissue?

C. ADVANCES IN COMPLEXITY SHOWN BY THE LIVERWORTS AND MOSSES

Complete Table 28-1. Indicate the advances shown by the mosses and liverworts over the algae with respect to morphology, habitat, and reproduction. The following suggested terms can be used.

- Morphology: unicellular, filamentous, spherical colonies, prostrate thallus, erect thallus, radial symmetry, bilateral symmetry.
- Habitat: fresh water, salt water, ditches, stagnant pools, damp woods, terrestrial.
- Asexual reproduction: fragmentation, cell division, zoospores, nonmotile spores, gemmae.
- Sexual reproduction: unicellular or multicellular reproductive structures, motile or nonmotile gametes, alternation of generations, isogametes or anisogametes, necessity of water for fertilization, dependency or nondependency of sporophyte, antheridia, archegonia.

REFERENCES

Bauer, L. 1962. On the Physiology of Sporangium Differentiation in Mosses. *J. Linn. Soc. Bot.*, **58**:343–351.

Bell, P. R., and C. Woodcock. 1968. *The Diversity of Green Plants*. Addison-Wesley, Reading, Massachusetts.

Smith, G. M. 1955. *Cryptogamic Botany*. 2d ed. Vol. 2. *Bryophytes and Pteridophytes*. McGraw-Hill, New York.

Watson, E. V. 1964. *The Structure and Life of Bryophytes*. Hutchinson University Press, London.

EXERCISE 29

Diversity in Plants: Vascular Plants

The members of this group represent an advancement in the plant kingdom. In these plants we find for the first time a system of specialized tissues that carry water, dissolved minerals, and other organic products throughout the plant. They are called **vascular tissues** and the plants that have them (ferns, pines, oaks) are called **vascular plants**, or **tracheophytes**. These plants are able to transport great quantities of water and minerals over long distances in relatively short periods of time. As a result, vascular plants grow much larger than the nonvascular mosses, algae, and fungi. In this exercise you will become familiar with the structure and reproductive patterns of ferns, conifers (which include pines, spruce and other evergreens), and the lily (a flowering plant).

A. FERNS

1. Sporophyte

Examine the mature sporophyte of *Polypodium*. Locate the horizontal stem, which may lie on, or just beneath, the surface of the soil. What name is given to this type of stem?

_____

Note the large, deeply lobed leaves, or fronds, of the plant (Fig. 29-1). On the undersurface of some of the leaves locate small yellowish brown spots or **sori**. Each sorus contains many **sporangia**. What are produced in the sporangia?

_____

What do we call a leaf that produces sporangia?

_____

Obtain part of a fern sporophyll that bears sori. Examine the sorus with a hand lens or dissecting microscope. Scrape the contents of a sorus into a drop of water on a slide. Examine under the low power of the microscope. Each sporangium is composed of a stalk and an enlarged capsule. Examine the capsule under high power. Note a ridge of cells (**annulus**) that extends around the capsule.

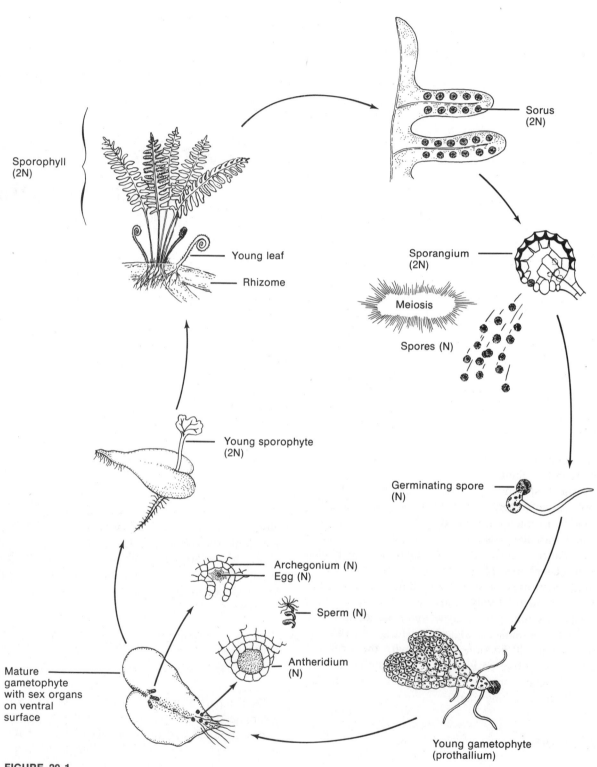

FIGURE 29-1
Life cycle of a fern.

DIVERSITY IN PLANTS: VASCULAR PLANTS

These cells have thickened inner and radial walls. As the capsule matures, the annular cells lose water and the thin outer walls tend to shrink. This results in a considerable amount of tension on the lateral, thin-walled "cheek cells," which then rupture, releasing the spores.

Remove the coverslip from the slide and blot the excess water with a piece of filter paper. Examine your preparation (without the coverslip) and observe the sporangia as the slide dries. Describe your observations.

_____

_____

2. Gametophyte

Upon germination the fern spore develops into a short filament of cells, which differentiates into a platelike structure called a **prothallium**. The mature prothallium or gametophyte resembles a heart-shaped structure with a notch, containing the growing point, at one end. With a dissecting microscope examine living fern gametophytes. Locate rhizoids on the ventral surface.

Among the rhizoids of mature gametophytes are found numerous antheridia (Fig. 29-1). Examine the gametophyte with a dissecting microscope. Locate archegonia on the ventral surface just posterior to the apical notch. Only the neck of the archegonium is visible. The venter is embedded within the gametophytic tissue. The sex organs are found on the ventral surface of the plant. Of what significance is this?

_____

Examine prepared slides of a mature gametophyte and locate the structures indicated above.

In the development of the new sporophyte, the zygote—following fertilization of the egg—undergoes a series of divisions and develops into an embryo. The young embryo, still held in the venter of the archegonium, differentiates into four lobes. One lobe develops into the "foot." A second lobe grows down into the soil and becomes a primary root. A third lobe develops into the primary leaf, and the last lobe becomes the stem. Until the primary root and leaf become functional the young sporophyte is dependent upon the gametophyte; but as soon as the root begins to absorb water and the leaf photosynthesizes, the sporophyte becomes independent. Soon after this, secondary leaves and adventitious roots are formed and the primary root, leaf, and gametophyte die.

B. CONIFERS

These plants are characterized by producing a seed in the life cycle. In the conifers the seeds are exposed (naked) on scalelike structures. Like the ferns, seed plants have a well-defined alternation of generations. In contrast to the ferns, the gametophyte is microscopic and completely dependent upon the large free-living sporophyte, which possesses stems, leaves, and roots. Although there are several living orders of conifers, your study will be confined to only one of these, the conifers, which have familiar representatives in the pines, spruces, cedars, firs, and others.

1. Pine Sporophyte

Study specimens of pine branches and note that two kinds of leaves are produced by pines. Locate needlelike, photosynthetic leaves borne on short spur shoots and inconspicuous scale leaves found at the bases of the spur shoots. How many needle leaves are borne on each spur shoot?

_____

Examine branches of other species of pine. Describe any variation in the number of leaves in a cluster.

_____

Would you consider this difference a means of separating one group of pine from another?

_____

Why are pines called "evergreens"?

_____

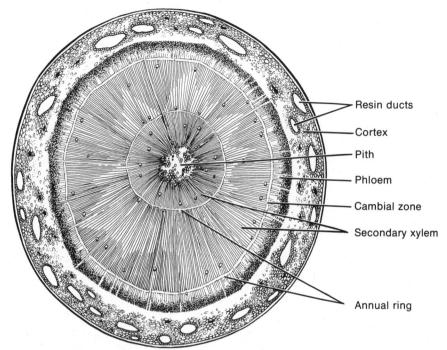

FIGURE 29-2
Diagram of an older pine stem.

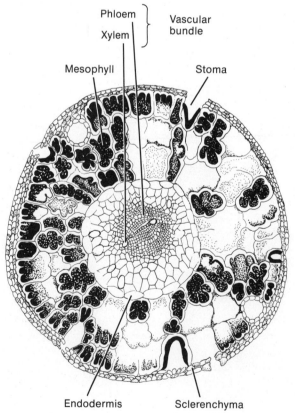

FIGURE 29-3
Pine needle leaf.

DIVERSITY IN PLANTS: VASCULAR PLANTS

Examine cross sections of young and old pine stems (Fig. 29-2). Identify the epidermis, cortex, and vascular tissue, and note the numerous resin canals. What economically important products come from pines?

In the older stems note the annual rings. How were these formed?

Now examine cross sections of a pine needle (Fig. 29-3). Note the heavily cutinized epidermal cells. Locate a layer of sclerenchyma just beneath the epidermis. Find depressions, or pits, in the epidermis. Where are the stomata located in the pine leaf?

How does the tree benefit by having the stomata located as they are?

Locate the central vascular cylinder surrounded by an endodermis. How many vascular bundles are present?

Examine cross sections of leaves of other species of pines. Describe any variation in structure among the various needles.

Examine the slides of the various leaves with a dissecting microscope. What correlation exists between the number of leaves in a cluster and the shape of an individual needle leaf?

---

2. Pine Gametophyte

The pine tree produces two kinds of cones on the same tree. Each cone (**strobilus**) is an auxiliary shoot composed of a central axis that bears spirally arranged sporophylls. Examine branches bearing **staminate** or **pollen cones**, usually in clusters at the ends. Dissect one of the staminate cones and locate the central axis to which are attached scalelike structures called **microsporophylls** (Fig. 29-4). Remove a microsporophyll and examine with a dissecting microscope. Note the two saclike **microsporangia**. Break the microsporangia open and examine the contents under the high power of your microscope. Depending on the stage of development, you will observe either young **microspores** (consisting of a single cell) or microgametophytes (consisting of two prominant cells and two degenerate cells). What is another name for the microgametophyte of pine?

Note the lateral, winglike appendages. How might these "wings" be advantageous?

Examine prepared slides of pine pollen.

A second kind of cone found on the tree is the **seed cone**, which is borne on short lateral branches near the apex of young branches. They are partially hidden by the terminal bud. Examine branches bearing seed cones and compare them to the staminate cones on the basis of size, shape, and location. Locate older seed cones that have opened and discharged their seeds. Dissect a seed cone and locate the overlapping, bractlike appendages (Fig. 29-4) attached to the central axis. On the upper surface of each bract, or **megasporophyll**, locate two small white structures, the ovules. Note that the ovules are not attached directly to the bract but are borne on paper-thin scales, which in turn are attached to each bract. Why are gymnosperms classed as having "naked" seeds?

Examine longitudinal sections of a young pine ovule with a dissecting microscope (Fig. 29-4). Locate an outer layer of tissue that almost encloses the inner tissues except for a small opening at one

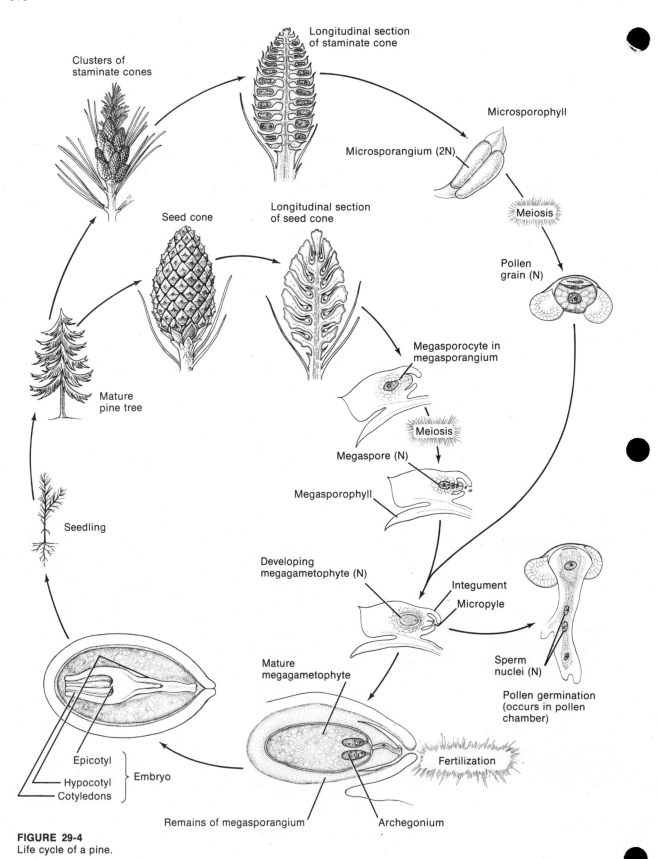

FIGURE 29-4
Life cycle of a pine.

DIVERSITY IN PLANTS: VASCULAR PLANTS

end. The outer layer is the **integument**; the opening is the **micropyle**. Within the integument locate the **megasporangium** in which the **megagametophyte** is found. At the micropylar end of the megagametophyte locate archegonia. From what cell did the megagametophyte arise?

_____

Note the pollen chamber just inside the micropyle.

3. Pollination and Fertilization

The pollen grains are released from the staminate cones toward the end of May. At about the same time, the young seed cones open and the pollen, by sifting down through the cone scales, comes to lie in the pollen chamber. The pollen grain germinates, produces a slender pollen tube containing the male gametes, and digests its way to the archegonia. Because the growth of the pollen tube and the development of the megagametophyte is quite slow, fertilization may not occur for as long as thirteen months following pollination.

Remove the seed coats from soaked pine seeds and cut through the gametophytic tissue to locate the embryo. Locate the epicotyl and hypocotyl. At the tip of the hypocotyl locate a coiled suspensor, which, by its growth, pushed the developing embryo deeply into the gametophytic tissue that contained large quantities of stored food. If available, examine stages in the germination of a pine seed. Note the peculiar manner in which the embryo comes out of the seed.

C. FLOWERING PLANTS

Sexual reproduction requires the union of gametes. The result of this union (fertilization) is a single cell, the zygote, which, by a series of morphogenetic changes, develops into a new individual. In the flowering plants the gametes are produced in modified branches that we call flowers. Examine a flower of a lily and locate the structures described below (Fig. 29-5).

1. **Sepals** are modified leaves and are the outermost structures of the flower; they are typically green, although they may be other colors and in some cases may be absent. Collectively, the sepals constitute the **calyx**.

2. **Petals** are internal to the sepals, and may be white or highly colored. Collectively, the petals make up the **corolla**.

3. **Stamens** are structures internal to the petals. Each stamen consists of a stalk (the filament) and a terminal capsule (the anther), which when mature contains the pollen grains. The pollen grains produce the male gametes.

4. **Pistils**—one or more—are found at the center of the flower. Pistils are made up of one or more carpels—leaflike structures bearing ovules. It is postulated that in evolution the carpels have rolled inward, enclosing the ovules. The pistil consists of three parts: an enlarged basal region (the ovary), a slender stalk (the style), and a somewhat flattened tip (the stigma). The ovary contains seedlike structures called ovules, which, when mature, contain the female gametes. In the process of pollination, the pollen grains are transported to the stigma, where they germinate and send a long tube down to the ovules. How does pollination in the flowering plants differ from that in conifers?

_____

_____

_____

_____

The pollen tube penetrates the ovule and releases two male gametes, one of which unites with the egg to form the zygote.

The lily plant is called a sporophyte, a stage in the life cycle of the plant during which spores are produced within specialized structures called sporangia. These are formed on or within modified leaves called **sporophylls**. What flower parts described above can be considered sporophylls?

_____

Certain cells—**spore mother cells** or **sporocytes**—undergo meiotic division within the sporangium to form these spores. The spore undergoes morphogenetic changes and develops into microscopic plants (gametophytes) that produce the gametes. In the flowering plants the gametophytic plant is parasitic upon the sporophyte. Since the male gamete is smaller than the female gamete, it is called a **microgamete** to distinguish it from the larger female **megagamete**. Similarly, the male gametophyte is called the microgametophyte as

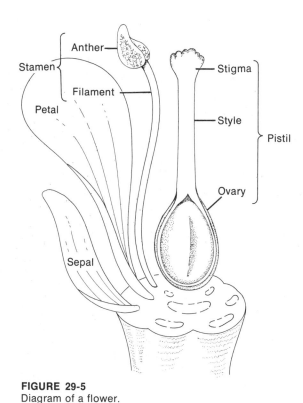

FIGURE 29-5
Diagram of a flower.

opposed to the female megagametophyte. This same terminology applies to all male and female structures. This scheme is summarized in Fig. 29-6.

In studying the stamens (microsporophylls), recall that the stamen consists of an anther and filament. The anther contains the microsporangium in which are found the microsporocytes. Examine slides of a young lily anther (Fig. 29-7). Locate the microsporocytes. What is the chromosomal condition of these cells (e.g., N or 2N)?

_____

What indication is there that some of the microsporocytes have divided?

_____

_____

What kind of division has occurred?

_____

The cells formed as a result of the division of the microsporocytes are called microspores. These will develop into mature microgametophytes or pollen grains. Remove an anther from your flower, crush it in a drop of water on a slide, and examine microscopically. For a more detailed examination of the pollen grain, obtain slides showing a section through an older lily anther. Locate the numerous pollen grains. The pollen grains are shed from the anther and are transferred by various means to the stigma of the same or a different flower. In this process of pollination a pollen tube is formed that is chemotropically oriented to grow down through the style to the ovules. During the growth of the pollen tube, gametes are formed as a result of the division of one of the nuclei in the pollen tube. Examine cultures of germinating pollen and observe the pollen tubes. If possible, locate the two male gametes. In Fig. 29-7 trace the stages in the development of the microgametophyte.

In the ovule, a centrally located mass of tissue enlarges to form the megasporangium or **nucellus** (Fig. 29-7). Concomitant with the development of the megasporangium, adjacent cells grow up and around the nucellus to form the integuments, which later develop into the outer coverings of the seed. The integuments do not fuse but leave an opening (the micropyle) at the apex of the nucellus, through which the pollen tube enters the ovule. As the nucellus grows, the nucleus of the centrally located megasporocyte undergoes meiosis to form four megaspore nuclei.

One of the megaspore nuclei migrates to the micropyle end; the other three migrate to the opposite (chalazel) end. This 3 + 1 arrangement represents the first 4-nucleate stage in the development of the embryo sac.

The three haploid nuclei then fuse to form a 3N nucleus. This stage is followed by a mitotic division of the 3N and N nuclei to form two triploid nuclei and two haploid nuclei. This is the second 4-nucleate stage. A final division occurs resulting in an 8-nucleate embryo sac consisting of four triploid and four haploid nuclei. At this point three of the triploid nuclei migrate to the chalazel end of the embryo sac where cell walls are formed around them. These cells, called antipodals, are probably nutritive in nature.

Three haploid nuclei migrate to the micropylar end of the embryo sac. The middle one becomes the egg. The role of the two lateral ones (called synergids) is not fully understood although recent work has shown that the pollen tube enters the synergid first and then the egg. Thus, the synergids might function in guiding the pollen tube. The two remaining nuclei migrate to the center of the em-

DIVERSITY IN PLANTS: VASCULAR PLANTS

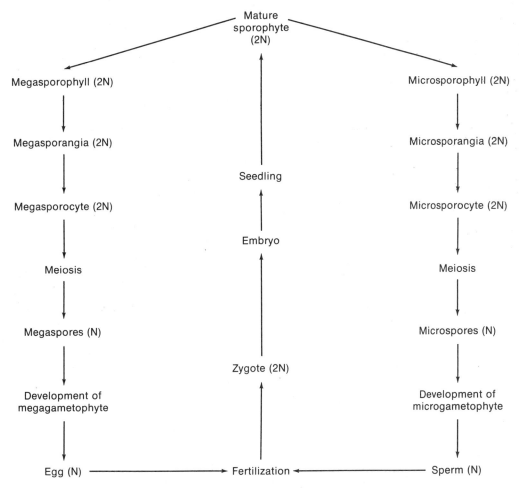

FIGURE 29-6
Generalized scheme of the reproductive cycle of higher plants.

bryo sac and form a fusion nucleus. What is the chromosome content of this nucleus?

___

Examine slides that show the developmental stages indicated above. As you study these slides, refer to Fig. 29-7.

Fertilization is complete when the pollen tube, having reached the micropyle of the ovule, digests its way through the nucellus, reaches the embryo sac, and releases both male gametes. One gamete unites with the egg to form the zygote, which subsequently develops into an embryo. The second gamete unites with the fusion nucleus to form the endosperm nucleus. What is the chromosome condition of the endosperm nucleus?

___

Subsequent division of the endosperm results in a mass of nutritive tissue that is used as a food source for the developing embryo. This double fertilization is a phenomenon peculiar to flowering plants and has no counterpart in the lower plants or in the animal kingdom.

D. ADVANCES IN COMPLEXITY SHOWN BY VASCULAR PLANTS

Complete Table 29-1. Indicate the increases in complexity shown by the tracheophytes over the bryophytes with respect to morphology and reproduction. The following are suggested terms to use.

- Morphology: true leaves present or absent, true stems and roots present or absent, vascular

TABLE 29-1
Advances in complexity shown by the vascular plants.

Character	Liverworts and mosses	Vascular plants
Morphology of sporophyte		
Morphology of gametophyte		
Asexual reproduction		
Sexual reproduction		

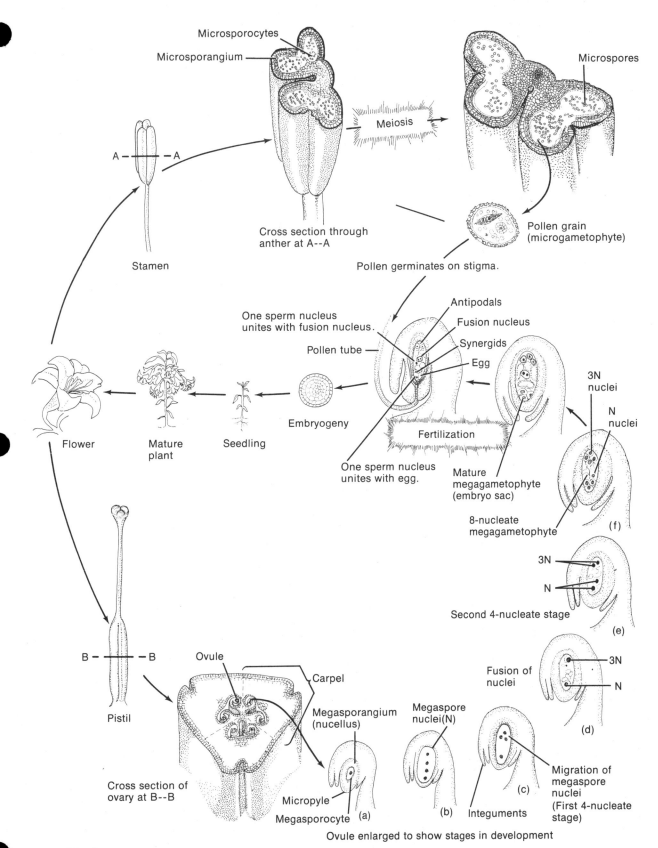

FIGURE 29-7
Life cycle of a lily.

tissue present or absent, erect or prostrate plant body.
- Asexual reproduction: spores, fragmentation, gemmae, death and decay of older parts, and so on.
- Sexual reproduction: unicellular or multicellular reproductive organs, motile or nonmotile gametes, alternation of generations, isogametes, anisogametes, water necessary or unnecessary for fertilization, antheridia, archegonia, dependency or interdependency of sporophyte, pollination.

REFERENCES

Delevoryas, T. 1966. *Plant Diversification*. Holt, Rinehart & Winston, New York.

Foster, A. S., and E. M. Gifford, Jr. 1959. *Comparative Morphology of Vascular Plants*. W. H. Freeman and Company, San Francisco.

Smith, G. M. 1955. *Cryptogamic Botany*. 2d ed. Vol. 2. *Bryophytes and Pteridophytes*. McGraw-Hill, New York.

Wallace, B., and A. M. Srb. 1964. *Adaptation*. 2d ed. Prentice-Hall, Englewood Cliffs, New Jersey.

EXERCISE 30

Diversity in Animals: Protozoa

The protozoa are a large and diverse group of one-celled organisms. They are widespread in nature, living in both fresh and salt water. Some forms live upon or within the soil. Others are parasites of certain higher animals and cause disease.

Four main groups of protozoa are recognized: ciliate, ameboid, spore-forming, and flagellate protozoa. The ciliate protozoa have many hairlike structures called **cilia** (singular, **cilium**) covering their cell surfaces. The flagellate protozoa have one or more structures called **flagella** (singular, **flagellum**), which are much like cilia except that they are considerably longer. Amoeboid protozoa move by the flow of cell contents in a given direction in response to a food source. These extensions of the cytoplasm are called **pseudopodia** (singular, **pseudopodium**). The spore-forming protozoa are made up entirely of parasitic forms that do not have locomotor structures.

A. AMOEBA

Amoeba proteus, one of the simplest protozoans available for laboratory study, lives in clean, fresh waters that contain green vegetation (Fig. 30-1).

Prepare a wet mount from the amoeba culture in your laboratory. This one-celled organism may be difficult to find unless the diaphragm is adjusted to cut down on the amount of light entering the microscope. After you have located an amoeba, watch it closely for a few minutes. Describe the shape of this organism. The amoeba moves by forming and extending fingerlike extensions, called pseudopodia, from any place on its body surface. This type of irregular flowing is called **amoeboid movement**; it occurs in many protozoa and also in

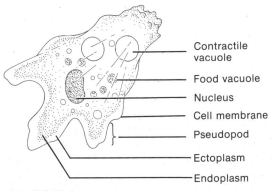

FIGURE 30-1
Amoeba.

the amoebocytes of sponges and the white blood cells of vertebrates.

The amoeba eats other protozoans, algae, and rotifers, preferring live flagellates and ciliates. It is attracted by the movements of its prey, or by substances diffusing from it. Food is taken in at any part of the cell surface by pseudopodia that encircle the food, which, with some water, is taken into the endoplasm as a food vacuole. The vacuoles are moved about by streaming movements in the endoplasm and are thus brought into contact with various parts of the cell. The food vacuoles decrease in size as digestion proceeds, and undigested residues are eliminated at any place on the cell surface.

The cytoplasm of the freshwater amoeba contains a higher salt concentration than the fluid in which the amoeba lives. Because of this difference, water tends to diffuse into the cell at a faster rate than it leaves the cell. Unless water is removed from the cell periodically, the amoeba will soon burst. A unique structure, the contractile vacuole, performs the function of "pumping out" excess water. Observe the surface of the amoeba closely and locate a **contractile vacuole**. How frequently does this organelle empty?

_____

Would you expect marine amoebas to have contractile vacuoles? Explain.

_____

_____

_____

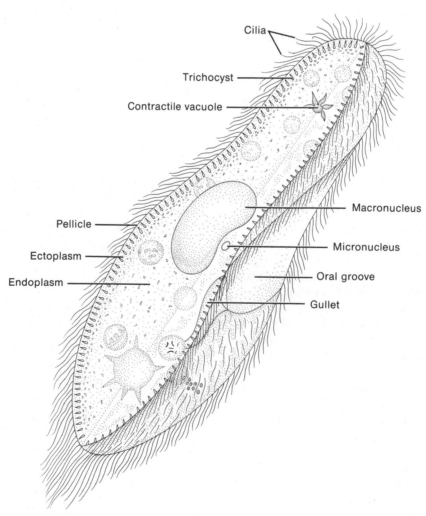

FIGURE 30-2
Paramecium.

DIVERSITY IN ANIMALS: PROTOZOA

● What environmental factors (for example, temperature) might affect the rate of contraction of the contractile vacuole?

_____

_____

Design an experiment to measure the effect of one factor.

B. PARAMECIUM

One of the most complex and extensively studied of all protozoa is *Paramecium* (Fig. 30-2). In nature it is commonly found in fresh waters that contain an abundance of decaying vegetation. These protozoa are readily cultured in the laboratory by growing them in a solution containing boiled hay, in which they feed on bacteria and smaller protozoa.

● Mount a drop of the culture on a slide. To slow down the movement of *Paramecium* add a drop of methyl cellulose before adding the cover slip. Observe the method of locomotion in these organisms. Note that they have three predominant locomotor patterns: forward, rotating, and swerving. Is the rotation during movement primarily clockwise or counterclockwise?

_____

Observe the manner in which this organism avoids obstacles in its path of motion.

Examine a single *Paramecium* under high power and observe the rhythmic pattern of motion of the cilia that cover the body. From the anterior end to about the middle of the organism is an oblique oral groove that empties into the mouth. Observe the manner in which food is carried into this groove by the beating cilia. What is the difference in the rate of beating of these cilia as compared with those on the body surface?

_____

● Note that the mouth opens into a narrow **gullet**, which in turn empties into a **food vacuole**. Numerous other food vacuoles can be seen in the endoplasm. Are these stationary or do they appear to be moving through the cytoplasm in some organized way?

At each end of the organism is a large contractile vacuole. Describe its appearance and activity.

_____

_____

In the middle of the organism observe a large, opaque, kidney-shaped structure, the **macronucleus**, that partially surrounds a much smaller **micronucleus**. The latter structure is best observed by examination of prepared slides. The other structures described in this part of the exercise should also be located on these slides.

1. Feeding and Digestion in *Paramecium*

Paramecium is a continuous feeder whose cilia constantly sweep food material into the mouth and gullet where it is collected in food vacuoles. To observe these characteristic feeding patterns, add a small drop of yeast stained with Congo red to a drop of thick suspension of paramecia. Stir the drop with a toothpick and add a cover glass. Observe the feeding paramecia under high power. Note the action of cilia in setting up currents that direct food into the oral groove. Time the formation of a food vacuole at the end of the gullet and observe what happens as it breaks away from the gullet. As you observe the movement of these food vacuoles through the endoplasm, note changes in color of the stained yeast cells. Congo red, a pH indicator, is bright orange-red above pH 5.0 and blue around pH 3.0. What changes in color occur in the food vacuole as it leaves the gullet?

_____

_____

From these observations, what can you deduce concerning changes in acidity and digestion in these vacuoles?

_____

_____

_____

Elimination of undigested materials in *Paramecium* occurs via an anal pore located midway between the mouth and the posterior end of the body. Since it is formed only when undigested materials are eliminated, it is not easily seen. In order to facilitate observation of this process, your instructor will prepare samples of *Paramecium* that have been fed on Congo red stained yeast cells and then concentrated by centrifugation. Observe the Congo red vacuoles moving about in the cytoplasm by **cyclosis** (streaming movements in the endoplasm). You should be able to see several food vacuoles empty through the anal pore in 10 or 15 minutes.

2. Reproduction

Paramecium reproduces by dividing into two equal daughter cells. This simple type of asexual reproduction, called **binary fission**, is characteristic of most protozoa. Examine prepared slides showing binary fission in *Paramecium*. The division of *Paramecium* occurs in a median plane.

Paramecium also reproduces sexually by a process called **conjugation**, in which two animals come together and exchange micronuclei. Conjugation can sometimes be observed in ordinary cultures of paramecia. Under certain conditions conjugation will occur when two different strains are mixed. If different strains are available, study the process as directed by your instructor. Also examine prepared slides showing *Paramecium* in conjugation.

C. SYMBIOSIS

Plants, because they contain chlorophyll, are generally considered to be the primary **producers** of the world's food supply. The largest amount of organic materials is produced by the photosynthetic activity of microscopic aquatic organisms. Other organisms are recognized as "**consumers**" because they do not have chlorophyll and thus are not able to synthesize the organic molecules necessary for metabolism. The consumers, therefore, depend on other organisms for the synthesis of organic compounds. Occasionally, however, there are established intimate associations between organisms that provide advantages to both parties. This relationship is called **symbiosis**. In this exercise you will study examples of this phenomenon in protozoans.

1. Symbiosis Between Two Animals

Trichonympha is a protozoan that inhabits the intestine of wood-eating termites. Although termites ingest bits of wood, they are unable to digest cellulose, the chief constituent of wood. *Trichonympha* forms pseudopodia that engulf the wood fragments ingested by the termite, digesting the cellulose and transforming it into soluble carbohydrates that can then be utilized by the termite (Fig. 30-3). Neither the termite nor *Trichonympha*

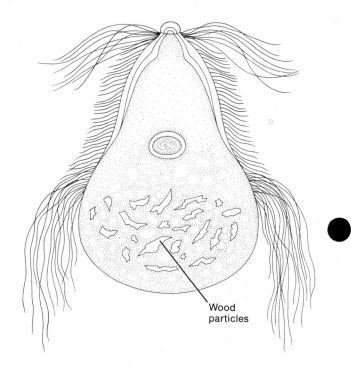

FIGURE 30-3
Trichonympha.

can live without the other. Examine slides of this organism. Note the large number of flagella that cover the upper part of the animal and the small fragments of wood in the cytoplasm.

2. Symbiosis Between an Alga and a Protozoan

Prepare a wet mount of the protozoan *Paramecium caudatum*. What color is this organism?

_____

Prepare a second slide of *Paramecium bursaria*.

DIVERSITY IN ANIMALS: PROTOZOA

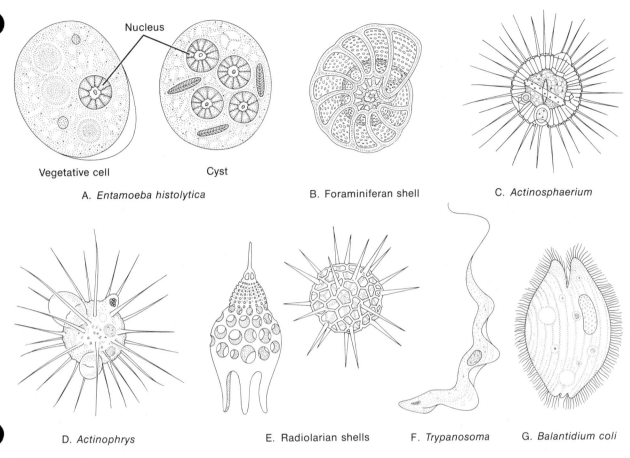

FIGURE 30-4
Protozoans.

How does this animal differ from *Paramecium caudatum*?

_____

What are the "green bodies" in this single-celled organism?

_____

How does the "consumer" in this symbiotic relationship benefit?

_____

_____

How does the "producer" benefit?

_____

_____

D. OTHER PROTOZOA

1. *Entamoeba histolytica* (Fig. 30-4A)

This is the causative organism of amoebic dysentery. Examine prepared slides showing both the active (vegetative) stages and the cysts (resistant, infective stages of the organism). The vegetative stages have a single nucleus; the cysts have four nuclei, as well as a heavy wall surrounding the entire cell.

2. Foraminiferans (Fig. 30-4B)

This is a large group of marine amoebas that secrete limy, snaillike shells about themselves. Long, delicate feeding pseudopodia stream out of minute pores that perforate the surface of the shells. Examine prepared slides showing a variety of these foraminiferan shells.

3. *Actinosphaerium* and *Actinophrys* (Fig. 30-4C and D)

These two protozoans are members of a group of spherical amoebas that have delicate pseudopodia that are supported by axial rods. Most of the species of this group live in fresh waters. Examine slides of *Actinosphaerium*. Note in particular the delicate pseudopodia, which have suggested the name "sun animalcules" for these organisms.

4. Radiolarians (Fig. 30-4E)

This is a group of amoebas that have silicious skeletons secreted by the cytoplasm, usually taking the form of an intricate latticework, through which extend stiff radiating spines. Examine prepared slides of radiolarian skeletons.

5. Trypanosomes (Fig. 30-4F)

These slender flagellates inhabit the blood of vertebrates and are carried from one host to another by bloodsucking invertebrates. The trypanosomes that cause human sleeping sickness in Africa are transmitted by the bloodsucking tsetse fly. Examine stained slides of human blood showing these parasitic flagellates.

6. *Balantidium coli* (Fig. 30-4G)

This is a causative agent of dysentery and is one of a very few large ciliated parasites inhabiting the human intestine. The most common source of infection is the pig. Examine prepared slides of *Balantidium*.

REFERENCES

Allen, R. D. 1962. Amoeboid Movement. *Scientific American*, February (Offprint No. 182). *Scientific American* Offprints are available from W. H. Freeman and Company, 660 Market Street, San Francisco 94104, and 58 Kings Road, Reading RG1 3AA, England. Please order by number.

Corliss, J. L. 1961. *The Ciliated Protozoa*. Pergamon, New York.

John, T. L. 1949. *How to Know the Protozoa*. Wm. C. Brown, Dubuque, Iowa.

Kudo, R. R. 1954. *Protozoology*. Charles C. Thomas, Springfield, Illinois.

MacKinnan, D., and R. Hawes. 1961. *An Introduction to the Study of Protozoa*. Oxford University Press, New York.

Needham, J. G. 1937. *Culture Methods for Invertebrate Animals*. Comstock, Ithaca, New York.

Sonneborn, T. M. 1955. Protozoa in the General Biology or Zoology Course. *Am. Biol. Teacher* 17:187–190.

Wichterman, R. 1952. *The Biology of Paramecium*. McGraw-Hill, New York.

EXERCISE 31

Diversity in Animals: Sponges and Coelenterates

The sponges and coelenterates are the first multicellular animals exhibiting evidence of tissue formation and are collectively called metazoans. However, the sponges differ from the colenterates in having intracellular digestion and certain peculiarities in embryonic development.

A. SPONGES

Most sponges are marine, their habitats ranging from the low-tide line to depths of $3\frac{1}{2}$ miles; one group is widespread in fresh waters. The scientific name Porifera (pore bearer) refers to the porous structure of the body with its many surface openings. Despite their simplicity, many of the more familiar species form extensive colonies on the floor of the sea.

Using a dissecting microscope, examine specimens of *Scypha* and note its slender, vase-shaped appearance. Clusters of this sponge, which rarely reaches over an inch in height, are found adhering to objects in shallow sea waters. Note the basal end by which this organism attaches itself. At the free end note the large single excurrent opening, the **osculum**. Minute pores (**ostia**) along the body wall allow water to enter into the single tubular central canal, or **spongocoel**. Note the bristlelike structures (**spicules**) along the body surfaces, which give rigidity to the body. The spicules are chiefly of four kinds: (1) long straight spicules around the osculum that protrude beyond the body surface, producing a bristly appearance; (2) short straight spicules surrounding the ostia; (3) T-shaped spicules lining the spongocoel; and (4) triple-branched spicules embedded in the body wall. Examine slides showing isolated spicules.

Study a stained cross section of *Scypha* under low and high magnification (Fig. 31-1). Note that the body wall is thick and folded to form many short radial canals. **Incurrent canals** open from the exterior through small pores, or ostia, and end blindly near the wall of the **spongocoel**—the digestive cavity that connects with the outside through

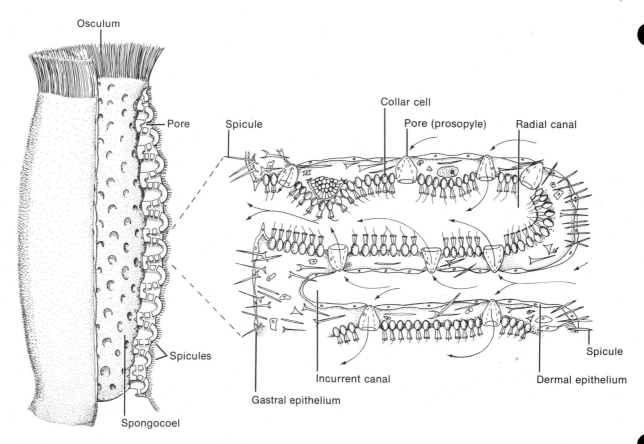

FIGURE 31-1
Canal system of *Scypha*.

a mouth. The **excurrent canals** open into the spongocoel by minute pores and end blindly near the outer surface. The incurrent and excurrent canals are connected by a series of small pores. The exterior body surface is covered by a thin dermal epithelium; the spongocoel is lined with gastral epithelium; and the radial canals are lined with small flagellated cells called **collar cells**. These cells bear a resemblance to some of the flagellated protozoa—a fact that indicates a probable evolution of the sponges from the Protozoa. Water, taken into the body through the ostia, is passed over the collar cells, which remove tiny bits of food, and then is extruded through the osculum.

Examine the sponges on demonstration in your laboratory. What you are observing is merely the skeleton of the living sponge, since all of the protoplasm has dried and decayed away. As already noted, the skeleton of *Scypha* consists of a loosely woven network of spicules of calcium carbonate and thus is called a calcareous sponge. Examine specimens of the sponge *Euplectella* ("Venus flower basket") and note that the skeleton is composed of a network of silicious spicules. As a group these sponges are called glass sponges. The bath sponges, on the other hand, have no spicules but have a skeleton made up of an elastic substance called "spongin."

B. COELENTERATES

The coelenterates consist of a diverse group of aquatic animals, which are almost exclusively marine and exhibit definite tissues. The members of this phylum may occur in two forms—polyps and medusae. The **polyps** are tubular organisms closed at one end with a mouth surrounded by tentacles at the other end. The **medusae** are umbrella-shaped, jellylike, free-swimming organisms with a mouth at the end of a central projection called a **manubrium**.

The phylum is divided into three classes. The **Hydrozoa** have a predominant polyp stage that gives rise to small free-swimming medusae. The **Scyphozoa** include the larger jellyfish, all of which have a predominant medusa stage consisting largely of gelatinous **mesoglea** or "jelly." These can be roughly distinguished from the medusa of the Hydrozoa by their size, which ranges from 1 inch to 7 feet in diameter. The polyp stage is minute or lacking. The **Anthozoa** have no medusa stage.

1. Class Hydrozoa

a. With a hand lens or a dissecting microscope examine a living hydra in a Syracuse dish or deep well slide containing pond water. These animals are slender organisms that vary from $1\frac{1}{2}$ to 1 inch in length. At the free end of the animal observe several elongate, actively moving tentacles, which are used to capture food. At the center of this circle of tentacles is the mouth. Tap the edge of your slide or dish and observe the extreme contractility of this organism. Observe that the animal expands after contraction. Compare the rate of elongation with that of contraction.

_____

_____

In order to observe feeding in hydras your instructor will add several brine shrimp that have been thoroughly rinsed with tap water to your slide or dish. Why are the brine shrimp washed before being fed to the hydra?

_____

_____

With a dissecting microscope observe the organism as it feeds. Describe the feeding process beginning with the first contact of the hydra's tentacles with the shrimp.

_____

_____

It is not unusual for a single hydra to capture and ingest 10 or more shrimp in the course of a 2-hour feeding period.

During the feeding reaction you may have noticed that as the tentacles of the hydra came in contact with the shrimp the latter was seen to jerk for several minutes and finally become quiet. This reaction occurred because the shrimp was stung by numerous stinging cells released from the tentacles of the hydra. Close examination of the tentacles of your specimen will reveal the presence of many wartlike structures called **nematocysts**. Each of these nematocysts is enclosed in a cnidoblast cell before discharge, with the coiled thread inside it. During discharge the coiled thread is turned inside out as it is released from the cnidoblast. The release of these nematocysts is best observed by placing a hydra in a drop of water and then adding a drop of dilute acetic acid to the water. Observe the tentacles as the acid diffuses toward them. Note the reaction as the nematocysts are discharged. The structure of the various types of nematocysts may be observed by adding a drop of dilute methylene blue or other suitable stain at the edge of the slide.

Hydras reproduce asexually by simple budding or sexually by the production of sperm and ova. Budding occurs by a simple outgrowth from the body wall so the two gastrovascular cavities are continuous. Upon maturity the bud separates from the parent and becomes a separate individual. Examine specimens of budding hydras.

The gametes are produced in organs called spermaries and ovaries, which appear as protuberances along the body wall. Both sex organs may occur simultaneously in a single organism.

To study the histology of *Hydra*, examine stained cross sections and note that the body wall surrounding the **gastrovascular cavity** consists of two distinct layers of cells—an outer layer of ectoderm and an inner layer of endoderm (Fig. 31-2). These two layers are held together by a thin, noncellular layer—the mesoglea. In the ectoderm are two principal cell types—the larger **epitheliomuscular cells** containing contractile fibrils at their base and the smaller **interstitial cells** located between them. What might you expect the function of the epitheliomuscular cells to be?

_____

_____

The ectoderm of the tentacles also contains nematocysts.

Locate epitheliomuscular and interstitial cells in the endoderm. Many of the large endodermal

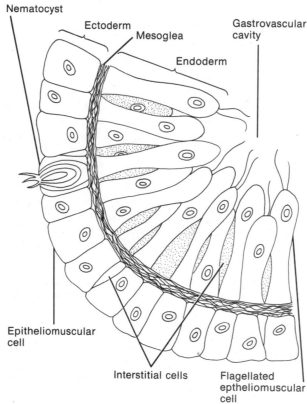

FIGURE 31-2
Cross section of *Hydra*.

cells contain food vacuoles where intracellular digestion occurs. These food vacuoles are formed as a result of the engulfing of food particles by pseudopodia that are extended into the gastrovascular cavity. Many of the digestive cells also contain flagellae, which are used to beat food in the direction of the pseudopodia. Extracellular digestion also occurs in hydra by the secretion of enzymes directly into the gastrovascular cavity, followed by absorption of the digested food material.

b. Place a small piece of *Obelia*, a colonial marine hydroid, in a Syracuse dish and examine under a dissecting microscope. The plantlike colony consists of numerous branches, which terminate in two kinds of polyps (Fig. 31-3). The feeding or **nutritive polyps**, possess tentacles and resemble *Hydra*. The **reproductive polyps** are club-shaped and lack tentacles.

Examine a prepared slide to study the components of this colonial hydroid. Locate a nutritive polyp. Note that the branch supporting it has a hollow internal cavity that is continuous with the nutritive polyp. This is the gastrovascular cavity, which serves as the digestive and circulatory organ of the colony. Food material is captured by the tentacles of the nutritive polyps and enters the common gastrovascular cavity through the mouths of the nutritive polyps. Each polyp, like that of *Hydra*, has an outer ectodermal and an inner endodermal layer.

Examine a reproductive polyp and note that it consists of a club-shaped stalk to which are attached a number of medusoid buds. These buds eventually break away from the stalk, escape through the terminal opening of the polyp, and swim away to develop into **medusae**. When these free-swimming medusae mature they produce eggs or sperm, which are released into the sea where fertilization occurs. The zygote develops into a free-swimming, ciliated **planula larva** that eventually attaches to a rock and develops into the colonial hydroid.

c. Since the medusae of *Obelia* are too small to study in detail we shall examine the large medusa stage of *Gonionemus* for details of its anatomy.

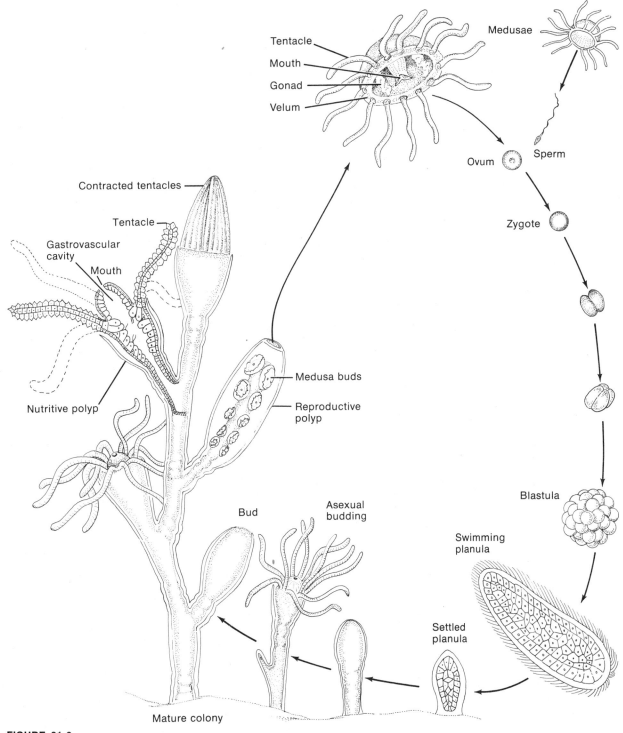

FIGURE 31-3
Obelia life cycle.

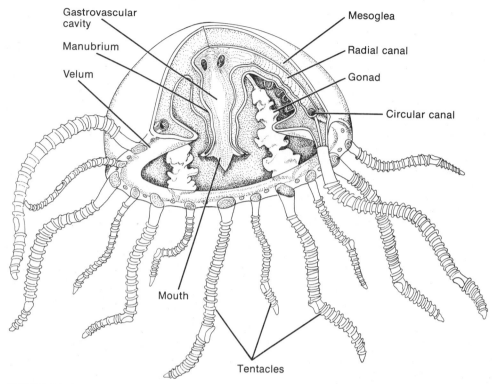

FIGURE 31-4
Internal anatomy of *Gonionemus*.

Using a hand lens or dissecting microscope, examine a specimen in a Syracuse dish (Fig. 31-4). Note that the medusa is umbrella-shaped. From the margin of the "umbrella" extending inward is a muscular ring of tissue called the **velum**. Suggest a function of the velum?

_____

_____

Extending into the cavity of the medusa is the **manubrium**, which contains a mouth at its tip. Extending from the manubrium along the inner surfaces are four **radial canals** that join with the circular canal at the margin and connect with the cavities of the tentacles. What is the function of these canals?

_____

_____

The reproductive organs are attached to the radial canals.

d. Examine demonstration specimens of other hydrozoa that show the great diversity of structure among the individuals forming a colony. In this respect, pay particular attention to *Physalia*, the Portuguese man-of-war, which shows a highly specialized form of **polymorphism** (many forms). In addition to a modified medusa that forms the gas-filled bloat, there are several kinds of tentacles that serve to paralyze and capture food for the colony.

2. Class Scyphozoa

Aurelia, one of the commonest of the scyphozoan jellyfish, occurs throughout the world. They are often seen in large shoals, drifting along together or swimming slowly by rhythmic contractions of the shallow, almost saucer-shaped bell. Great numbers are sometimes cast on shore during storms. They range in size from less than 3 inches to about 12 inches across the bell.

The body is shallowly convex above and concave below and is fringed by a row of closely spaced

DIVERSITY IN ANIMALS: SPONGES AND COELENTERATES

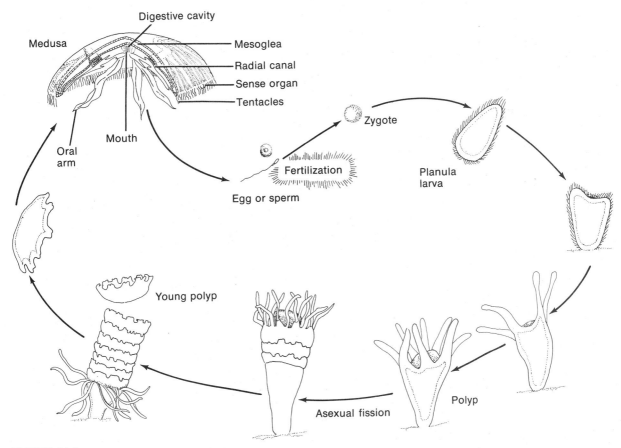

FIGURE 31-5
Life cycle of *Aurelia*.

marginal tentacles (Fig. 31-5). These are interrupted by eight equally spaced indentations, each with a sense organ between them. The sense organs consist of a pigmented eyespot, sensitive to light; a hollow statocyst, containing minute calcareous particles whose movements set up stimuli that direct the swimming movements; and two sense pits, lined with cells that are thought to be sensitive to food or to other chemicals in the water. Many circular muscle fibers are present in the margin. What is the function of these muscle fibers?

_____

_____

The mouth is located in the center of the oral (concave) surface at the end of a very short manubrium. The manubrium lies between four tapering oral arms, each of which has a ciliated groove. Stinging capsules in the lobes paralyze and entangle small animals, which are then swept up the grooves, through the mouth, and into a digestive cavity that extends into four pouches, in which are tentaclelike projections. These are covered with nematocysts, which paralyze prey that arrives in the pouches still alive and struggling. Many radial canals extend through the mesoglea from the pouches to a ring canal in the bell margin. Flagella lining the entire gastrovascular cavity maintain a steady current of water, which brings a constant supply of food and oxygen to, and removes waste from, the internal parts of this large animal.

Examine a specimen of *Aurelia* in a dish of water. Identify as many of the structural characteristics of this jellyfish as you can.

The four horseshoe-shaped, colored bodies by which *Aurelia* is recognized are the testes or ovaries; they occur on the floor of the large central part of the gastrovascular cavity. In a male medusa the sperm cells are discharged into the gastrovascular cavity and are shed to the outside through the mouth. The eggs of the female are shed into the cavity where they are fertilized by sperm cells entering the mouth along with incoming food.

The zygotes emerge and lodge on the oral lobes where each develops into a ciliated planula larva. This larva soon escapes, swims about for a while, and then settles and attaches to the sea bottom. Losing the cilia, it becomes a minute trumpet-shaped polyp with a basal disc, mouth, and tentacles. This polyp feeds and stores food and may survive in this stage for many months, meanwhile budding off other small polyps like itself.

Usually in the fall and winter, the polyp develops a series of horizontal constrictions, which resemble a pile of minute "saucers" with fluted borders. One by one the "saucers" pinch off from the parent and swim away as little medusas, gradually developing into adult jellyfish.

Examine demonstration specimens and slides showing the stages in the life cycle of *Aurelia*.

3. Class Anthozoa

Anthozoa are marine polyps that have no medusa stage. They are distinguished from other hydrozoan polyps by a gastrovascular cavity that is divided by a series of vertical partitions; the surface ectoderm turns in at the mouth to line the gullet. Externally there is no difficulty in distinguishing the large sea anemones or the stony corals from most of the small fragile hydrozoan polyps. Besides the familiar sea anemones and stony corals, this class includes the soft, horny, and black corals, the colonial sea pens and sea pansies, and others. They are abundant in warm shallow waters, but some inhabit polar seas, while other species inhabit areas ranging from the tide lines to depths of 17,000 feet.

a. Sea anemones are among the most highly specialized of the polyp types of coelenterates. They have a well-developed nerve net, mesenchyme cells between ectoderm and endoderm, and several sets of specialized muscles.

Examine the common sea anemone *Metridium*, and note the stout cylindrical body, expanded at its upper end into an oral disc around a slitlike mouth surrounded by several rows of tentacles. When undisturbed and covered by water, the body and tentacles are widely extended. If irritated or exposed by a receding tide, the oral disc may be completely inturned and the body closely contracted. Cilia on the tentacles and disc beat to keep these surfaces free of debris. The basal end forms a smooth, muscular, slimy disc on which the anemone can slide about very slowly and by which it holds to rocks so tenaciously that one is likely to tear the animal in trying to pry it loose. They commonly attach themselves to the shells of crabs or other shelled animals in the ocean. In this way the anemones may become widely dispersed. Many anemones are exquisitely colored. Examine a preserved specimen and identify the external and internal features described below.

From the mouth a muscular gullet hangs down into the gastrovascular cavity (Fig. 31-6). The gullet is not cylindrical but is flattened and lined with cilia that beat downward, thus drawing a current of water into the gastrovascular cavity and providing the internal parts with a steady supply of oxygen. At the same time other cilia lining the gullet beat upward, creating an outgoing current of water that takes with it carbon dioxide and other wastes. When small animals touch the tentacles, the cilia of the gullet reverse their beat, and the food is swept down the gullet into the digestive cavity. The gullet is connected with the body wall by a series of vertical partitions. What is the function of these partitions?

_____

_____

The free, inner margin of each partition bears nematocysts for paralyzing their prey and gland cells for the secretion of enzymes needed in digestion.

Anemones sometimes reproduce asexually by pulling apart into two longitudinal halves; each half then regenerates into a new adult. Sexual reproduction is accomplished by the small rounded gonads located at the edges of the partitions. The sexes are separate. Eggs and sperm leave the gonads through the mouth, and fertilization occurs in the water. The fertilized egg develops into a ciliated planula larva, which finally settles down in some rocky crevice and grows into an adult anemone.

b. Corals are small, fragile polyps that may form huge colonies consisting of millions of individuals. Each member of a colony secretes a protective skeleton of limestone, containing a pocket into which the polyp partly withdraws when attacked. New individuals build their skeletons on the skeletons of dead ones, and thus, over many years, by the action of these minute animals, huge undersea ledges known as coral reefs are built up. Three

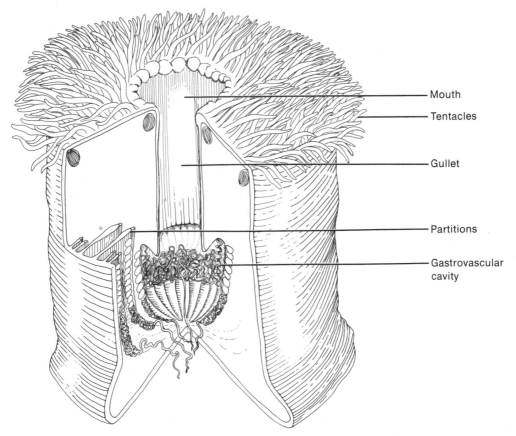

FIGURE 31-6
Internal anatomy of *Metridium*.

main types of coral reefs are recognized. **Fringing reefs** grow in shallow water and closely border the coasts or are separated from them at the most by a narrow stretch of water that can be waded when the tide is out. **Barrier reefs** also parallel coasts but are separated from them by a channel (lagoon) deep enough to accommodate large ships. The most famous of these is the Great Barrier Reef along the northwest coast of Australia, which is about 1,200 miles long and from a few to 90 miles off shore. **Atolls** are ring-shaped islands enclosing central lagoons and not enclosing an island. They are hundreds or thousands of miles from the nearest land, and slope off into the depths of the ocean.

Examine samples of coral on demonstration in the laboratory and note the variety of different forms. Locate and examine the small pockets that contained the polyps.

REFERENCES

Buchsbaum, R. 1948. *Animals Without Backbones*. University of Chicago Press, Chicago.

Meglitsch, P. A. 1967. *Invertebrate Zoology*. Oxford University Press, New York.

EXERCISE 32

Diversity in Animals: Flatworms, Roundworms, and Parasites of the Frog

A. FREE-LIVING FLATWORMS

The free-living flatworms are best exemplified by a group of small freshwater organisms collectively called the planarians. The genus *Dugesia* is commonly found in slow streams or ponds of cool, fresh water, either adhering to or crawling over sticks, stones, leaves, and other debris.

Examine a living planarian with a hand lens or dissecting microscope. Observe its general shape and size. From its movements determine its anterior, posterior, dorsal, and ventral sides. What type of symmetry is evident?

What is its coloration?

What are its reactions to such stimuli as jarring, contact with a toothpick, or being turned on its dorsal surface?

Observe the protrusible pharynx on the ventral surface. What is the advantage of such a structure?

Note the large eyespots on the anterior dorsal surface.

With a dissecting microscope examine a prepared slide of a whole mount of a planarian cleared to show the internal anatomy. Observe the branched digestive tract leading from the pharynx. What is the function of the numerous lateral projections

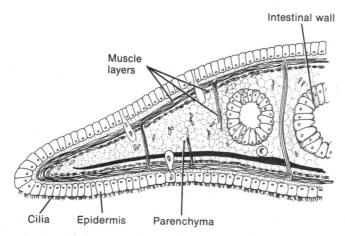
FIGURE 32-1
Cross section of *Dugesia*.

from each of the three main branches of the digestive tract?

_____

_____

Excretory tubes and nervous elements are difficult to observe because of the thickness of the animal.

Examine cross sections of a planarian cut through the pharynx and through regions anterior and posterior to the pharynx. Observe the various portions of the digestive tract in each of these regions (Fig. 32-1). There is no body cavity. The space between the surface ectodermal and the endodermal linings of the digestive tract are filled with a mass of cells of mesodermal origin (parenchyma). Observe cilia on some of the ectodermal surface cells. Where are they most prominent?

_____

What is the arrangement of the muscle layers?

_____

How does such an arrangement of muscles give the planarian flexibility?

_____

_____

Locate the two nerve cords just inside the ventral epidermis, one on either side.

Planarians have been used extensively in studies of **regeneration**. These regenerative powers may be readily demonstrated. Add several drops of the anesthetic tricain methanesulfonate or magnesium sulfate to a Syracuse dish and allow 15 minutes for the planarian to become completely anesthetized. Carefully cut the planarian with a sharp razor blade wiped clean of any oil. Diagram your cut in Fig. 32-2. (Your instructor will indicate different cuts to be made.) Add fresh pond water and place your dish in a cool dark spot in the laboratory. Cover to prevent evaporation. Observe the planarian during the next two weeks and make sketches of the regeneration.

B. PARASITIC FLATWORMS

1. Flukes

All members of this class are parasitic. Many show structural modifications due to their parasitic existence. Determine how their structural organization has adapted them to the parasitic life.

a. Chinese Liver Fluke (*Chlonorchis sinensis*)

This fluke lives in man and has a life cycle involving two intermediate hosts. It is a common parasite in China, Japan, and Korea, where it is estimated that there are approximately 19 million human infections.

The adult fluke, about ½ inch long, lives in the bile ducts of the human liver where it remains firmly attached by means of a pair of suckers. The

FIGURE 32-2
Regeneration in planaria.

fluke is hermaphroditic and capable of self-fertilization. What advantage is gained from possessing the capability for self-fertilization?

_____

_____

_____

The fertilized eggs that are released into the bile ducts are eventually voided in the feces of the host. If the feces get into water, the eggs are eaten by snails (Fig. 32-3). Within the digestive system of the snail the egg opens, and a larval form, called a **miracidium**, emerges and makes its way through the tissues of the snail, where it becomes asexually transformed into other larval forms. It has been estimated that a single miracidium can ultimately give rise to 250,000 infective larvae called **cercariae**. Why is it essential that one miracidium give rise to such a large number of cercaria in the life cycle of the fluke?

_____

_____

_____

The cerariae escape from the snail and swim about until they come in contact with a second intermediate host, a particular species of fish. They burrow through the skin of the fish, lose their tails, and encyst to form **metacercariae**. When raw or partially cooked fish is eaten by man, the cyst walls are digested away in the stomach and metacercariae are released. They then migrate to the bile ducts and mature into adult flukes to complete the life cycle. In light of what you know about the life cycle of the fluke, what would be the most effective methods of controlling the infection in man?

_____

_____

_____

_____

Examine a prepared slide of the Chinese liver fluke and note the flattened, leaflike shape of the worm. Observe an anterior (oral) sucker, the center of which is the mouth, and a ventral sucker about one-third of the way back. Observe the bilobed intestine. Note that the intestine has no opening

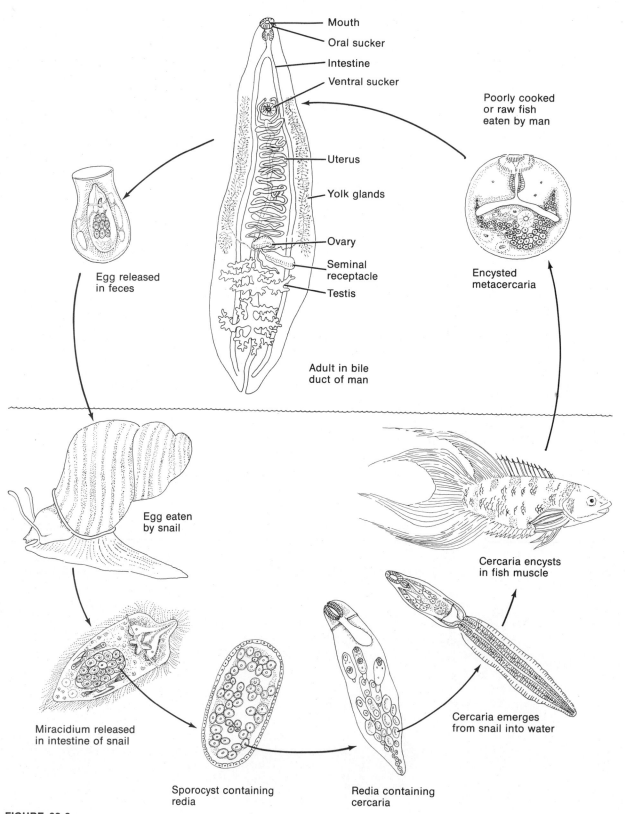

FIGURE 32-3
Life cycle of the Chinese liver fluke.

other than the mouth. A pair of irregularly branched testes occupies the posterior third of the body. The long, coiled uterus lies behind the ventral sucker. The ovary, a single body lying near the middle of the animal, is connected to a lighter-staining seminal receptacle used for the storage of sperm. Also leading into the ovary by means of two delicate tubules are the yolk glands, which consist of many small rounded bodies characteristically found in the lateral midportions of the body.

b. Sheep Liver Fluke (*Fasciola hepatica*)

This fluke infests the livers of sheep, causing a disease known as "liver rot." The adult parasite may also infect horses, rabbits, camels, pigs, and goats. The adult fluke is a flat, leaflike worm ranging in length from $\frac{1}{2}$ to 2 inches. Examine demonstration slides of this fluke. How does it differ morphologically from the Chinese liver fluke?

_____

_____

_____

_____

c. Giant Intestinal Fluke (*Fasciolopsis buski*)

This organism, measuring approximately 3 inches in length, is probably the largest of the flukes found in man. It is common in southeast Asia where it is estimated that there are 10 million human infections. Except for an unbranched intestine, it resembles the Chinese liver fluke in general arrangement of organs. Examine demonstration slides of this fluke.

From your study of the life cycle and morphology of the flukes, explain how they are adapted to their parasitic life.

_____

_____

_____

_____

2. Tapeworms

Tapeworms are usually long, flat, ribbonlike organisms. The tapeworm's adaptation to a parasitic existence is so extreme that the digestive and nervous systems have become degenerate, while the reproductive system is highly developed. The flatness of these worms enhances diffusional exchanges across cells so that there is no need for respiratory and circulatory systems.

a. Dog Tapeworm (*Dipylidium caninum*)

In this exercise the common cat or dog tapeworm will be studied as a typical example of a tapeworm. Children are sometimes infected by this tapeworm as a result of accidentally swallowing fleas while playing with dogs or cats. The flea is the normal carrier of the larval stage of this species.

Examine a prepared slide showing characteristic segments of an adult tapeworm (Fig. 32-4). Note the head, or **scolex** with its four suckers and several rows of hooks used for attachment to the intestine of its host. Beyond the scolex is a short neck, followed by a row of segments called **proglottids**. As you move back from the scolex note that the proglottids show a high degree of internal organization. These mature proglottids contain a double set of reproductive organs, each of which leads to a lateral genital pore. The smaller duct leading from the genital pore is the oviduct which leads into the diffuse ovary. The testes are found distributed throughout the proglottid. The vas deferens, by which the sperm leave the proglottid, is the larger duct leading to the genital pore. Lateral and transverse excretory canals occur in these mature proglottids as well as thin, longitudinal lateral nerve cords.

The more posterior proglottids of these tapeworms become filled with many ovarian capsules, each of which contains a number of eggs. In these gravid segments the reproductive organs observed in the mature segments have atrophied and the uterus enlarged to fill the proglottid. The gravid segments break loose from the tapeworm and are released in the feces of the host. The embryos released from these segments are eaten by the larvae of the dog or cat flea and develop into an infective stage called a **cysticercus**. When swallowed by a dog or cat, such an infected flea will be digested by the host and the cysticercus released. It attaches to the intestinal wall and develops into an adult tapeworm.

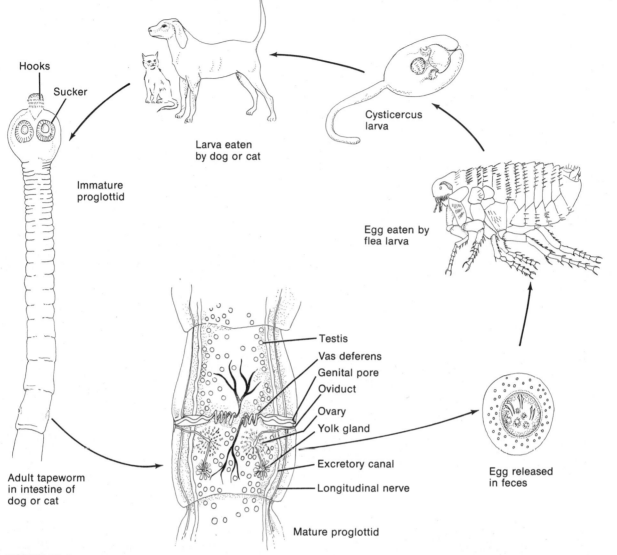

FIGURE 32-4
Life cycle of the dog tapeworm.

b. Beef Tapeworm (*Taenia saginata*)

The most common intestinal tapeworm of man is the "beef tapeworm," which ordinarily reaches a length of 15 to 20 feet. In addition to cattle, the buffalo, giraffe, and llama have been recorded as natural hosts of the cysticercus stage. Meat inspection has reduced the occurrence of this parasite to about 1% of the cattle in the United States. In countries where beef is prepared by broiling large pieces over open fires, the incidence of the parasite is quite high. Why?

_____

_____

Examine preserved specimens of this tapeworm, as well as slides showing scolices and mature and gravid proglottids.

c. Pork Tapeworm (*Taenia solium*)

The pork tapeworm closely resembles the beef tapeworm and has a similar life history, except that the larval stages, called **bladderworms**, develop in pigs. Meat inspection has made it uncommon in the United States.

d. Fish Tapeworm (*Dibothriocephalus latus*)

The "broad, fish tapeworm," which may reach a length of 60 feet, is the largest tapeworm that lives in man. This tapeworm requires two intermediate hosts to complete its life cycle. The eggs, when released into water, develop into larvae that are eaten by small crustaceans called copepods. These in turn are eaten by fish, in which the larval stage (**plerocercoid**), which is infective to man, develops. Man gets the parasite when he eats raw or partially cooked fish. This tapeworm has recently become well established in the western Great Lakes region of North America, where it commonly infects pike and pickerel.

Examine prepared slides of this tapeworm and note that the scolex has sucking grooves, as opposed to the suckers and hooks found in the other tapeworms you have studied. Also note that the proglottids are broader than they are long and that the reproductive organs are located in the center of the proglottid.

Do you consider tapeworms to be more, or less, specialized to parasitism than flukes? Give the reasons for your answer.

_____

_____

_____

_____

C. ROUNDWORMS (Nematodes)

Among multicellular animals, roundworms are probably second only to insects in number. The majority of the 500,000 species of nematodes are free-living. Some are found in fresh and salt water, and some in mud, field, or garden soils. Other nematodes are parasites of roots, stems, leaves, and even the seeds of many of our agricultural and horticultural plants where they cause inestimable damage to crops. Many thousands of species of roundworms are parasites of invertebrate and vertebrate animals. Among the vertebrates it is doubtful if any species is not parasitized by one or more of the roundworms. Although over 50 species have been known to occur in man, only about a dozen are important human parasites. Most of the species that are either free-living or parasitic in plants or invertebrates are barely visible to the eye, are transparent, and have simple life cycles. The species that are parasitic in vertebrates, on the other hand, are often several feet in length and have more complicated life cycles.

1. Pig Roundworm (*Ascaris lumbricoides*)

Ascaris is a roundworm commonly parasitic in the intestine of hogs. The sexes are separate, with the male worm shorter and more slender than the female and readily distinguished from her by a sharply curved posterior end. In addition, the male has a pair of hairlike structures (spicules) extending from the anal opening, which are used during copulation. Both worms are several inches long. Examine preserved specimens of males and females. Using a hand lens or dissecting microscope, observe the terminal mouth surrounded by three lobelike lips. Locate the excretory pore just behind the mouth. Note the external chitinous cuticle.

With a scalpel carefully slit a female worm along the whole length of the body. Expose the internal organs by pinning back the body wall in a dissecting pan (Fig. 32-5). Cover the worm with water. Locate the long, flat digestive tract extending the length of the worm. The Y-shaped female reproductive system occupies most of the body cavity. Identify the single vagina, which divides into a pair of large, straight, and uncoiled uterine tubes. The uteri narrow down into the oviducts, which in turn continue into the long, thin, extremely coiled ovaries.

The male reproductive tract consists of a single, long, very coiled tube, which is divided into a long slender testis, sperm duct, enlarged seminal vesicle, and a small ejaculatory duct opening into the terminal end of the digestive tract.

Examine a prepared slide of a cross section of the female (Fig. 32-6). The body wall consists of an outer, noncellular cuticle and an inner hypodermis. What is the function of the cuticle?

_____

_____

The hypodermis has four extensions into the body cavity. Two of these extensions contain the dorsal and ventral nerve cords, while the other two contain the excretory tubes. The greater part of the body wall is composed of muscle tissue. The digestive tract is centrally located. Observe the two large cross sections of the uteri and note the shelled ova. Identify as many meiotic stages as are evident in your slide.

_____

_____

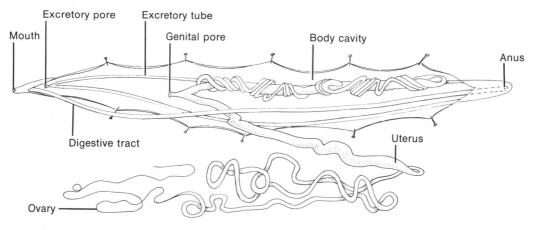

FIGURE 32-5
Internal anatomy of female *Ascaris*.

Sections through the oviducts and ovaries can be distinguished by their smaller sizes and by the absence of shells around the egg cells.

2. *Trichinella*

Infection by this parasitic roundworm, which causes trichinosis in man, results from eating raw or partially cooked pork containing the encysted larvae of this parasite. In the human intestine the larvae are released from their cysts and develop into adult worms. The mature female worm gives birth to larval worms, which then migrate through the body and encyst in the muscles of the diaphragm, ribs, and tongue. Greatest damage results during this migration phase. Why?

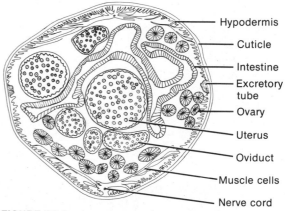

FIGURE 32-6
Cross section of female *Ascaris*.

The incidence of human infections in the United States is astonishingly high—approximately 17% of the population is infected.

Examine prepared slides of adult worms and encysted larvae.

3. Hookworm (*Necator americanus*)

No group of nematodes causes more injury to man or greater economic loss through attacks on domestic animals than the hookworms. These tiny parasites have highly developed mouth cavities that contain plates by which the worm holds onto the intestine while it sucks blood and tissue. Larvae develop from eggs on moist soil in the warmer climates of the south. Human infections occur by direct penetration of the skin by the larvae. Infection most often occurs from going barefoot in areas where this parasite is prevalent.

Examine prepared slides of hookworms. Locate the buccal capsule at the anterior end of this worm. Note the cutting plates inside this capsule. At the posterior ends of these worms locate the bursa, an umbrellalike expansion of the cuticle, which is used in copulation.

4. Pinworm (*Enterobius vermicularis*)

This roundworm is one of the most common parasites of children, with an estimated incidence of 30%–60% of the population in Canada and the United States. Infection occurs by the accidental swallowing of ova carried under finger nails, in bed clothing, or on toilet seats. Infections in children lead to severe itchiness in the anal region, insomnia, restlessness, loss of appetite, and irritability.

Examine slides of adult pinworms. These small, spindle-shaped worms are characterized by a lateral expansion of the cuticle at their anterior ends. The posterior end of the male is strongly curved.

5. Free-Living Nematodes

Free-living nematodes can be found in almost any loose soil containing organic matter. Several days before this laboratory period your instructor placed small piles of loose soil onto the surface of Petri dishes containing agar. Examine a dish under a dissecting microscope or the low power of a compound microscope. Scrape several of these worms onto a slide, add a drop of water, and examine under the low and high magnifications of your microscope. With the exception of the nervous system, these nematodes are transparent, and the various organ systems can be observed. If the movements of these worms are too rapid, add a drop of dilute hydrochloric acid to your slide. Describe and account for the type of movement exhibited by these worms.

_____

_____

_____

_____

Compare the structure of these roundworms with those previously studied.

A very common free-living nematode is the vinegar eel (*Turbatrix aceti*), which is frequently found at the bottom of vinegar barrels where it feeds on the bacteria and yeasts that have settled there. Mount a drop of vinegar eel culture on a slide and examine microscopically.

D. PARASITES OF THE FROG

Frogs are an excellent source of a number of animal parasites. Pith a frog supplied by your instructor and then cut through the ventral body wall to expose the internal organs. For the procedure of pithing, see Appendix D. Refer to Fig. 32-7 for some of the more common frog parasites.

1. Parasites of the Blood

Take blood from one of the larger blood vessels or the kidney, place a drop on a clean slide, and draw it out in a thin film, using the edge of another slide. Add a cover slip and examine for blood parasites. The most easily recognized forms are the flagellate trypanosomes. Less readily observed are the parasites of the red blood cells. Examine the red blood cells under high power and locate clear oval or spindle-shaped parasites in the cytoplasm (Fig. 32-7A).

2. Parasites of the Lung

Carefully remove the lungs by cutting them at their point of attachment to the larynx. Place them in a Syracuse dish containing warm Ringer's solution and examine for dark bodies. Cut open the lungs and observe the movements of these parasites. Among the species of trematodes commonly found in the lung is a relatively large one called *Haematoloechus* (Fig. 32-7B). Mount the parasites on a slide, add a drop of water, and examine for details of internal anatomy.

3. Parasites of the Digestive Tract

With a scissors, slit open the digestive tract from the rectum to the esophagus. Examine each region for the presence of roundworms and flatworms. Several species of parasites are commonly found in the frog digestive tract. Carefully examine the contents of the intestine for small trematodes. *Diplodiscus*, a relatively large trematode, lives in the rectum and is recognized by its large ventral sucker (Fig. 32-7C).

A protozoan common to the rectum of the frog is *Opalina*, which is a large, multinuclear, parasitic ciliate (Fig. 32-7D). Cut open the rectum and look for other protozoa (Fig. 32-7F).

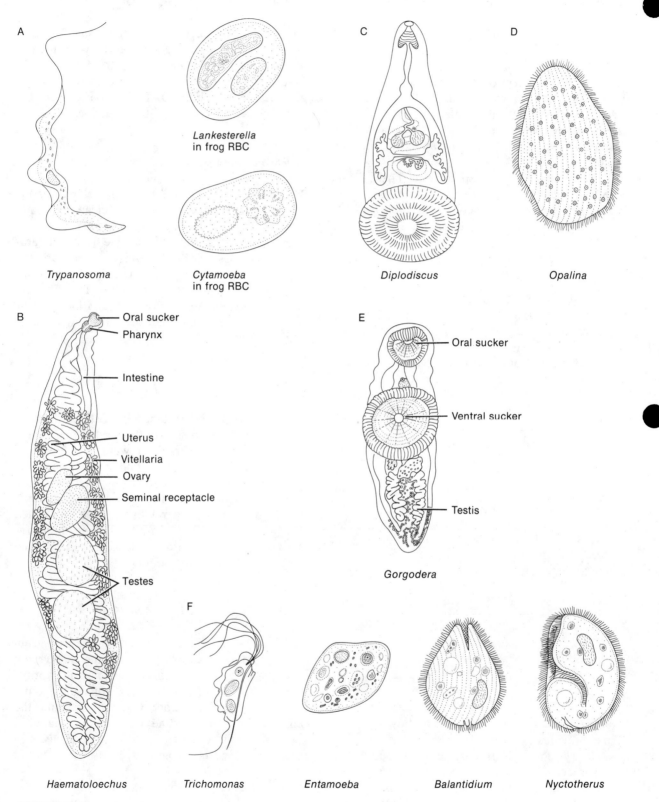

FIGURE 32-7
Common parasites of the frog.

4. Parasites of the Bladder

Cut open the urinary bladder and examine for the presence of parasites. At least six species of trematodes have been reported in the frog bladder. One of the more common of these trematodes is *Gorgodera* (Fig. 32-7E).

5. Other Parasites

Carefully examine all other regions of the frog body for parasites (e.g., lining of body cavity, mesenteries, mouth, and surfaces of internal organs). Identify and list the kinds of parasites found (i.e., roundworms, flukes, tapeworms, or protozoa).

REFERENCES

Buchsbaum, R. 1948. *Animals Without Backbones.* University of Chicago Press, Chicago.

Chandler, A. C., and C. P. Read. 1961. *Introduction to Parasitology.* 10th ed. Wiley, New York.

Meglitsch, P. A. 1967. *Invertebrate Zoology.* Oxford University Press, New York.

Noble, G. K. 1954. *The Biology of the Amphibia.* Dover, New York.

Russell-Hunter, W. D. 1968. *A Biology of the Lower Invertebrates.* Macmillan, New York.

EXERCISE 33

Diversity in Animals: Annelids

The annelids are a group of worms possessing a true coelom derived from mesoderm and bodies characteristically divided into a series of segments. The type of segmentation encountered in these organisms is said to be **homonomous**; all segments of the body are essentially alike. Most members of this phylum are marine worms, although the most commonly studied member is the earthworm. The members of this group show significant advances over the previous phyla. These should be noted during your study.

A. CLAMWORMS (CLASS POLYCHAETA)

These annelids are characterized by having a head with sensory appendages, and lateral body appendages called **parapodia** that bear many bristlelike structures called **setae**.

The clamworm [*Neanthes* (*Nereis*) *virens*] is a marine worm that lives in burrows in the sand at the tide level. It remains in its burrow during the day but crawls or swims about at night. Examine specimens and note the characteristic body segments, each of which bears a pair of lateral parapodia that terminate in bristlelike setae. Suggest a function for the setae.

_____

_____

Examine the head region and note the two large feelers, a pair of smaller tentacles, and four eyes.

B. SEGMENTED WORMS (CLASS OLIGOCHAETA)

1. Earthworm

The familiar earthworm (*Lumbricus*) is studied as a representative of this group of annelids because of its universal availability and distribution. Other members are found in a variety of terrestrial, freshwater, and marine habitats. Some species live in burrows as do the earthworms, while others live in special tubes.

The earthworm moves by means of coordinated peristaltic waves of muscle contraction passing backward along the body. Each wave consists of an initial elongation and thinning, followed by a shortening and thickening of the body. Observe an earthworm crawling on wet paper toweling or other rough surface. Examine the ventral side of the worm and note that in every segment, except the most anterior and most posterior, there are four pairs of setae. Stroke the body and note the direction in which they slant.

Suggest a function for the setae in locomotion?

With a sharp scalpel, or razor blade, make a short longitudinal slit in the dorsal body wall of the worm and observe it as the animal crawls. Describe the opening and closing of the slit in relation to the passage of peristaltic waves along the earthworm.

What kind of orientation of muscle fibers must be present in order to account for the opening and closing of the slit during locomotion?

Repeat these observations on a transverse slit.

What kind of orientation of muscle fibers is responsible for these movements?

While the worm is crawling forward, stimulate the anterior end in various ways (i.e., stroking gently, jarring the table, applying heat, poking, and so on). What happens to the peristaltic waves after each kind of stimulation?

What is the difference in the reactions to gentle and strong stimulations?

Although earthworms lack definite eyes, they do possess light-sensitive cells (**photoreceptors**) distributed over the body. These can best be studied by using a "light pencil," which can direct a point of light to small areas of the earthworm's body. Such a light pencil is easily constructed by drawing out a glass rod to a tapered point, cutting off the cone that is formed and cementing the base to a pocket "pen light." The glass cone is then painted black, except for the tip, to permit the emission of a fine beam of light.

Using your light pencil, test the light sensitivity of various regions of dark-adapted worms (worms kept in a dark, cool place for several hours before the laboratory period). How does the sensitivity of the head and tail regions compare with that of the middle segments?

Why would you expect a difference in sensitivity of various regions of the body?

DIVERSITY IN ANIMALS: ANNELIDS

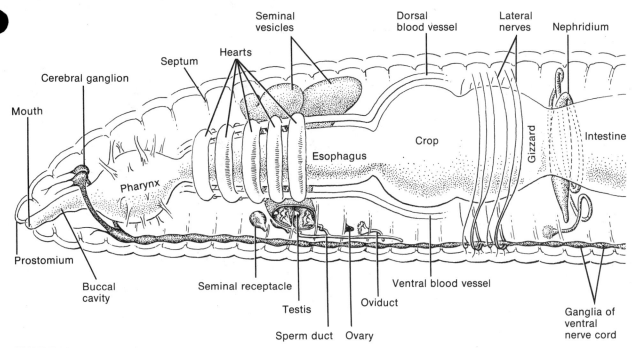

FIGURE 33-1
Internal anatomy of the earthworm.

Examine preserved specimens of this large annelid and note the many regular segments. These are usually numbered, beginning at the anterior end. Count the number of segments.

The anterior end is the most pointed and contains the mouth. Note the **prostomium**, the dorsal, lobe-like projection of the segments above the mouth. The anal opening is in the terminal segment. Approximately one-third of the distance from the anterior end certain segments are larger than the others. These constitute the **clitellum**, which is important in reproduction.

On the ventral surface of the 15th body segment are two lateral swollen areas, each with a central opening. These are the openings of the **sperm ducts** through which the sperm are discharged during copulation. On the 14th segment are the smaller openings of the **oviducts**.

To examine the internal anatomy, carefully slit the body wall along the middorsal line, using a sharp scissors or scalpel. Carefully cut the internal partitions and pin back the body wall in your dissecting pan. Cover the worm with water and, using a hand lens or dissecting microscope, study the internal organs.

The digestive tract is the straight tube running the length of the body (Fig. 33-1). The mouth, which opens into a small buccal cavity, is followed by the pharynx, which is connected to the body wall by numerous muscle fibers. Behind the **pharynx** is the long, narrow esophagus. It is partially hidden by the three large, light-colored pairs of seminal vesicles. The **crop**, a thin-walled expanded portion of the digestive tract, is followed by the heavy, muscular gizzard. The crop apparently serves only for storage. The **gizzard**, with the aid of small soil particles taken in during feeding, grinds the food thoroughly before passing it on to the **intestine** for final digestion and absorption. Solid waste products of digestion are passed to the exterior through the **anus**.

The earthworm possesses a closed circulatory system—that is, it consists of a system of tubes, or vessels, in which a circulating fluid is continually kept in motion by the pumping action of the dorsal blood vessel. What is the advantage of such a system over the gastrovascular cavities found in the coelenterates and flatworms?

_____

_____

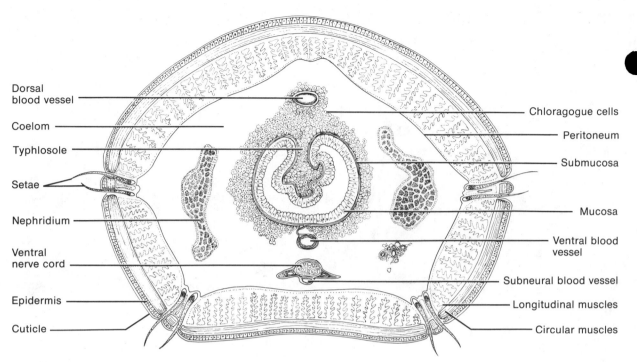

FIGURE 33-2
Cross section of an earthworm.

Locate the dorsal **longitudinal blood vessel** that runs along the dorsal midline of the digestive tract and carries blood forward from the intestine, where digested food material has been absorbed. The **ventral blood vessel**, suspended below the digestive tract, carries blood backward and distributes it to various organs and tissues of the body. In the region of the esophagus, the dorsal and ventral vessels are connected by five pairs of enlarged vessels, the "hearts," which pump blood from the dorsal to the ventral blood vessel. Carefully examine your dissected earthworm and trace the blood flow as described.

Push aside the intestine and note the white ventral nerve cord extending the length of the earthworm. Follow the nerve cord to the anterior end of the worm. Note that it is connected with the **cerebral ganglia** on the dorsal side of the pharynx by a pair of nerve cords that pass on either side of the pharynx.

Using a dissecting microscope, observe the pair of delicate, white, closely coiled tubes lying against the body wall in each segment. These structures are the **metanephridia** or excretory organs. Liquid wastes, which collect in the body cavity, are excreted by these organs.

Although earthworms are **monoecious** (possess a complete set of male and female reproductive organs) they nevertheless undergo cross-fertilization during copulation. Two worms become temporarily joined along their ventral sides by the secretion of a "slime tube." During copulation sperm cells, released from the **seminal vesicles** of each worm, are stored in the **seminal receptacles** of the opposite earthworm. After the animals separate, the **clitellum** secretes a mucus ring that slides over the anterior segments and picks up eggs from the oviducts in the 14th segment and sperm as it slides over the 9th and 10th segments. The mucus ring eventually slips over the anterior end of the worm to form the egg cocoon from which the young eventually hatch. Examine the anterior segments of your earthworm and locate the various reproductive structures described above. The seminal vesicles have been described as lying alongside of the crop and gizzard. The testes are inside of these bodies and cannot be seen. Near the seminal vesicles, in segments 9 and 10, are the four small, spherical **seminal receptacles**. By careful examination with a hand lens or dissecting microscope you may be able to locate the single pair of ovaries in segment 13.

Study a prepared slide of a cross section taken from the intestinal region. Examine under low power for general features and under high power for structural details. The body wall can be seen to be made up of four layers (Fig. 33-2). The outermost layer is the thin, noncellular cuticle secreted

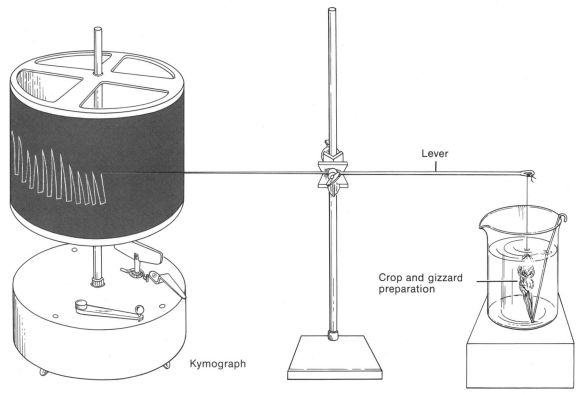

FIGURE 33-3
Kymograph apparatus for measuring muscle contractions.

by the epidermis, which consists of a layer of columnar epithelial cells, among which are many glandular cells. Beneath the epidermis is a layer of circular muscle fibers. The next and thickest layer of the body wall is made up of longitudinal muscle fibers arranged in a featherlike pattern. The innermost layer of the body wall consists of a thin layer of flattened cells composing the **peritoneum**. The coelom is the space between the body wall and the digestive cavity. In the coelomic cavity locate the dorsal and ventral blood vessels and the ventral nerve cord. Nephridia may be observed in the coelomic cavity on either side of the intestine. Setae may be observed in some of the sections.

The thin-walled intestine bulges laterally in each somite; its dorsal wall has an infolded **typhlosole**. The bulges and typhlosole provide increased surface area for digestion and absorption of food. The wall of the intestine is composed of three layers. The innermost layer is the mucosa, a layer of narrow, ciliated, columnar cells. Beneath this layer lies the submucosa, which contains circular and longitudinal muscle fibers and small blood vessels. The outermost layer is made up of slender columnar epithelial cells containing numerous granules and **chloragogue cells**, which aid in the elimination of wastes from the blood.

One of the significant advancements of the annelids over the roundworms and flatworms is their highly developed muscular system. In many respects the physiology of annelid muscle tissue is remarkably similar to that of vertebrate smooth muscle.

The crop and gizzard of the earthworm exhibit considerable rhythmicity when isolated. Anesthetize an earthworm in 5% alcohol. Dissect out the crop and gizzard along with a short segment of the pharynx and intestine. Tie a thread to each end of the dissected tissue. Attach one end to a glass rod and suspend in a bath of Ringer's solution (Fig. 33-3). The other end is attached to the lever of a **kymograph**. If placed under slight tension, regular contractions should commence within a few minutes. Does the preparation contract as a whole or does one region start first and initiate a wave of contraction through the rest of the preparation?

TABLE 33-1
Effects of chemicals on contraction of an earthworm crop-gizzard preparation.

Chemical treatment	Observations
Adrenaline	
Acetylcholine	
Eserine	
Conclusions	

Determine the effect of adding several drops of adrenaline (1:100,000 dilution), acetylcholine (1:100,000), and eserine (1:100,000) on the rhythmic contractions of your preparation. Record your observations and conclusions in Table 33-1. The crop-gizzard preparation should be thoroughly washed with warm Ringer's solution between each of the tests. Fresh solution should be added to the bath for each test.

2. Tubifex

Tubifex, small, reddish brown, freshwater annelids commonly found on the bottom of lakes and streams, live in tubes from which they protrude their posterior ends. Since this worm is semitransparent, its internal anatomy can easily be studied. Prepare a wet mount of this worm and examine under the microscope. Note the dark, tubular, digestive tract, which occasionally shows waves of contraction along its length. The pulsating dorsal blood vessel with its reddish color can also be seen. A single aortic loop may be evident when the worm straightens out. The anterior segments each possess paired groups of setae, while the posterior segments have only two pair. Describe the motion of these setae.

C. LEECHES (CLASS HIRUDINEA)

Leeches are found in some lakes, ponds, and streams. The body of a leech is dorsoventrally compressed, lacks a well-defined head, and has anterior and posterior suckers for attachment. The anterior sucker is smaller and surrounds the mouth, which contains three teeth. Suggest a function for the teeth?

Although the body is divided by many transverse divisions, these are not considered to be true segments since they are superficial and in fact exceed the number of true segments found internally.

REFERENCES

Barrett, T. J. 1947. *Harnessing the Earthworm.* Humphries, Boston.

Hess, W. N. 1924. Reactions to Light in the Earthworm, *Lumbricus terrestris. J. Morphol.* **39**:515–542.

Laverack, M. S. 1963. *The Physiology of Earthworms.* Macmillan, New York.

Meglitsch, P. A. 1967. *Invertebrate Zoology.* Oxford University Press, New York.

Ratner, S. C., and D. C. Stein. 1965. Responses of Worms to Light as a Function of Interatrial Interval and Ganglia Removal. *J. Comp. Physiol. Psychol.* **59**:301–305.

Rounds, H. D. 1968. *Invertebrates*. Reinhold, New York.

Russell-Hunter, W. D. 1968. *A Biology of Lower Invertebrates*. Macmillan, New York.

Wells, M. 1968. *Lower Animals*. McGraw-Hill, New York.

EXERCISE 34

Diversity in Animals: Arthropods

This group of organisms constitutes the largest of all plant and animal phyla. Of the million or so known species of animals, over three-fourths are arthropods. The arthropods are considered to have attained the greatest "biological success": they have the largest number of species and individuals, occupy the greatest variety of habitats, consume the largest amounts and kinds of food, and are capable of defending themselves against their enemies.

A. PERIPATUS (CLASS ONYCOPHORA)

Peripatus, an animal having characteristics common to both the annelids and arthropods, comes closer than any other to being the "missing link" between any two phyla. *Peripatus* resembles the annelids in the structure of the eyes, the nephridia, the ciliated reproductive ducts, and the simple gut. It resembles the arthropods in having a hemocoel (body cavity that functions as part of a blood-vascular system), a dorsal heart, and the general structure of the reproductive organs.

Examine a preserved specimen of *Peripatus*. Note the absence of external segmentation, though there is a pair of legs for each internal segment of the body. The legs terminate in claws, which superficially resemble those of arthropods but differ in not being jointed. The head has three segments, each of which bears three pairs of appendages: two short antennae, two blunt oral papillae, and two small horny jaws.

B. CENTIPEDES (CLASS CHILOPODA) AND MILLIPEDES (CLASS DIPLOPODA)

These classes include a group of slender, dorsoventrally flattened animals, having a distinct head and numerous body segments, with either a single pair of walking legs (centipedes) or a double pair of walking legs (millipedes) per segment.

Examine specimens of centipedes and millipedes.

C. SPIDERS, SCORPIONS, AND HORSESHOE CRABS (CLASS ARACHNIDA)

The body segments of this class are fused into two main regions: a cephalothorax, bearing four pairs of walking legs, and an abdomen. Arachnids differ from the other arthropods by the absence of compound eyes, antennae, or true jaws.

Examine specimens of spiders, scorpions, ticks, mites, and horseshoe crabs.

D. INSECTS (CLASS INSECTA)

The insects—with some 675,000 species—are the most successful and abundant of all land animals. Insects are the principal invertebrates that can live in dry environments and the only ones able to fly. These habits are made possible by the chitinous body coating (**exoskeleton**) that protects the internal organs against injury and loss of moisture and by the system of tracheal tubes that enable insects to breathe air. The ability to fly permits them to readily find food. Under favorable conditions insects can multiply rapidly.

1. Grasshopper (*Romalea microptera*)

Grasshoppers live all over the world, mainly in open grasslands, weed patches, and roadside growth, where they eat grasses and other leafy vegetation. The grasshopper has three distinct body parts and chewing mouth parts. It undergoes a gradual **metamorphosis** (definite series of changes during development) from the young (**nymph**) stages to the adult. The following description, although primarily pertaining to the short-winged lubber grasshopper (*Romalea microptera*), will serve for any common species.

The body of the grasshopper is divided into a head consisting of six fused segments (**somites**); a thorax of three somites to which are attached the legs and wings; and a long segmented abdomen that terminates with the reproductive organs (Fig. 34-1). The outer **exoskeleton** consists largely of chitin, which is secreted by the epidermis. In order to grow, the grasshopper periodically sheds this exoskeleton (**molts**); adults do not molt. Pigment in and under the cuticle provides a protective coloration pattern by which the grasshopper resembles its surroundings.

The **head** (Fig. 34-2) has one pair of slender, jointed **antennae**, two **compound eyes**, and three simple eyes or **ocelli**. The mouth parts are of the chewing type and include (1) a broad upper lip or **labrum**; (2) a tonguelike **hypopharynx**; (3) two heavy blackish lateral jaws or **mandibles**, each with teeth along the inner lateral margin for chewing food; (4) a pair of **maxillae** of several parts, including palps (sensory appendages) at the side; and (5) a broad lower lip or **labium**, with two short palps.

The **thorax** consists of three parts, a large anterior **prothorax**, the **mesothorax**, and the posterior **metathorax**. Each part bears a pair of jointed legs, and the meso- and metathorax each a pair of wings. Identify the leg segments indicated in Fig. 34-1. The anterior wings of the grasshopper are thickened, acting as shields for the larger pair of flight wings. The wings are derived from the cuticle and have thickened portions (veins) that strengthen the wings. Stretch out one pair of wings and examine the anterior protective wings and the flight wings.

The slender **abdomen** consists of 11 somites, the terminal ones being modified for reproduction. The male has a blunt terminal segment while the female has four sharp conical prongs, the **ovipositors**, which are used in egg laying (Fig. 34-1). Along the lower sides of the thorax and abdomen are 10 pairs of small openings, the **spiracles**. These connect to a system of elastic air tubes, or **tracheae**, that branch to all parts of the body and constitute the respiratory (tracheal) system of the grasshopper. This is a system of air tubes that brings atmospheric oxygen directly to the cells of the body. This system also offers an enormous area for water uptake. The spiracles may be actively opened and closed to regulate the flow of air. The three most anterior pairs of spiracles are inhalatory, piping air directly to all body tissues. The other spiracles are exhalatory.

It is difficult to preserve the internal organs of the grasshopper because the preservative often fails to penetrate the exoskeleton. Careful dissection is necessary to study the internal anatomy.

After removing the wings start at the posterior end and make two lateral cuts towards the head with a pair of scissors or fine scalpel as indicated

DIVERSITY IN ANIMALS: ARTHROPODS

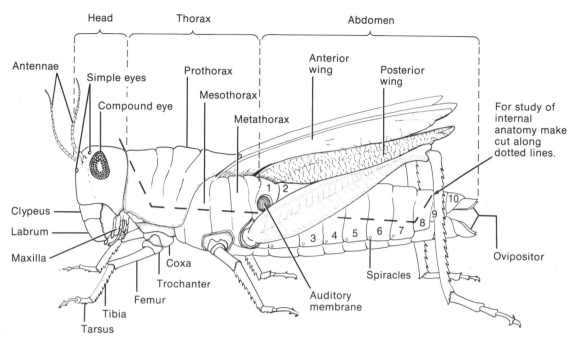

FIGURE 34-1
External anatomy of a female grasshopper.

in Fig. 34-1. Remove the dorsal wall. On the inside of the body wall locate the muscles noting their arrangement. What is their function?

_____

_____

There is a space between the body wall and digestive tract, the **haemocoel**, which is a blood space filled with colorless blood.

Study the digestive tract and identify its parts (Fig. 34-3). Beginning at the anterior end is the **mouth**, which opens between the mandibles and continues through a short **esophagus** into the **crop**. This is followed by the **stomach** to which are attached six double finger-shaped digestive glands (**gastric caeca**). The latter produce enzymes that are secreted into the stomach to aid digestion. The digestive tract continues as the **intestine**, which consists of a tapered anterior part, a slender middle portion, and an enlarged **rectum** that opens at the **anus**. During feeding, food held by the forelegs, labium, and labrum is lubricated by secretions from the salivary glands and chewed by the mandibles and maxillae. Chewed food is stored in the crop, digested in the stomach, and absorbed in the intestine. In the rectum, excess water is absorbed from any undigested food.

The excretory system is made up of numerous tiny tubules, the excretory or **Malpighian tubules**, which empty their products into the anterior end of the intestine. These tubules remove urea and salts from the blood.

The sexes are separate and show sexual characteristics in the terminal abdominal segments as previously described. In the male, each of the two **testes** is composed of a series of slender tubules or follicles above the intestine that are joined to a lengthwise **vas deferens**. These are joined to a common ejaculatory duct to which are attached **accessory glands**. In the female, each **ovary** is composed of several tapering egg tubes (**ovarioles**), which produce the ova. These are joined to an **oviduct** leading to a common **vagina** to which is attached a pair of accessory glands and a single **spermatheca**. The latter organ is used to store sperm received at copulation from the male.

The insect circulatory system may be studied by examining the wing veins of a living adult grasshopper or cricket. This can be done by preparing a plasticene or wax cell large enough to hold the insect on a microscope slide. Pin the animal down with two strips of paper, one across the thorax, the

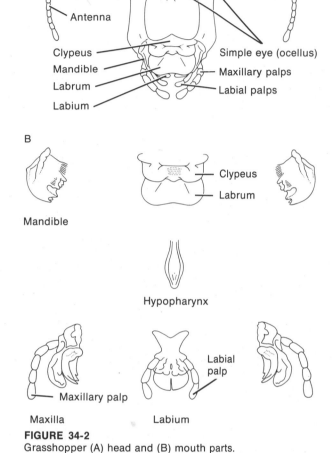

FIGURE 34-2
Grasshopper (A) head and (B) mouth parts.

other across the body beneath the wings. Slip a piece of tinfoil or glazed white paper beneath the anterior wing and examine under a dissecting microscope. Note the direction of flow of the hemolymph in the various parts of the wing, and the moving blood cells. In order to observe the beating heart, pin the insect's wings out to the side, and focus light through the body with the concave mirror. What is the direction of blood flow in the dorsal tubular heart?

_____

2. Insect Metamorphosis

Most insects undergo a definite series of changes during their development (**metamorphosis**). These changes are of several types:

- In the group without metamorphosis, the eggs hatch into young that, except for size, resemble the adults.
- In those having gradual metamorphosis, there is a development of wings as the insects mature. Except for the size and rudimentary condition of the genital appendages, the young (called **nymphs**) resemble the adults. Furthermore, the young and the adults occupy the same habitat and feed on the same food.
- In those insects having complete metamorphosis, the young and adults are totally unlike in appearance. For example, caterpillars develop into

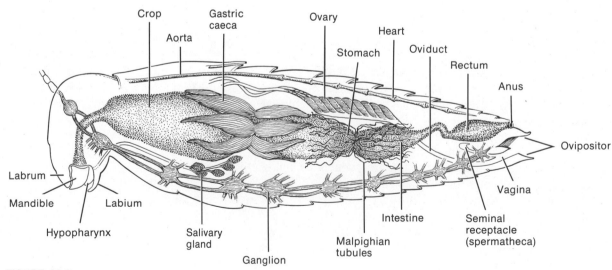

FIGURE 34-3
Internal anatomy of a female grasshopper.

DIVERSITY IN ANIMALS: ARTHROPODS

TABLE 34-1
Types of insect metamorphosis.

Insect	Type of stage present					Type of metamorphosis
	Egg	Nymph	Larva	Pupa	Adult	
Grasshopper						
Housefly						
Bee						
Butterfly						
Beetle						
Damsel fly						
Wasp						
Dragonfly						
Common roach						

moths or butterflies, maggots become flies, and grubs transform into beetles. The young stage, known as the **larva,** occupies an entirely different habitat and has different food requirements from those of the adult. When the larva has reached maturity, it ceases to feed, settles down in one place, forms a **pupa** within which it undergoes a series of morphogenetic changes, and emerges as an entirely new form, the adult.

Examine the various insect life cycles on demonstration. Complete Table 34-1 for each of the insects listed. Based on the presence of the stages you have indicated, determine the type of metamorphosis exhibited by each insect.

E. CRUSTACEANS

This class of arthropods includes the fairy shrimps, water fleas, barnacles, crayfish, and crabs. Most of them are marine, but many others inhabit inland waters, and a few, like the sow bugs, live in moist places on land. Many crustacea are microscopic while others, like the giant crabs, may reach a length of 12 feet.

Crustaceans have three important arthropod features. The body is segmented and covered by a chitinous exoskeleton, and the appendages are jointed. However, the segmentation of the crustaceans is more specialized than that of the annelids. In the latter the segments are nearly identical, whereas in the crustaceans and most arthropods there is considerable regional specialization including a well-developed head.

1. Crayfish (*Cambarus*)

The crayfish is a cannibalistic scavenger that lives on the muddy bottoms of streams and ponds. It emerges at night to feed on dead and decaying matter, insect larvae, and worms. It is up to six inches long, with appendages differentiated to serve special purposes.

Observe living specimens of a freshwater crayfish in shallow pans of water or in aquaria. Study their manner of walking and swimming. In which direction does a crayfish normally crawl?

Hold a crayfish firmly against the bottom of the container and then introduce a drop of India ink at the posterior end of the abdomen. From your observations, describe the direction of water flow and the mechanism used by the crayfish to set up such water currents.

_____

_____

Feed a crayfish a live insect (cockroach or cricket) and note the coordinated activity of the mouth parts.

Examine a preserved specimen of the crayfish. The body is divided into an anterior **cephalothorax** (fused head and thorax) and a posterior **abdomen**. The chitinous exoskeleton effectively protects the crayfish from predators. In order to grow, however, the crayfish periodically **molts** (i.e., sheds its exoskeleton). During the period that a new skeleton is being formed, the crayfish is defenseless; it hides to escape its enemies.

The **carapace** is a saddlelike covering over the cephalothorax. A transverse groove separates the fused head from the thoracic region. Laterally the carapace covers the gills. The **rostrum** is an anterior, pointed extension of the head. The eyes are located on either side of the rostrum.

The **abdomen** consists of several segments and is terminated by the **telson**, an extension of the last abdominal segment. To escape its enemies, the crayfish spreads the telson like a fan, rapidly drawing it forward under its body. This motion causes the crayfish to dart backwards into the muddy bottom, which is further clouded by the flipping movements. Each abdominal segment contains a dorsal plate, a ventral plate, and two lateral plates. Examine the appendages on the lower side of the abdomen. These are the **swimmerets** and are composed of a basal segment, the **protopodite**, to which is attached a pair of elongate appendages, the **endopodite** and **exopodite**. What is the function of these appendages?

_____

_____

_____

You will note that all other appendages are modifications of this three-part plan. Although the number of swimmerets is the same in the male and female, in the male those adjoining the thorax have been modified. The exopodites and endopodites are elongate and can be brought together to form a troughlike channel, used for the transfer of sperm from the male to the seminal receptacles of the female. In the female, eggs are attached to the abdominal swimmerets and are aerated by gentle, waving movements of the swimmerets.

Examine the five pairs of walking legs and count the number of segments in each. Note that there is no pincer on the fifth pair of legs. Locate the male genital pore, which opens into the base of the fifth pair of walking legs. The openings of the oviducts in the female are located at the base of each third walking leg. Carefully cut the membrane at the base of each leg and remove the right walking legs, starting with the most posterior one and working forward. Arrange these appendages in order on a piece of paper. How many of these legs have feathery gills attached to them?

_____

What is the advantage of the featherlike structures in these gills?

_____

_____

_____

The first pair of walking legs, the **chelipeds**, are much larger than the rest. Cut the exoskeleton from the dorsal surface of the claw and study the action of the muscles in opening and closing the pincer.

The appendages of the crayfish and other crustaceans are homologous organs; that is, they are all fundamentally similar in structure and arise from a similar embryonic rudiment. The basic structural plan is evident in the simple abdominal appendages consisting of a single protopodite and the exopodite and endopodite. The structural adaptations of the appendages in different regions are correlated with the special function they perform. When corresponding structures in different segments of the same animal exhibit homology, it is called **serial homology**. On the other hand, analogous organs are similar in function but not necessarily in structure. Cite examples of analogous organs.

_____

_____

_____

With a sharp scalpel and scissors, carefully dissect

DIVERSITY IN ANIMALS: ARTHROPODS

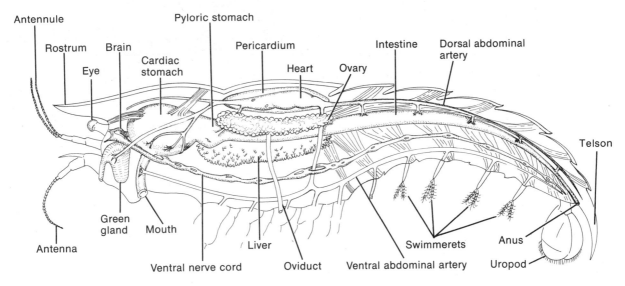

FIGURE 34-4
Internal anatomy of a female crayfish.

away the dorsal half of the exoskeleton to expose the internal organs (Fig. 34-4). In the middorsal region locate the diamond-shaped heart. Remove the heart and locate the gonads, found anterior to the heart in both sexes. In the female crayfish they appear as a pair of tubular structures bilaterally arranged in front of the heart and continuing behind it as a single mass. In the male the testes are highly coiled white tubes.

Locate the mouth. This leads to a short tubular esophagus, which in turn leads to the large saclike stomach. The stomach is made up of two parts: the large anterior **cardiac stomach**, in which food is stored, and the small, posterior **pyloric stomach**; it is in the pyloric stomach where most of the digestion occurs, as a result of the grinding action of the gastric mill and the enzymes that are secreted into the region by the digestive glands. The remaining part of the digestive system consists of the intestine, a small straight tube leading to the anus. In the intestine, digestion is completed and absorption takes place. Anterior to the stomach and just behind each antenna are the excretory structures, commonly called the **green glands**. Find their external openings at the bases of the antennae.

The nervous system is similar to that of annelids, but it exhibits fusion of several originally separated nervous elements, which is not evident in annelids. Remove the main organs from the thoracic and abdominal regions of your crayfish and locate the ganglion or brain in front of the esophagus. Expose the brain by careful dissection and find the nerves passing to the eyes, antennae, and antennules. The brain is connected to the ventral nerve cord by a pair of nerves that pass around the esophagus. Follow the ventral nerve cord as it passes through the thorax and abdomen and count the number of ganglia. Each ganglion gives off pairs of nerves to the appendages and internal organs of the segment in which it lies.

2. Physiological Behavior Among Crustaceans

a. In this experiment a syringe-type, constant-pressure respirometer will be used to measure the change in oxygen consumption of crayfish subjected to different salt concentrations (see Appendix G for instructions in the use of this respirometer). Place two small crayfish in each of the five respiration chambers, containing sufficient quantities of the following room temperature solutions: (1) 3.5% NaCl (equivalent to NaCl concentration of full-strength sea water), (2) 2.5% NaCl, (3) 1.75% NaCl, (4) pond water, and (5) distilled water. Which of these is the control for this experiment?

TABLE 34-2
Effect of salinity on respiration of crayfish.

Time (minutes)	Oxygen consumption (ml)				
	Distilled water	Pond water	1.75% NaCl	2.5% NaCl	3.5% NaCl
0					
15					
30					
45					
60					

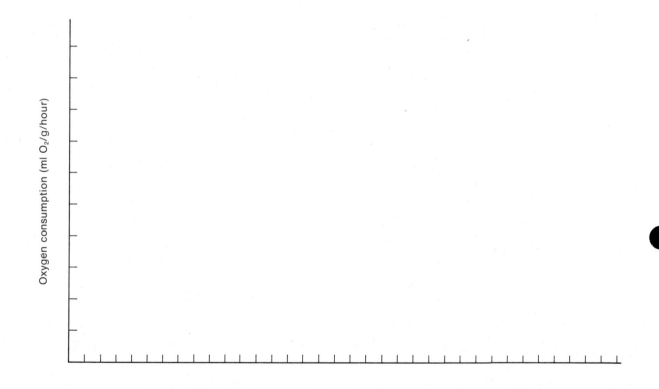

FIGURE 34-5
Effect of salinity on oxygen consumption of crayfish.

TABLE 34-3
Effect of temperature on heart rate in *Daphnia*.

Temperature (°C)	Heart rate (beats/minute)
0	
10	
20	
30	
40	

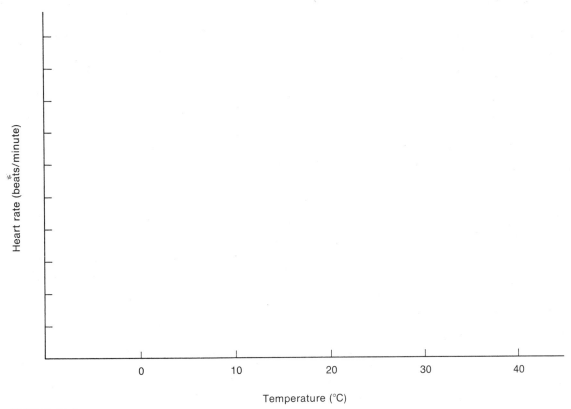

FIGURE 34-6
Effect of temperature on heart rate in *Daphnia*.

Measure the rate of oxygen consumption at 15-minute intervals during the next hour. Record your data in Table 34-2. At the conclusion of the experiment, weigh the crayfish and calculate the respiratory rate in milliliters of oxygen consumed per gram of body weight per hour. Plot the total volume against the total time for each experimental condition in Fig. 34-5. From these results, what can you conclude about oxygen consumption in relation to salinity?

_____

_____

_____

b. The common freshwater crustacean *Daphnia* will be used to determine the effect of temperature on heart rate. Put a small dab of vaseline on the bottom of a clean Syracuse dish, place a *Daphnia* in it, and add pond water. Determine the heart rate (beats/minute) and respiratory movements at room temperature. Now pipet out the pond water and add water that has been chilled with ice to 0°C. Allow 2–3 minutes for the animal to adapt to the change in temperature and then recheck the heart rate. Repeat this experiment by adding water at 10°, 20°, 30°, and 40°C. Make at least two determinations at each temperature and average and record your data in Table 34-3. Plot the average rate against temperature in Fig. 34-6. What conclusions can you make about the effect of temperature on the heart rate in *Daphnia*?

_____

_____

_____

REFERENCES

Barnes, R. D. 1963. *Invertebrate Zoology*. Saunders, Philadelphia.

Buchsbaum, R. 1948. *Animals Without Backbones*. University of Chicago Press, Chicago.

Meglitsch, P. A. 1967. *Invertebrate Zoology*. Oxford University Press, New York.

Rau, P. 1940. The Life History of the American Cockroach, *Periplaneta americana* Linn. (Orthop.: Blattidae). *Entomol. News* 51:121–124, 151–155, 186–189, 223–227, 273–278.

Russell-Hunter, W. D. 1968. *A Biology of Lower Invertebrates*. Macmillan, New York.

Wells, M. 1968. *Lower Animals*. McGraw-Hill, New York.

EXERCISE 35

Diversity in Animals: Molluscs and Echinoderms

A. MOLLUSCS

One of the largest and most familiar groups of invertebrates is the molluscs. This group of organisms is characterized by having a ventral muscular foot, a dorsal fleshy mass containing the viscera, and a protective shell secreted by an underlying mantle layer. The ventral foot and the fleshy parts of the body show extreme modifications among the members of this phylum. Because of their soft, fleshy bodies, the molluscs—more than any other invertebrates—are widely used as food by man. Included in this phylum are the snails, slugs, clams, chitons, squids, nautili, and octopi.

1. Clams (Class Pelecypoda)

Clams have a shell of two valves. They include the common freshwater clams and marine varieties of clams and oysters. Freshwater clams are abundantly distributed and live on the bottoms of lakes, rivers, ponds, and streams, where they feed on small plant and animal life (Fig. 35-1E).

Study the shell of the freshwater clam *Anodonta* or the marine clam *Venus* and note that it consists of two valves hinged together along the dorsal side. On the anterior part of each valve is a swollen region, the **umbo**. Concentric lines extend outward from the umbo and represent lines of growth.

The valves are held together by two large muscles. Cut these muscles by inserting a scalpel between the shells and cutting in the direction where the shells are held together. Open the valves and observe that they are lined by a membranous structure, the **mantle**. At the posterior margin the mantle is thickened and comes together to form two openings, the **siphons**. Water enters the clam through the ventral incurrent siphon, circulates through the mantle cavity and over the gills, and leaves through the dorsal excurrent siphon.

Remove one of the valves and its mantle, exposing the mantle cavity and the internal organs (Fig. 35-2). Observe the large muscular foot extending

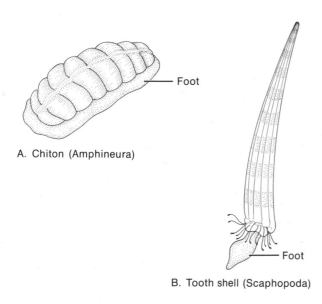
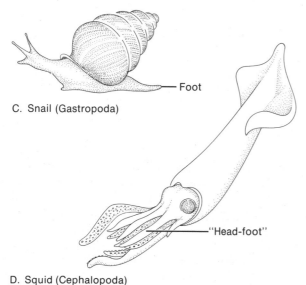
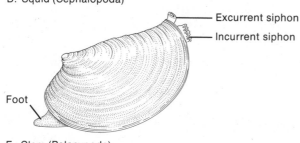

FIGURE 35-1
Classes of molluscs.

down from the visceral mass. Locate the two pair of **gills**, hanging down into the mantle cavity. Locate the two pair of flaplike **labial palps** that surround the mouth near the anterior muscle on the edge of the visceral mass. Dorsal to the gills is the pericardial sac that encloses the heart. Carefully cut open the pericardium and note that the heart consists of three chambers: two lateral auricles and one ventricle. Carefully cut away the gills and locate the **nephridium**, which appears as a dark-colored organ, lying near the base of the gills and just below the pericardial cavity. These organs remove organic wastes from the blood and pericardial fluid.

Most of the digestive system is located in the visceral mass. With a sharp scalpel or razor blade carefully cut the visceral mass into left and right halves. The mouth, located between the labial palps, leads by way of a short esophagus, into an expanded stomach, which is flanked on either side by two large digestive glands. Carefully remove these glands and observe that the stomach leads into an intestine that winds through the visceral mass and then passes through the pericardial sac as the rectum. The rectum empties into the excurrent siphon via the anus.

2. Chitons (Class Amphineura)

All members of this primitive class of molluscs are marine. They are characterized by having a shell composed of eight plates (Fig. 35-1A). A large, broad, flat muscular foot occupies the greater part of the ventral surface. Examine specimens of chitons on demonstration.

3. Tooth Shells (Class Scaphopoda)

The visceral mass of these marine molluscs is enclosed in a long, tubular, toothlike shell (Fig. 35-1B). The foot is typically cone-shaped and is used for burrowing in the sand. Examine the tooth shell on demonstration.

4. Snails and Slugs (Class Gastropoda)

This is by far the largest class of molluscs (Fig. 35-1C). In all forms the visceral mass is enclosed in a coiled shell during the early stages of development. In most gastropods the coiled shell is re-

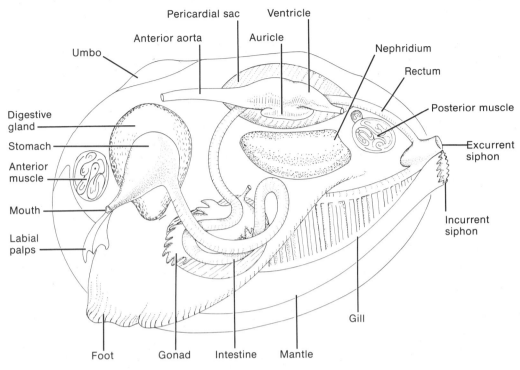

FIGURE 35-2
Diagrammatic representation of the internal organs of the clam.

tained in the adult, but in the slugs not only has the shell been lost but some species have become land dwellers. Examine the gastropods on demonstration and observe the modifications of the shell.

5. Squid and Octopi (Class Cephalopoda)

This class contains a diverse group of well-developed marine molluscs having an external or internal shell, a foot modified in the form of tentacles, and a head with prominent, highly developed eyes. In contrast to most other molluscs, the cephalopods are highly active, free-swimming animals.

Examine a squid. What morphological characteristics can you see on this organism that are adaptations to a predatory existence (Fig. 35-1D)?

_____
_____
_____
_____

Since the squid relies on its ability to swim rapidly for protection, it has no need of a cumbersome external shell. Consequently the shell is vestigial, represented by a horny plate buried in the visceral mass.

Next examine an octopus. How are the squid and octopus similar?

_____
_____
_____

How are they different?

_____
_____

B. ECHINODERMS

The echinoderms are exclusively marine organisms possessing spiny skin, radial symmetry (although the larva is bilaterally symmetrical), and true coeloms arising as mesodermal out-pouchings from

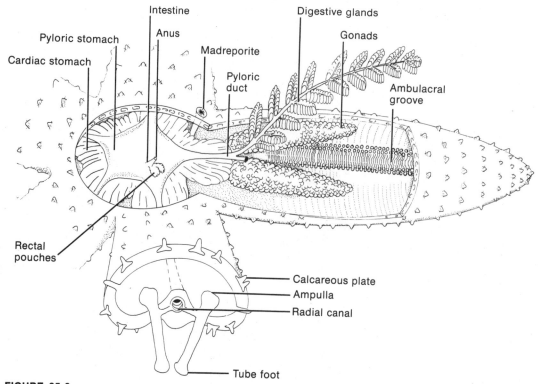

FIGURE 35-3
Diagrammatic representation of the internal organs of the starfish.

embryonic mesoderm. It is on the basis of this last characteristic, and because the bipinnaria larva of echinoderms more closely resembles the chordate larva, that the echinoderms are said to be related to the chordates.

The phylum is divided into five classes, the most common of which is represented by the starfish.

1. Starfish (Class Asteroidea)

The common starfish, *Asterias*, found along the Atlantic coast of North America, is a typical example of this class. Starfish crawl over the shallow bottom or live in tide pools among the rocks and sand of the seashore and coral reefs. They have been serious predators of oyster beds. At one time oyster fishermen caught starfish, cut them up and threw them back into the ocean. Then, it was discovered that each of the pieces could regenerate, growing into another starfish. Today, "sea mops" made of cloth are dragged over the oyster beds to entrap the starfish. They are then exposed to the sun to dry.

Examine a preserved specimen and note that the body is composed of a central disc from which radiate five arms (Fig. 35-3). Some of the specimens may have fewer arms, but this is usually because they have been broken off in handling. Some starfish have more than five arms; in rare cases specimens with up to 25 have been found. The oral, or ventral, surface of each arm contains grooves extending outward from the centrally located mouth. The opposite spiny surface is the aboral, or dorsal, surface. Locate the **madreporite**, a small porous, buttonlike structure, at one side of the aboral surface. The spines of the aboral surface are part of a small calcareous plate that lies buried beneath the integument. These plates form the exoskeleton. The grooves in the oral surface of each arm are called the **ambulacral grooves**. Located in these grooves are rows of small, fingerlike projections, **tube feet**, that act as organs of locomotion. At the tip of each arm locate the small light-sensitive eye spot. These light-sensitive tips are thrust upward during locomotion.

Cut off the tip of one of the arms and then make longitudinal cuts on one side and then on the other side of the arm to the central disc. Carefully remove the aboral surface to expose the internal organs. Note that most of the space in each arm is taken up by two highly branched **digestive glands**. Examine the glands with a hand lens or dissecting microscope and note the numerous lobes that secrete digestive enzymes. The two main ducts of these glands join at the base of the arm to form the pyloric duct, which enters the centrally located saclike stomach. A ventral mouth and a short esophagus lead directly into the stomach, which consists of a lobed, lower cardiac region and an upper pyloric region. During feeding the cardiac region is everted through the mouth. The food is partially digested and passed into the pyloric region, which empties into the anus located in the center of the aboral disc. Two small **rectal pouches** can usually be found near the anus. These are thought to function as temporary storage areas for waste products.

Cut the pyloric duct where it enters the stomach and remove the digestive glands. If the starfish was caught during the breeding season, the rays will be filled with the **gonads** (reproductive organs). At other times the gonads are normally very small.

The male and female gonads look alike. To determine the sex of the starfish, the gonads must be examined microscopically. To do this, remove a small piece of the reproductive organ and mince it in a drop of water on a slide. Add a cover slip and examine under the low and high power of the microscope. The testes of the male will have flagellated sperm. The ovaries of the female produce spherical eggs that are considerably larger than the sperm. Eggs and sperm are discharged into the water where fertilization occurs.

The **water vascular system** of the starfish consists of a series of seawater-filled ducts that function in locomotion and feeding. To study this system, carefully remove the reproductive organs and the remaining parts of the digestive system, the stomach and anus. Be careful not to damage the sieve plate.

Water enters this system through the **sieve plate**, which is connected to a circular **ring canal** by the **stone canal**. The water is then distributed to the radial canals that pass into each of the rays. Lining the ridge through which each radial canal passes is a double row of bulblike structures called **ampullae**. These are connected to the **tube feet**, which project from the **ambulacral groove** on the under surface of each ray. Water from the radial canal collects in the ampullae. Contraction of the ampullae causes the tube feet to elongate as water is forced into them. Expansion of the ampullae results in shortening of the tube feet. Thus, through the use of small suction discs at the end of each tube foot and the alternate expansion and contraction of the ampullae, the starfish is able to move.

2. Other Classes

Examine representatives of the other four classes of echinoderms (sea urchins, sand dollars, sea cucumbers, brittle stars and sea lilies), which you will find on demonstration in your laboratory.

REFERENCES

Buchsbaum, R. 1948. *Animals Without Backbones.* University of Chicago Press, Chicago.

Meglitsch, P. A. 1967. *Invertebrate Zoology.* Oxford University Press, New York.

Rounds, H. D. 1968. *Invertebrates.* Reinhold, New York.

Russell-Hunter, W. D. 1968. *A Biology of Lower Invertebrates.* Macmillan, New York.

Smith, J. E. 1945. The Nervous System of Starfish. *Biol. Rev.* **20**:29.

Wells, M. 1968. *Lower Animals.* McGraw-Hill, New York.

Wilbur, K. M., and C. M. Yonge, Eds. 1964. *The Physiology of Molluscs.* Academic Press, New York.

EXERCISE 36

Diversity in Animals: Chordates

Characteristically, the chordates possess a notochord (a longitudinal elastic rod that serves as the internal skeleton); **pharyngeal gill slits**; and a **dorsal (tubular) nerve cord**. Few of these basic structures persist into adulthood among the largest group of chordates, the vertebrates (those animals with a vertebral column).

Chordates are classified into three subphyla: the headless and unsegmented urochordates; the headless and segmented cephalochordates; and the head-possessing and segmented vertebrates.

A. TUNICATES (UROCHORDATA)

This subphylum is characterized by having an outer body layer, or **tunic**, composed of translucent, celluloselike material. Examine preserved specimens of *Molgula* or *Corella*. Note that these animals have two anterior openings: a mouth, through which water is taken in, and an **atrium**, through which the water is ejected after the food material has been removed in the ciliated pharynx (Fig. 36-1). In the adult the only chordate characteristic that persists is the pharyngeal gill slits.

Study slides of the larval stages of a tunicate and identify the notochord, nerve cord, and pharyngeal gill slits. The larval stage, which resembles a tadpole, swims about at first, but finally settles down, attaches to a rock, and undergoes metamorphosis into the adult form. During metamorphosis the notochord and nerve cord degenerate.

B. AMPHIOXUS (CEPHALOCHORDATA)

Amphioxus, the most commonly studied member of this group, is a small, fishlike animal found in shallow marine waters in many parts of the world. Although the animal can swim with lateral, undulating movements of its body, it spends most of the time buried in the sandy bottom with its anterior end projected.

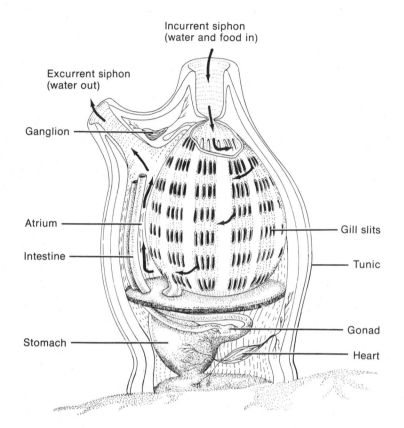

FIGURE 36-1
Tunicate.

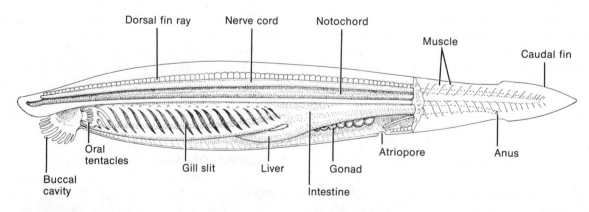

FIGURE 36-2
Longitudinal section of amphioxus.

DIVERSITY IN ANIMALS: CHORDATES

Examine a specimen of amphioxus. Although this organism is like a fish in appearance, it lacks a true head. Note the enlarged tail (**caudal fin**) used in locomotion. There is also a **dorsal fin** running the length of the body. Observe the V-shaped muscles on the sides of the animal. At the anterior end of the body is the funnel-shaped **buccal cavity** surrounded by a circle of oral tentacles. Approximately two-thirds of the way back from the anterior end, on the ventral side, locate the **atriopore**, through which water leaves after having been pumped through the pharynx.

Study the internal anatomy of amphioxus by examining preserved or stained specimens under the dissecting microscope (Fig. 36-2). Identify the nerve cord, which extends nearly the entire length of the animal's body. Above the nerve cord are short rods of connective tissue, called **fin rays**, that strengthen the dorsal fin. Locate the **notochord**, a cartilagelike rod that lies ventral to the nerve cord and extends the length of the body. The notochord is a longitudinal elastic rod of cells serving as an internal skeleton. Ventral to the notochord is the digestive tract. The mouth, located in the buccal cavity, leads into the relatively large pharynx. Observe the gill slits in the wall of the pharynx. The pharynx is lined with cilia that beat inward to produce a steady current of water that enters the mouth, passes over the gill slits (leaving behind suspended food particles), and is then eliminated through the atriopore. Posteriorly, the pharynx joins the intestine. Close to the point where the pharynx joins the intestine is a ventral outgrowth, the liver, which secretes digestive enzymes into the intestine.

Examine a cross section of amphioxus taken from the region of the pharynx and locate the structures outlined in Fig. 36-3. On either side of the pharynx locate the gonads, paired bodies containing the sex cells. Where would you expect fertilization to occur in these animals?

_____

The sex of your specimen can be determined by examining the cells of the gonads. The testes are made up of a large number of small, dense cells. The ovaries contain fewer, larger cells with vesicular nuclei. What is the sex of your specimen?

_____
_____

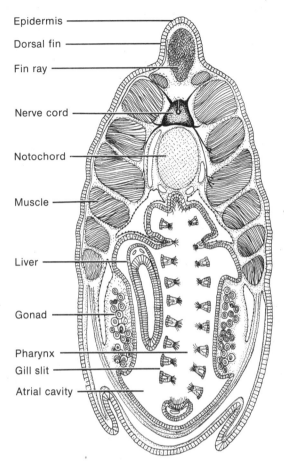

FIGURE 36-3
Cross section of amphioxus.

C. VERTEBRATES

This is the largest group of chordates and consists of animals in which the notochord is replaced by a **vertebral column** as the primary supporting structure. Examine representatives of each of the seven classes that make up this subphylum. Identify those external characteristics which serve to distinguish each class. In your study of the various classes, note the evolutionary changes that have occurred in the transition from invertebrates to vertebrates to the land-dwelling adaptations of the higher vertebrates.

1. Lampreys and Hagfish (Class Agnatha)

These eellike vertebrates lack the paired fins, scales, and jaws characteristic of sharks and bony fish but do possess seven or more paired gill slits.

Note the round mouth surrounded by horny teeth that are used to attach to fish on which these animals feed. The teeth are specialized structures of epidermal origin and are not homologous with teeth of higher vertebrates, which are of mesodermal origin. Examine the ammocoete larva of the lamprey. Note the resemblance of this larva to the adult amphioxus previously studied. What is the significance of this similarity?

_____

_____

2. Sharks, Skates, and Rays (Class Chondrichthyes)

These vertebrates possess skeletons that remain cartilaginous throughout their life. They also have five to seven pairs of gill slits, paired fins, and jaws containing teeth that are modified scales similar to those covering this animal. Examine specimens of the dogfish shark, a small species of shark that abounds in the shallow coastal waters of the United States. The appendages are of two types—median fins and paired lateral fins. The paired fins are of two types: two large pectoral fins in the anterior region and two smaller pelvic fins in the posterior region. In the male the pelvic fins possess **claspers**, which are used in copulation. Locate the lidless eyes, nasal openings of the olfactory organs, spiracles, and cloacal opening.

3. Bony Fish (Class Osteichthyes)

Bony fishes are characterized by scales, jaws, a bony skeleton, and an **operculum** covering the four pairs of gills. Examine specimens of the common freshwater perch. Compare the arrangements of the fins with those of the dogfish shark. Remove a scale and examine under low power. The scales grow throughout life, increasing in size with the fish. There is no molt of the body covering, although occasional scales may be lost and replaced. On many species (salmon, trout, bass, perch) growth of the scale results in a series of fine concentric ridges. After growth ceases during the winter months, the first ridges of the next season's growth form parallel to the scale margin, making a definite "winter line." Thus, age determination is possible in many fish species.

4. Frogs, Toads, and Salamanders (Class Amphibia)

These four-legged vertebrates are characterized by their larval forms; the larvae breathe by means of gills, whereas the adults lose their gills and become lung-breathers. What adaptations are demonstrated by this group in their transition from an aquatic to a terrestrial environment?

_____

_____

_____

_____

5. Turtles, Snakes, and Lizards (Class Reptilia)

The members of this class are covered by scales or bony plates, breathe by means of lungs, spend most of their lives on land, and lay eggs. How has the evolution of the land egg contributed to the ability of these organisms to occupy diverse habitats on land?

_____

_____

_____

Study representatives of the various groups of reptiles on demonstration, noting their great diversification in form.

6. Birds (Class Aves)

In these warm-blooded vertebrates the forelimbs have been modified as wings and the legs are covered with horny scales. If you have made a previous study of chick development, what chordate characteristics were present in the 72-hour chick?

_____

_____

DIVERSITY IN ANIMALS: CHORDATES

● What chordate characteristics persist in the adult?

_____

_____

7. Mammals (Class Mammalia)

The warm-blooded animals that constitute this class possess bodies covered with hair and mammary glands that secrete milk for the nourishment of their young. The mammals have an interesting evolution with respect to the development and care of the young. Primitive mammals, such as the duck-bill platypus, resemble the reptiles in that they are egg-layers. More advanced mammals, such as the marsupials (opossum and kangaroo), have limited uterine development of the young; they are born in a very immature condition and transferred to a pouch where they are suckled until they are more mature. In the most advanced mammals the young are retained in the uterus until they are in an advanced stage. What is the advantage of the retention of the developing embryo in the uterus for such a long period of time?

_____

_____

_____

_____

D. MORPHOLOGICAL ADAPTATIONS OF VERTEBRATES TO THE ENVIRONMENT

The present study will be concerned with an examination of the morphological adaptations that enable vertebrates to survive in a given habitat. Special attention will be given to structural changes that have evolved with respect to the food habits of these animals.

The specialized structural characteristics of animals are related to their physical and biotic environments. For an adaptation to be favored in the processes of natural selection, it must be of some benefit to the animal. The structural characteristics of an animal are, therefore, often indices to its mode of life—its ecological niche.

Most of the specialized structural characteristics of mammals are related either directly or indirectly to their high rate of living. Striking skeletal specializations are found particularly in the limbs as they are modified for specialized locomotion and in the jaw region, where modifications (particularly in the dentition) reflect the nature of the food habits of these animals.

1. Adaptation to Walking and Running

The reptilian (crocodile), feline (cat), and human skeletons displayed in the laboratory serve to show the evolutionary differences acquired by the mammals in evolving from the reptiles. The pectoral and pelvic girdles of the crocodile are so constructed that the fore and hind limbs project laterally, thus giving the animal its characteristic crawling position (Fig. 36-4A). In mammals, on the other hand, there has been a change in position of the limbs from a laterally to ventrally projecting position, thus raising the animal well off the ground (Fig. 36-4B). Mammalian appendages are not only rotated but also lengthened in proportion to the body, as can be seen by comparison of the crocodile and cat skeletons. The pectoral girdle of mammals consists chiefly of the scapula, which has become fused to form a large, flat bone for the support of the strong pectoral muscles. In the mammals the pelvic girdle has been replaced by a single larger and stronger pelvic bone. Note that the mammalian vertebral column has been modified to give less movement than is possible in the reptiles. Mammalian vertebrae allow little lateral movement and the number of cervical vertebrae has been reduced from 9 in the reptile to 7 in the mammal.

Few events in evolution are more remarkable than the adaptation of the vertebrate skeleton to the erect posture of the human race. The changes that occurred are largely limited to the position of the skull, the form of the chest, the curvature of the spine, the shape of the pelvis, and the structure of the feet.

From the point of view of adaptation to erect posture, the most important change is the effective balancing of the skull atop the backbone, instead of being slung in front of it. The double curvature of the spine—in place of the single arching curvature of the quadruped back—keeps the head and shoulders above the hips; they are thus balanced, rather than drooping in front. In all four-footed forms the

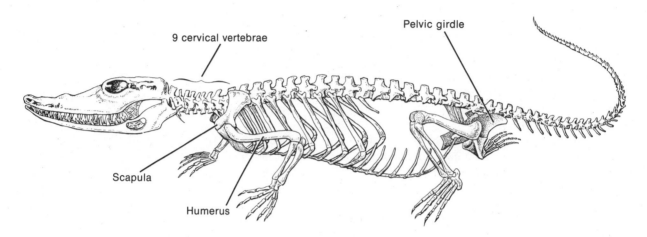

A. Reptilian (crocodile) skeleton

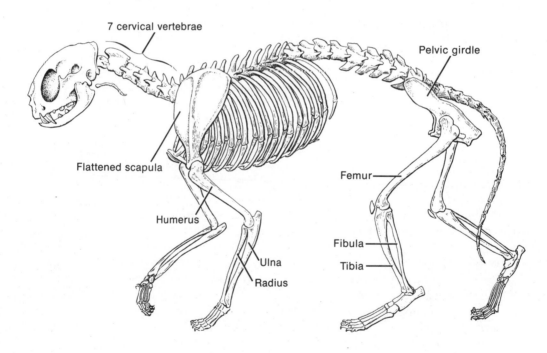

B. Mammalian (cat) skeleton

FIGURE 36-4
Typical (A) reptilian and (B) mammalian skeletons.

DIVERSITY IN ANIMALS: CHORDATES

hip bones are long and narrow; but in humans the hip bones have become short and broad so as to adequately support the internal viscera.

Despite the overwhelming importance of hands in the development of human culture, our hands have undergone surprisingly little modification from the primitive paws of ancient amphibians. The human foot—with its unique arch—has been far more radically altered by evolution. Whereas all other vertebrates stand on either the toes or the flattened sole, man stands with his heels, toes, and outer border of the foot pressed to the ground. The arch formed by the rest of the foot provides a means of distributing the weight over a triangular area much larger than the base of the leg bones. Examine the various mammalian skeletons on display in the laboratory and note their adaptations to running and walking.

2. Adaptation to Flying

Adaptations for flight are apparent in the skeleton of birds. The hollow, thin bones are light in weight. The skeleton of the frigate bird, with a wingspan of 7 feet, weighs only 4 ounces. This is less than the weight of its feathers. This is an extreme example, but the skeleton of all birds weighs less, relative to their body weight, than the skeleton of mammals. The bones are very strong because most of the bony substance is located at the periphery of the bone, where it gives better structural support. Indeed, bird bone may be compared to a metal tube, which is more resistant to certain types of stress than a metal rod of equal weight. Many bird bones are further strengthened by internal bony struts, arranged similarly to the struts inside the wing of an airplane.

The necks of birds are long with the cervical vertebrae articulated in such a way that the head and neck are highly mobile. Since the bird's bill is used for feeding, nest building, and defense, freedom of head movement is important. In contrast the trunk region is shortened, with the trunk vertebrae firmly united to form a strong fulcrum for the action of the wings and a strong point of attachment for the pelvic girdle and hindlimbs, which bear the entire weight of the body when the bird is on the ground. The sternum, with its large midventral keel, increases the area available for the attachment of the flight muscles. The bones of the wing are homologous to those of the pectoral appendage of the frog and other tetrapods. The fifth toe of all birds has been lost, as well as the fourth toe in some species. In most species the first toe is turned back to serve as a prop that increases the grasping action of the foot when the bird perches. The various fusions of the limb bones reduce the chance of dislocation and injury, for birds' legs must act as shock absorbers when they land.

3. Adaptation to Burrowing

The mole is a good example of an animal adapted to a burrowing existence. Examine a mole and note that it is more or less spindle-shaped with a long, pointed snout. What is the significance of the reduction in the eyes and ears of this animal?

_____

_____

_____

Note the size and position of the limbs. The forefeet are laterally placed for maximum efficiency in digging. Internally, the pelvic region is modified by a reduction and fusion of the bones, since the hind limbs function merely to push the animal through the earth. The hair of the mole is short, dense, and capable of lying both forward and backward, to offer the least resistance to the animal's passage through a burrow.

4. Adaptation to an Aquatic Existence

Modifications for living in the water may be observed in many mammals from those that spend their entire existence in the water to those making occasional sojourns into the water. Between these two extremes is a host of species that illustrate diverse adaptive development for a partial aquatic existence.

The muskrat is an example of an animal adapted for aquatic living. Swimming is accomplished by means of a laterally compressed tail and a stiff fringe of hairs on the outer edge of the broadly webbed feet. Suggest a function for the extremely dense underfur.

_____

_____

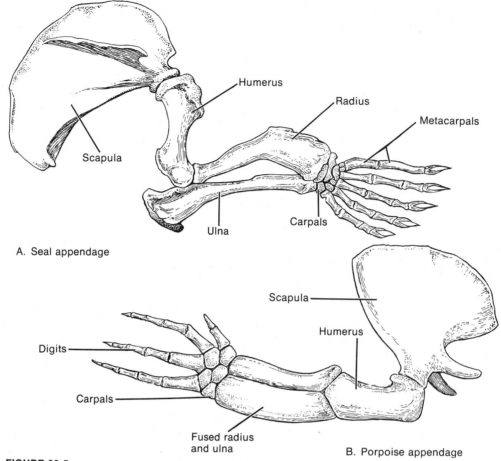

FIGURE 36-5
(A) Seal and (B) porpoise appendages.

Note the short powerful limbs, which provide maximum efficiency for swimming.

In the seal and porpoise the anterior appendage is especially well adapted for swimming. The humerus, radius, and ulna in the seal flipper are short, heavy bones that are specially adapted for power (Fig. 36-5A). The digits are covered with skin to form paddlelike fins. In the porpoise the humerus is short, broad, and fused with the radius and ulna (also fused), thus providing a powerful swimming structure (Fig. 36-5B).

5. Adaptation for Flight

Bats are the only mammals that are adapted for true flight (Fig. 36-6). They possess modified forelimbs with greatly elongated fingers that are joined together by a membrane that extends to the sides of the body and legs as well as between the legs and tail. The shoulder girdle is better developed than the pelvis; the sternum is usually keeled.

6. Adaptation for Climbing

A great many small mammals have become adapted to climbing trees for food and to escape predators that would readily overcome them had they not evolved climbing habits. All of these mammals have handlike feet. The toes, which are long and flexible, are supplemented by sharp, strong claws adapted for clutching the rough bark of trees.

Examine preserved specimens and/or skeletons of the mammals listed in Table 36-1. Enter the information asked for in the table and then decide the habitat in which each animal might be found.

7. Adaptation for Feeding

Adaptations of the mammalian skull are not necessarily correlated with adaptations in the feet and limbs, but have evolved separately; these changes reflect the varied food habits of mammals. In general there has been a reduction in the number

DIVERSITY IN ANIMALS: CHORDATES

TABLE 36-1
Morphological adaptations among vertebrates.

Species	Body shape	Size and shape of limbs	Position of limbs	Special characteristics of limbs	Nature of sense organs	Ecological habitat
Gopher						
Beaver						
Racoon						
Shrew						
Opossum						
Woodchuck						
Rabbit						
Squirrel						
Porcupine						
Ground squirrel						
Muskrat						
Bat						

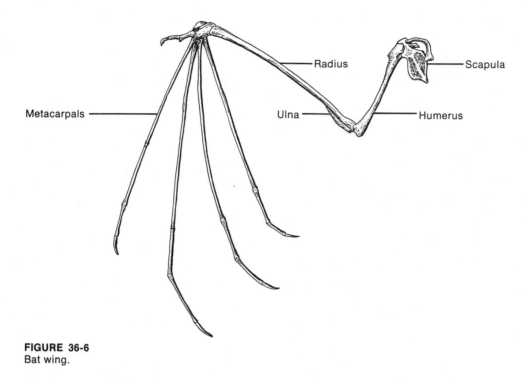

FIGURE 36-6
Bat wing.

of bones in the skull and an expansion dorsally and laterally of the walls of the cranium. The lower jaw is represented by a single fused bone, the **dentary**. In this study particular attention will be given to modifications in dentition—the **zygomatic arch**, the **dentary**, and the **diastema**—as related to the food habits of various groups of mammals.

a. The teeth of **insectivores** (insect-eating mammals) are low-crowned, with sharply pointed cusps (Fig. 36-7A). The canines, while always present, are simple and shaped like the nearby incisors. The skull has a small brain case much reduced between the orbits, while the zygomatic arch is greatly reduced or lacking. The dental formula in the mole, a common insectivore, is:

$$\frac{\text{Upper jaw} | \text{3 incisors, 1 canine, 4 premolars, 3 molars}}{\text{Lower jaw} | \text{3 incisors, 1 canine, 4 premolars, 3 molars}} \times 2 = 44$$

b. In the **carnivores** or flesh eaters, the dentition is specially adapted to the diet. Examine the skull of a cat. Note that the incisors and the well-developed canines are sharp-pointed for seizing, while the premolars and molars are flat and sharp-edged for cutting (Fig. 36-7B). One of the molars is larger than the rest and takes the form of a compressed blade having two or three roots. The lower jaw is hinged to the skull by a long straight half-cylindrical joint, which prevents any considerable lateral movements of the jaws; it thus requires the teeth to shut past one another in close proximity, like the blades of a pair of shears. What is the function of the large, prominent zygomatic arch in the carnivores?

_____

_____

_____

The dental formula of a typical carnivore, such as the cat, is:

$$\frac{\text{Upper jaw} | \text{3 incisors, 1 canine, 4 premolars, 2 molars}}{\text{Lower jaw} | \text{3 incisors, 1 canine, 4 premolars, 2 molars}} \times 2 = 40$$

c. Examine the skull of a beaver (a **rodent**). The most prominent characteristics of these gnawing mammals are the chisellike incisors, the absence of canines, and the wide diastema separating the incisors from the cheek teeth (Fig. 36-7C). The outer side of the incisors is of hard enamel; the inner surface is of softer cement, which wears

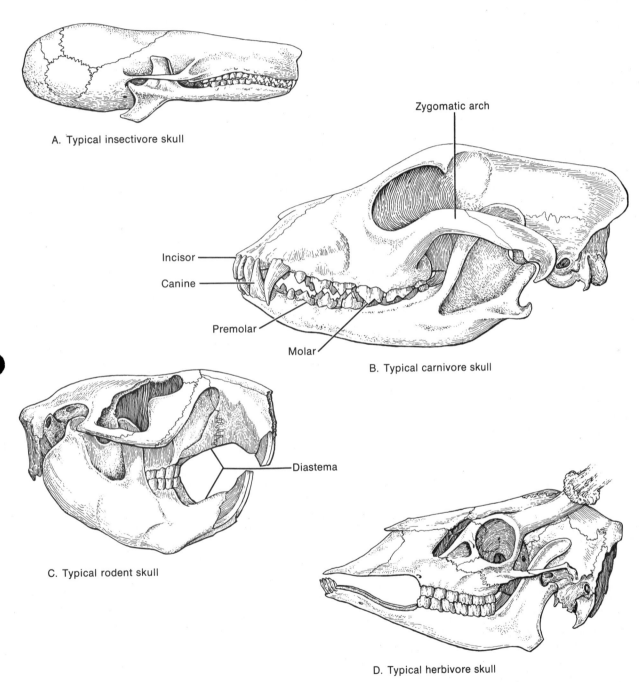

FIGURE 36-7
Types of mammalian skulls.

TABLE 36-2
Morphological characteristics of unknown skulls.

Skull number	Size and shape of cranium	Size of zygomatic arch	Dental formula	Food habit
1				
2				
3				
4				
5				
6				
7				
8				

away rapidly, thus maintaining a chisellike form for gnawing stems, seeds, or other foods. Gnawing teeth keep growing throughout life, so that the worn edges are constantly replaced. The zygomatic arch is also well developed in the rodents. In these mammals the lower jaw glides in a slot on either side of the skull in such a way that the jaw is moved backward and forward instead of from side to side. The dental formula of a typical rodent, such as the beaver, is:

$$\frac{\text{Upper jaw}}{\text{Lower jaw}} \begin{vmatrix} 1 \text{ incisor, 0 canines, 0 premolars, 5 molars} \\ 1 \text{ incisor, 0 canines, 0 premolars, 5 molars} \end{vmatrix} \times 2 = 24$$

d. The **herbivores** include such animals as the horse, which grazes on grasses and herbs, and the deer, which browses on leaves and twigs of shrubs and trees. Examine the skull of a deer or horse. Note that they lack upper incisors and canines and have a large diastema. The incisors are flat and suited to biting off herbage (Fig. 36-7D). The cheek teeth are broad-crowned, with folds of hard enamel imbedded in a cement that wears faster and keeps the tooth surface properly roughened for grinding. Note that the bony attachment of the lower jaw on each side is so rounded that the jaw can readily swing from side to side in a grinding motion. Why is this type of dentition essential to the herbivore?

_____

_____

Since these animals are vegetarians, the zygomatic arch is characteristically reduced because of the weaker jaw muscles. The dental formula of a typical herbivore is:

$$\frac{\text{Upper jaw}}{\text{Lower jaw}} \begin{vmatrix} 0 \text{ incisors, 0 canines, 3 premolars, 3 molars} \\ 4 \text{ incisors, 0 canines, 3 premolars, 3 molars} \end{vmatrix} \times 2 = 32$$

Examine the unknown skulls available in the laboratory and enter their characteristics in Table 36-2; the dental formula can be abbreviated. On the basis of these characteristics, determine the food habits of these specimens.

REFERENCES

Hamilton, T. H. 1967. *Process and Pattern in Evolution.* Macmillan, New York.

McCoy, C. J., Jr. 1968. *Vertebrates.* Reinhold, New York.

Wallace, B., and A. M. Srb. 1970. *Adaptation.* 3d ed. Prentice-Hall, Englewood Cliffs, New Jersey.

Appendices

APPENDIX A

Use of the Bausch & Lomb Spectronic 20 Colorimeter

The Bausch & Lomb Spectronic 20 colorimeter is an extremely versatile instrument useful for spectrophotometric or colorimetric determinations of solutions.

The optical system is shown in Fig. A-1. White light is focused by a lens (1) onto an entrance slit (2), where it is collected by a second lens (3) and refocused on the exit slit (4) after being reflected and dispersed by a diffraction grating (5). Rotation of this grating by a cam (6) enables one to select various wavelengths of light in a range from 375 nm to 625 nm. The addition of a filter can extend the usable wavelength to 950 nm. After the light passes the exit slit, it goes through the sample being measured (7) and is picked up by a phototube (8). The amount of light absorbed by the sample may then be read directly on a dial.

A. COLORIMETRY

Directions for colorimetric use are as follows:

1. Rotate the wavelength control (1) shown in Fig. A-2, until the desired wavelength is shown on the wavelength scale (2). The wavelength for a given substance can be found by referring to the literature or by determining it experimentally.

2. Turn the instrument on by rotating the "0" control (3) in a clockwise direction. Allow 5 minutes for the instrument to warm up.

3. Adjust the "0" control with the sample holder cover closed (5) until the meter needle reads 0 on the Percent Transmittance scale (4).

4. A colorimeter tube containing water or other solvent is placed in the sample holder and the cover is closed.

5. Rotate the light control (6) so that the meter reads 100 on the Percent Transmittance scale. This control regulates the amount of light passing through the second slit to the phototube.

6. The unknown samples may then be placed in the tube holder and the percentage of transmittance read on the dial. The needle should always return to zero when the tube is removed. Check the "0" and 100% transmittance occasionally with the solvent tube to make certain the unit is calibrated.

(Note: Always check the wavelength scale to be certain that the desired wavelength is being used.)

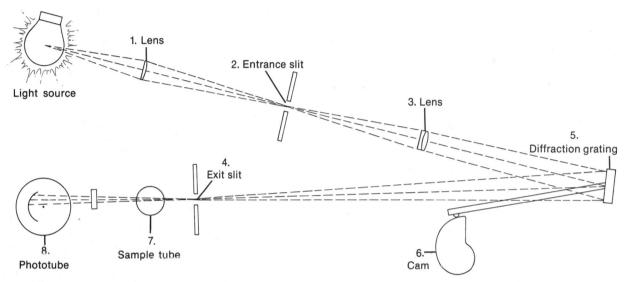

FIGURE A-1
Optical system of the Spectronic 20.

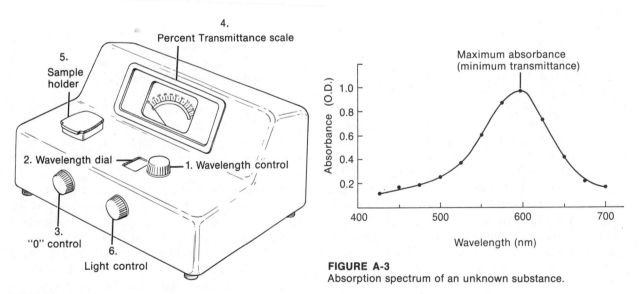

FIGURE A-2
Controls on the Spectronic 20.

FIGURE A-3
Absorption spectrum of an unknown substance.

The colorimetric measurements made with this apparatus employ standard matched tubes. These are selected so that variation in light transmission of the tubes due to slight variations in diameter and wall thickness are minimal. You will be issued a set of such matched tubes. *They are to be used only for colorimetry.* The matched tubes must be handled carefully to avoid etching or scratching of the surfaces exposed to the light beam. Obviously, the tubes will no longer be "matched" if scratched or etched, since such defects will cause the absorption and scattering of light.

B. SPECTROPHOTOMETRY

The method of operation for spectrophotometry is essentially the same as for colorimetry. The main difference is that the wavelength is reset for each reading and thus the blank, or solvent, control must be readjusted at each new wavelength setting.

This procedure can be used when no information is available to determine the proper operating wavelength for an unknown substance. To do this plot an absorption curve [absorbance (or optical density) versus different wavelengths] of the unknown substance (Fig. A-3). An operating wavelength may then be chosen according to the following:

1. Choose the wavelength at which the substance maximally absorbs the light (the minimum transmittance), since the greatest sensitivity will be obtained at this wavelength.

2. Do not choose wavelengths on the slope since a small error in wavelength will cause a large error in reading.

REFERENCES

Newman, D. W. 1964. *Instrumental Methods of Experimental Biology.* Macmillan, New York.

Umbreit, W. W., R. H. Burris, and J. F. Stauffer. 1964. *Manometric Methods.* 4th ed. Burgess, Minneapolis.

Van Norman, R. W. 1971. *Experimental Biology.* 2d ed. Prentice-Hall, Englewood Cliffs, New Jersey.

APPENDIX B

Colorimetry

If white light is passed through a solution containing a colored compound, certain wavelengths of light are selectively absorbed. The resultant intensity of light is due to the transmitted light. A number of biochemical methods for quantifying substances employ the measurement of color intensity by means of suitable photoelectric devices for measuring the intensity of the light transmitted through a colored solution. These methods require that the substance being determined is either colored itself or can be converted into a colored compound. With modern photoelectric colorimeters such measurements are made rapidly, are reasonably accurate, and, where speed is desirable and extremely high accuracy not essential, are supplanting the more tedious gravimetric and volumetric measurements.

The transmittance of light through a solution containing light-absorbing material depends upon (1) the nature of the substance, (2) the wavelength of the light, and (3) the amount of light-absorbing material in the path of the light. The third condition depends upon the concentration of the substance and the length of the light path through the solution.

The **Lambert-Beer law** states that the relationship between the concentration of a colored substance and the amount of light transmitted through the solution can be expressed

$$\log_{10} \frac{I_0}{I} = Ecl$$

where I_0 = intensity of the incident light,
I = intensity of the transmitted light,
l = length of the light path through the solution in centimeters,
E = molar extinction coefficient, a constant which is characteristic of each colored substance,
c = concentration of solution in moles.

$\log_{10} (I_0/I)$ is termed the **optical density (O.D.)** or the **absorbance** (A). It is the most immediately useful term, since it is seen from the equation that the absorbance is directly proportional to

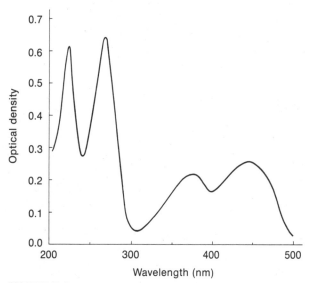

FIGURE B-1
Absorption spectrum of riboflavin. (From *Experimental Biochemistry* edited by John M. Clark, Jr. W. H. Freeman and Company. Copyright © 1964.)

the concentration of the colored substance. This relationship holds most accurately when monochromatic light is used and when the wavelength of the light beam coincides with the wavelength of maximum light absorption by the compound. The performance of a colorimetric analysis, assuming that the substance in question can be converted quantitatively to a colored derivative, requires that we know, first, what wavelengths of light are most strongly absorbed by the colored solution and, second, the quantitative relationship between color intensity and concentration of the substance (that is, the value of E).

The first requirement can be met by determining the absorption spectrum of the substance in a spectrophotometer. This is an optical device that determines the absorbance at different wavelengths of light, which are selected by use of a monochromator. An **absorption spectrum** is a plot of absorbance versus wavelength. The wavelength at which absorbance is greatest is the wavelength at which colorimetric measurement is most satisfactory. For example, the absorption spectrum of riboflavin shown in Fig. B-1 shows that it has a maximum absorption at a wavelength of 450 nm in the visible range. The ultraviolet absorptions having maxima at 220 and 270 nm are not visually recorded, but can be recorded with special instruments.

The second requirement deals with establishing the quantitative relationship between absorbance and concentration of the colored substance—that is, evaluating the value of the constant E in the basic expression $\log_{10}(I_0/I) = Ecl$. This is done by preparing a series of standards of the substance analyzed in graded known concentrations ("color standards"). Since absorbance is *directly* proportional to concentration, a plot of absorbance versus concentration of the standard yields a straight line, the slope of which represents the constant E, which varies with the compound, wavelength, light path, and other characteristics of the measuring device. Such a plot is called a **calibration curve** or **standard curve** (see Fig. B-2 for standard curve of riboflavin). Unknown solutions can be analyzed by determining the absorbance of the solution and reading off the concentration from such a plot. It is customary to set up a limited series of standards with each batch of determinations, since the slope of the plot will vary slightly, depending on the conditions of analysis and other factors. An unknown (X) plotted in Fig. B-2 has a concentration of 0.4 micromoles. This is determined by dropping a line from its absorbance on the curve to the base line.

The principle of the common photoelectric colorimeter such as the Bausch & Lomb Spectronic 20 that will be used in your laboratory is simply diagrammed in Fig. B-3. (If this is unavailable, the Klett-Summerson colorimeter may be effectively substituted.)

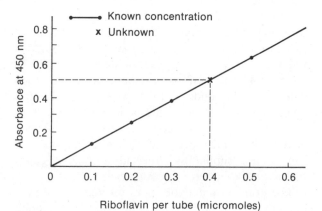

FIGURE B-2
Standard curve for riboflavin. (Adapted from *Experimental Biochemistry* edited by John M. Clark, Jr. W. H. Freeman and Company. Copyright © 1964.)

COLORIMETRY

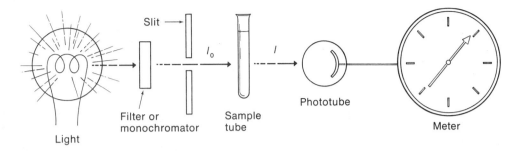

FIGURE B-3
Typical photoelectric colorimeter. (From *Experimental Biochemistry* edited by John M. Clark, Jr. W. H. Freeman and Company. Copyright © 1964.).

The light source of such a colorimeter must be quite constant in intensity, so as not to vary between readings. This is ensured by voltage regulation devices. In some colorimeters the chosen wavelengths of light are provided by a monochromator and an optical slit that can be adjusted. In others, light filters of colored glasses may be placed in the light path. The former arrangement is more satisfactory and is used in the colorimeters in your laboratory. The optical cell, for very precise work, has plane parallel windows and the length of the light path is accurately known. In routine work, however, test tubes are commonly used. These must be matched to give the same "blank" light transmission. The current developed in the photoelectric cell activates a galvanometer needle. The galvanometer scale of the colorimeter is usually graduated in one way or another suitable for easy calculation of data. The Bausch & Lomb Spectronic 20 that you will use is graduated in optical density (O.D.) and in percentage of transmittance (%T).

Colorimetric methods consist of a direct or indirect comparison of the absorption of light by a standard solution containing the same colored solute in known concentration. Therefore, at the same wavelength and depth of solution, the log (I_0/I) becomes the same for both solutions and the relation between their respective densities, D_1 and D_2, and concentrations, C_1 and C_2, is

$$\frac{C_1}{C_2} = \frac{D_1}{D_2}$$

If one of these two solutions is a standard of known concentration and the other is of unknown concentration, then, if their respective densities are measured, the concentration of the unknown is

$$\text{concentration of unknown} = \frac{\text{O.D. of unknown}}{\text{O.D. of standard}} \times \text{concentration of standard}$$

REFERENCES

Bauman, R. P. 1962. *Absorption Spectroscopy.* Wiley, New York.

Clark, G. L. 1960. *The Encyclopedia of Spectroscopy.* Reinhold, New York.

Newman, D. W., Ed. 1964. *Instrumental Methods of Experimental Biology.* Macmillan, New York.

Van Norman, R. W. 1971. *Experimental Biology.* 2d ed. Prentice-Hall, Englewood Cliffs, New Jersey.

APPENDIX C

Chromatography

Chromatographic techniques are among the most useful methods available for the resolution of complex mixtures. Chromatography is an analytical technique for the separation of solutes based on their selective adsorption as they are passed over adsorbants such as charcoal, starch, cellulose powders, ion-exchange resins, and filter paper, either as strips or in compressed piles. The most widely used chromatographic techniques are column chromatography on ion-exchange resins, paper partition chromatography on filter paper, and thin-layer methods using an adsorbant bound to some supporting material such as glass. These techniques are usually employed in the isolation of proteins, enzymes, lipids, hormones, plant growth substances, pigments, and other naturally occurring organic materials.

A. PAPER CHROMATOGRAPHY

Paper chromatography has revolutionized the art of detecting and identifying small amounts of organic and inorganic substances. It permits the separation of mixtures on a very small scale that no other simple method affords. The technique has found widespread application in many areas of biology. It has been used to separate and identify such substances as amino acids, carbohydrates, proteins and peptides, sterols and steroid hormones, fatty acids, antibiotics, and many other naturally occurring substances.

A small spot of the substance to be chromatographed is placed near one end of a length of filter paper. This end of the paper is then immersed in a solvent system that is usually composed of two or more miscible substances. In **descending paper chromatography**, the solvent is contained in a trough near the top of the chromatographic chamber and is allowed to irrigate the paper by a downward flow via capillary action and gravity (Fig. C-1). In **ascending chromatography** the solvent is placed at the bottom of the chamber and is allowed to rise upward through the length of the paper by capillary action. Although ascending chromatography is often preferred because of the simplicity of the setup, the flow of solvent is faster in the descending technique, and, if necessary, the solvent may be allowed to run off the end of the descending paper to elute the solute or enhance

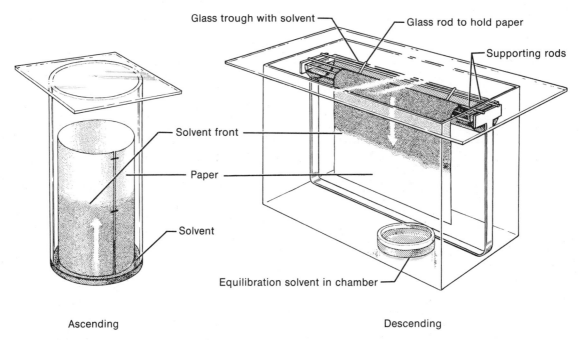

FIGURE C-1
Apparatus for paper chromatography. (From *Experimental Biochemistry* edited by John M. Clark, Jr. W. H. Freeman and Company. Copyright © 1964.)

resolution. As the solvent travels the length of the paper, the substance chromatographed partitions itself between the paper and the moving solvent in such a way that it moves at a characteristic rate in the solvent system used. After drying, the position of the material is identified by spraying with a reagent that combines with the substance to give a color.

The distance travelled by each compound from the origin, or base line, relative to the solvent front is defined as the R_f.

$$R_f = \frac{\text{distance from base line travelled by compound}}{\text{distance from base line travelled by solvent}}$$

The R_f value is a characteristic of the particular compound measured under specified conditions: solvent system, temperature, manner of development (whether ascending or descending), type of paper, and grain or "machine direction" of the paper. Since it is affected by so many variables, the R_f of a substance in a known system is only a rough indication of identity. Therefore, it is common practice to chromatograph a sample of a known material—the material presumed to be identical to the unknown—along with the unknown. It should also be noted that two different compounds may have the same R_f in one solvent system. Therefore, results as to the identity of a component obtained by paper chromatography (even though confirmed in several different solvent systems) should be verified by other means.

For increased resolution, the technique of *two-dimensional paper chromatography* is used (Fig. C-2). The material to be chromatographed is applied near one corner of the paper. The sheets are developed in one direction as above, allowed to dry, and then developed with another solvent system in a direction at right angles to the first development.

If the materials being separated on paper chromatograms absorb ultraviolet, or if they fluoresce, the spots may be detected readily by examining the chromatogram under a strong ultraviolet light (for example Mineralite). Other techniques for locating the compounds on paper include color tests, frequently carried out by spraying the chromatogram with appropriate reagents.

Quantitative analysis of a compound separated by paper chromatography can be accomplished in one of two ways: (1) the intensity of colored or ultraviolet-absorbing spots may be measured directly on the paper, or (2) the compound may be eluted from the paper and analyzed by any suitable method.

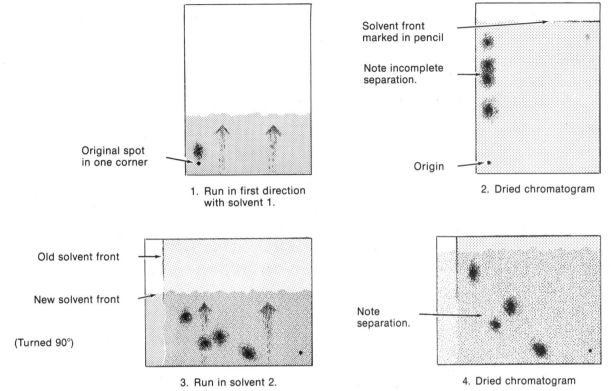

FIGURE C-2
Two-dimensional paper chromatography. (From *Experimental Biochemistry* edited by John M. Clark, Jr. W. H. Freeman and Company. Copyright © 1964.)

B. THIN-LAYER CHROMATOGRAPHY

In thin-layer chromatography an adsorbant is applied to a supporting material, frequently glass or aluminum, in a very thin layer. Generally a binding agent is used to adhere the adsorbant to the supporting material.

In one of the more common procedures, a slurry of silica-gel adsorbant, with calcium sulfate as the binder, is spread as a thin layer on a 20-cm² glass plate. The plates, dried to remove excess moisture, are then handled in much the same way as paper in ascending chromatography. After separation, the spots can be scraped from the glass for detailed analysis or eluted and rechromatographed.

Separation is accomplished very quickly on thin-layer plates. Using paper chromatographic methods, separation of some mixtures may take 24 hours. The same separation may take less than an hour when using thin-layer procedures.

REFERENCES

Alexander, P., and R. J. Block, Eds. 1960. *A Laboratory Manual of Analytical Methods of Protein Chemistry.* Vol. 1. *The Separation and Isolation of Proteins.* Pergamon, New York.

Block, R. J., E. L. Durrum, and G. Zweig. 1958. *A Manual of Paper Chromatography and Paper Electrophoresis.* 2d ed. Academic Press, New York.

Cassidy, H. G. 1957. *Fundamentals of Chromatography.* Interscience, New York.

Kirchner, J. G. 1967. *Thin-Layer Chromatography.* Interscience, New York.

Kunin, R. 1958. *Ion Exchange Resins.* 2d ed. Wiley, New York.

Lederer, E., and M. Lederer. 1957. *Chromatography.* American Elsevier, New York.

APPENDIX D

Use of Live Animals in the Laboratory

A. GENERAL PROCEDURES

If this is the first time living vertebrate animals are being used in your class, it is important that you understand the purpose in bringing these animals into the laboratory. First, there must be a genuine reason for using live animals. Second, live animals must always be treated in a humane way; never cause them unnecessary irritation or injury. Accordingly, if any organ or tissue damage may result from legitimate experimentation, the animal is first put under an anesthetic, or the nervous system is treated to make the organism insensible to pain. Avoid injuring the animal's tissues or making it bleed; such damage makes the animal less capable of "normal" reactions.

In handling organs or tissues that have been exposed or removed, use a glass hook or a small camel-hair brush moistened with Ringer's saline solution. Never handle tissues with your fingers. When you need to lay down an excised organ or part, never place it on the table, or on dry paper; place it in a watch glass or other dish and keep it moistened. Apply Ringer's solution as often as necessary while working, to keep the tissue from drying out.

B. PITHING PROCEDURE

A spinal frog (one whose entire brain has been destroyed) is prepared by the procedure of **pithing**. Grasp the animal as indicated in Fig. D-1, using the thumb and fingers to secure the limbs. With the index finger, depress the snout so that the head is at a sharp angle to the body. Run the tip of the dissecting needle down the midline of the head. At a point 2–3 mm behind the posterior border of the eardrums the needle should dip. This marks the location of the **foramen magnum**, a large opening in the skull through which the spinal cord emerges from the **cranium**. Place the point of a dissecting needle in this groove, and with a sharp movement force it through the skin and foramen

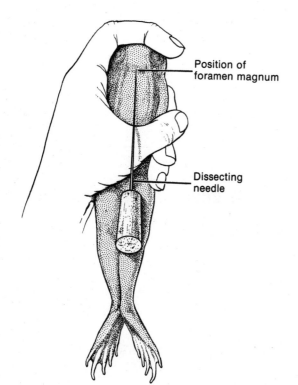

FIGURE D-1
Pithing the brain of the frog.

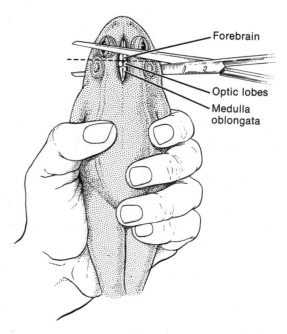

FIGURE D-2
Decerebration of the frog.

magnum into the brain. Twist and turn the needle to destroy the brain. Halt any bleeding and treat the pithed frog in the same way as the decerebrate animal.

C. DECEREBRATION PROCEDURE

A decerebrate frog is one in which the forebrain is no longer functional. Select a large, active frog and grasp it firmly in the left hand, pinioning the forelimbs and hindlimbs. Position the blades of a pair of *sharp* scissors as indicated in Fig. D-2, just behind the posterior margins of the eyes. Cut quickly and cleanly to avoid excessive nerve shock. Stop any blood flow with cotton or gauze pads. Lay the animal ventral side down on a damp paper towel. Be sure to moisten the skin from time to time, as the bulk of the respiration of such animals takes place through the skin.

APPENDIX E

Radioisotopes

The nuclei of atoms are composed of neutrons and protons. In any single element the nuclei contain the same number of protons. Some nuclei, however, may contain a *different number of neutrons* and thus have a different atomic mass. Atoms having nuclei with the same number of protons but different numbers of neutrons are called **isotopes**. Most isotopes are stable. However, others have certain combinations of protons and neutrons that make them unstable. Such isotopes tend to reach a stable state and in the process emit a distinctive type of radiation. During the disintegration of such isotopes, called **radioisotopes**, three types of radiation may be emitted:

- **Alpha (α)** rays are particles consisting of two neutrons and two protons each. Thus, alpha rays are basically charged helium atoms.
- **Beta (β)** rays are electrons emitted by some nuclei. The energy levels of these electrons may vary and produce hard beta radiation (highly charged electrons) or soft beta radiation (low-charged electrons). Phosphorus-32 is an example of a hard beta emitter; carbon-14 is a soft beta emitter.
- **Gamma (γ)** rays are electromagnetic radiation and closely resemble x-rays. Cobalt-60 is an example of a gamma emitter.

Because of the constant loss of particles, radioisotopes are said to "decay." The amount of radioactive material that remains will therefore be reduced with time. The rate of decay is known for many radioisotopes and is expressed as the **half-life** of the isotope. Half-life is the length of time it takes for a radioactive substance to lose half of its radioactivity. For example, the half-life of uranium-238 is 4,500,000,000 years; phosphorus-32 is 8.0 days.

Because radiation is damaging to human tissues, any radioactive materials used in the laboratory must be handled with extreme caution. Radioisotopes should be handled with the same care as harmful bacteria or strong acids. Remember, your "senses" cannot detect radiation. You *must* use proper techniques and follow certain rules.

SAFETY PRECAUTIONS IN THE USE OF RADIOISOTOPES

1. *Do not* eat, drink, smoke, chew gum, or apply cosmetics in the laboratory.
2. Wear an apron or laboratory coat.
3. To guard against contamination, place some disposable material (such as aluminum) over the table top on which you are working.
4. *Never* pipet any radioactive material by mouth. Use the special pipets provided by the instructor.
5. Discard paper, and other materials used, in the container designated for this purpose.
6. Never pour radioactive liquids into the standard drain. Pour them into the "waste" jar provided by the instructor.
7. With the instructor's permission, trace amounts of activity (such as that found when washing glassware) may be discarded into the sink. To avoid splashing, pour these *directly* into the drain.
8. Wash hands thoroughly before leaving the laboratory and after any suspected contact with radioactive material.
9. If you should spill radioactive material, do these things:
    a. Notify your instructor immediately.
    b. Wipe up liquids, using soap, water, and paper toweling. Give the wet towels to the instructor for disposal.
10. Clean up immediately after the experiment. Wash all contaminated equipment and set it aside to be checked by the instructor for radioactivity.

REFERENCES

Chase, G. D. and J. L. Rabinowitz. 1967. *Principles of Radioisotope Methodology.* 3d ed. Burgess, Minneapolis.

Van Norman, R. W. 1971. *Experimental Biology.* 2d ed. Prentice-Hall, Englewood Cliffs, New Jersey.

Wang, C. H. and D. L. Willis. 1965. *Radiotracer Methodology.* Prentice-Hall, Englewood Cliffs, New Jersey.

APPENDIX F

Elementary Statistical Analysis

The inherent variability of organisms, coupled with the degree of error normally encountered in most measuring systems, demands the use of some form of data evaluation before valid conclusions can be made or inferences drawn. It is particularly important that students of the biological sciences become familiar with statistical methods of handling data. The following section briefly outlines some of the basic methods used in this form of analysis.

The procedures fall into two groups. The first is concerned with those statistics that define the nature and distribution of the data. The second outlines three procedures that may be used to compare two or more sets of data. This introduction to statistical methods is not to be regarded as complete, nor is it expected to substitute for formal training in this area. Wherever possible, the student should supplement the information given by use of the references listed at the end of this appendix.

A. DEFINING THE DATA

In order to define a sample of data, one must have some knowledge of the central tendencies and degree of dispersion of the data. The statistics usually used for this purpose are the arithmetic mean, the standard deviation, and the confidence interval of the mean.

1. Arithmetic Mean

The mean ($\bar{X}$) is computed by summing (Σ) the individual sample measurements (x_i) and dividing by the total number of measurements (N):

$$\bar{X} = \frac{\Sigma x_i}{N}$$

2. Standard Deviation

The mean of a group of data gives little information concerning the distribution of the data about the mean. Obviously, different numerical values can give the same mean value. For example:

SET 1: 32, 32, 36, 40, 40 $\bar{X} = 36$
SET 2: 2, 18, 24, 36, 100 $\bar{X} = 36$

The means for both sets of figures are identical. Yet the variation from the mean in Set 2 is so great as to make the average meaningless. The standard deviation is a measure of data dispersion about the mean. The range covered by the mean plus or minus (±) one standard deviation includes about 68% of the data upon the basis of which the mean was calculated. The range covered by the mean ± two standard deviations will include approximately 95% of the data. The calculation of the standard deviation is summarized below.

a. Compute the arithmetical mean ($\bar{X}$) by summing (Σ) the individual measurements (x_i) and dividing by the total number of measurements (N): $\bar{X} = \Sigma x_i / N$.

b. Calculate the deviation from the mean for each measurement: ($x_i - \bar{X}$).

c. Square each of the individual deviations from the mean: $(x_i - \bar{X})^2$. This allows one to deal with positive values.

d. Determine the sum of the squared deviations: $\Sigma(x_i - \bar{X})^2$.

e. Calculate the standard deviation (s) of the sample, using the formula

$$s = \sqrt{\frac{\Sigma(x_i - \bar{X})^2}{N - 1}}$$

A sample set of standard deviation measurements is shown in Table F-1, where:

Number of individual measurements (N) = 10
Arithmetical means ($\bar{X}$) = 220/10 = 22
Degrees of freedom ($N - 1$) = 9

$$s = \sqrt{\frac{\Sigma(x_i - \bar{X})^2}{N - 1}} = \sqrt{\frac{42}{9}} = \sqrt{4.66} = 2.16$$

3. Confidence Interval for a Sample Mean

The confidence interval (C) for a sample mean equals the standard deviation of the sample divided by the square root of the sample number (N) and multiplied by a factor (t) that is determined by the probability level desired and the value of the sample number:

$$C = \pm t \left(\frac{s}{\sqrt{N}} \right)$$

It is highly improbable that a sample mean, based upon a relatively small series of data, will correspond exactly to the true mean calculated from an infinitely large sample of the population. It is necessary, therefore, to define a range within which the true mean might be expected to lie. To define this range—the confidence interval—the standard error of the mean (S.E.$_{\bar{x}}$) must be known. The standard error equals the standard deviation divided by the square root of the number in the sample:

$$\text{S.E.}_{\bar{x}} = \frac{s}{\sqrt{N}}$$

The confidence interval is then calculated by multiplying the S.E.$_{\bar{x}}$ by t, whose value depends on the number in the sample (N), and the level of probability selected. Normally, a probability or significance level of 0.05 is accepted in biological studies. This implies that in only 5% of the samples taken separately from a given population would the parameters defined by the sample *fail* to have significance.

The t values may be obtained from the t table (Table F-2). These values are listed in columns for 0.10, 0.05, and 0.01 probability levels. Note that it is necessary to have a value for the number of degrees of freedom (D.F.) of the sample. In this case, the D.F. for the sample equals the number in the sample minus one ($N - 1$).

B. COMPARISON OF DATA

1. Standard Error of the Difference of Means

The standard error of the difference of means is computed by using the formula

$$\text{S.E.} \bar{X}_1 - \bar{X}_2 = \sqrt{\frac{s_1}{N_1} + \frac{s_2}{N_2}}$$

where s_1 and s_2 represent the standard deviations of two different groups, N_1 and N_2 represent the number of individuals in each group, (preferably at least 20) and $\bar{X}_1$ and $\bar{X}_2$ are the respective means. Using this formula, if the difference between the two means is *larger than two times* the standard error of the difference, it can be concluded that the difference between the groups was not due to chance alone, but was due to the treatment given. It can be further concluded that similar plants or animals under similar treatment could be expected to respond in a similar manner.

ELEMENTARY STATISTICAL ANALYSIS

TABLE F-1
Sample calculation of the standard deviation.

Observation number	Individual measurement (x_i) of stem length (mm)	Deviation from mean ($x_i - \bar{X}$)	$(x_i - \bar{X})^2$
x_1	20	−2	4
x_2	24	+2	4
x_3	22	0	0
x_4	19	−3	9
x_5	26	+4	16
x_6	22	0	0
x_7	24	+2	4
x_8	20	−2	4
x_9	22	0	0
x_{10}	21	−1	1
	TOTAL = 220 = Σx_i	0	42 = $\Sigma(x_i - \bar{X})^2$

TABLE F-2
Significance limits of student's *t*-distribution.

(n) degrees of freedom	Confidence levels		
	0.10	0.05	0.01
1	6.314	12.706	63.657
2	2.920	4.303	9.925
3	2.353	3.182	5.841
4	2.132	2.776	4.604
5	2.015	2.571	4.032
6	1.943	2.447	3.707
7	1.895	2.365	3.499
8	1.860	2.306	3.355
9	1.833	2.262	3.250
10	1.812	2.228	3.169
11	1.796	2.201	3.106
12	1.782	2.179	3.055
13	1.771	2.160	3.012
14	1.761	2.145	2.977
15	1.753	2.131	2.947
16	1.746	2.120	2.921
17	1.740	2.110	2.898
18	1.734	2.101	2.878
19	1.729	2.093	2.861
20	1.725	2.086	2.845
21	1.721	2.080	2.831
22	1.717	2.074	2.819
23	1.714	2.069	2.807
24	1.711	2.064	2.797
25	1.708	2.060	2.787
26	1.706	2.056	2.779
27	1.703	2.052	2.771
28	1.701	2.048	2.763
29	1.699	2.045	2.756
30	1.697	2.042	2.750
40	1.684	2.021	2.704
60	1.671	2.000	2.660
120	1.658	1.980	2.617
∞	1.645	1.960	2.576

Adapted from Table III of Fisher and Yates, *Statistical Tables for Biological, Agricultural, and Medical Research*, Oliver & Boyd, Edinburgh.

2. Student's t Test

The student's t test is used to determine whether, within a selected degree of probability, two groups of data represent samples taken from the same or different populations of data. In other words, it is used to determine if two groups of data are significantly different. This test uses both the means and standard deviations of the two samples. It is calculated as

$$t = \frac{(\bar{X}_1 - \bar{X}_2)\left(\sqrt{\frac{N_1 N_2}{N_1 + N_2}}\right)}{\sqrt{\frac{(N_1 - 1)(s_1^2) + (N_2 - 1)(s_2^2)}{N_1 + N_2 - 2}}}$$

where s_1 and s_2 represent standard deviations of two different groups, N_1 and N_2 represent the number of individuals in each group, and $\bar{X}_1$ and $\bar{X}_2$ are the respective means.

The calculated t value is then compared to the value in the t table (Table F-2) at the probability level chosen (usually 0.05) and at the combined degrees of freedom of the two samples ($N_1 + N_2 - 2$). If the value for t is *less* than that found in the table, then the two groups of data are not considered significantly different at the chosen level of probability. If the t value *exceeds* that in the table, then the two groups of data may be considered significantly different.

3. Wilcoxon Test

This test, although mechanically easier to carry out than the student's t test, should only be used on small amounts of data — such as in a preliminary study — to help determine if a particular experiment is worth pursuing.

Basically, this test employs a ranking system. A simple example, given below, is a comparison of the heart rate between normal male and females to test for sexual differences.

The hypothesis to be tested (H_0–The Null Hypothesis) is

$$PR\,♀ = PR\,♂$$

that is, there is no difference in heart rate (in terms of pulse beat) between males and females.

a. Raw Data

Sex	Pulse Rate
♂	74 77 78 75 72 71
♀	80 83 73 84 82 79

b. Calculation of Means

$$\Sigma\,♂ = 447;\ \bar{X} = 74.5$$
$$\Sigma\,♀ = 481;\ \bar{X} = 81.7$$

c. Ranking of Data

(1) Rank the data in an order of increasing magnitude. Underline all values from the group having the smaller mean. In this example it is the males.

(2) Also write the rank number under the data beginning with 1 for the smallest value and proceeding to the largest value. Again, underline the rank number corresponding to the male values.

<u>71</u> <u>72</u> 73 <u>74</u> <u>75</u> <u>77</u> <u>78</u> 79 80 82 83 84
<u>1</u>   <u>2</u>   3   <u>4</u>   <u>5</u>   <u>6</u>   <u>7</u>   8   9   10 11 12

(3) Next, reorder the rank numbers by sex and add them up. If there is no real difference in pulse rate between sexes, then the sum of rank numbers should be equal.

♂ $\underline{1} + \underline{2} + \underline{4} + \underline{5} + \underline{6} + \underline{7} = 25$
♀ $3 + 8 + 9 + 10 + 11 + 12 = 53$

d. Computing U

In this case, the sums of the rank numbers are different. But is there a significant difference between males and females? To determine this the statistical value U must be determined:

$$U = W_1 - \tfrac{1}{2}n_1(n_1 + 1)$$

where W_1 = the total of the rank numbers belonging to the group with the smallest mean, i.e., the males ($W_1 = 25$).

n_1 = the number of individuals (or measurements) in the group having the smallest mean, i.e., the males ($n_1 = 6$).

Thus:

$$U = 25 - \tfrac{1}{2}\cdot 6\,(6 + 1)$$
$$= 25 - 21$$
$$= 4$$

ELEMENTARY STATISTICAL ANALYSIS

TABLE F-3
Wilcoxon distribution (with no pairing). The numbers given in this table are the number of cases for which the sum of the ranks of the sample of size n_1 is less than or equal to W_1.

n_1	n_2	$C_{n_1 n_2}$	\multicolumn{21}{c}{Values of U, where $U = W_1 - \frac{1}{2}n_1(n_1 + 1)$}
			0	1	2	3	4	5	6	7	8	9	10	11	12	13	14	15	16	17	18	19	20
3	3	20	1	2	4	7	10	13	16	18	19	20											
3	4	35	1	2	4	7	11	15	20	24	28	31	33	34	35								
4	4	70	1	2	4	7	12	17	24	31	39	46	53	58	63	66	68	69	70				
3	5	56	1	2	4	7	11	16	22	28	34	40	45	49	52	54	55	56					
4	5	126	1	2	4	7	12	18	26	35	46	57	69	80	91	100	108	114	119	122	124	125	126
5	5	252	1	2	4	7	12	19	28	39	53	69	87	106	126	146	165	183	199	213	224	233	240
3	6	84	1	2	4	7	11	16	23	30	38	46	54	61	68	73	77	80	82	83	84		
4	6	210	1	2	4	7	12	18	27	37	50	64	80	96	114	130	146	160	173	183	192	198	203
5	6	462	1	2	4	7	12	19	29	41	57	76	99	124	153	183	215	247	279	309	338	363	386
6	6	924	1	2	4	7	12	19	30	43	61	83	111	143	182	224	272	323	378	433	491	546	601
3	7	120	1	2	4	7	11	16	23	31	40	50	60	70	80	89	97	104	109	113	116	118	119
4	7	330	1	2	4	7	12	18	27	38	52	68	87	107	130	153	177	200	223	243	262	278	292
5	7	792	1	2	4	7	12	19	29	42	59	80	106	136	171	210	253	299	347	396	445	493	539
6	7	1716	1	2	4	7	12	19	30	44	63	87	118	155	201	253	314	382	458	539	627	717	811
7	7	3432	1	2	4	7	12	19	30	45	65	91	125	167	220	283	358	445	545	657	782	918	1064
3	8	165	1	2	4	7	11	16	23	31	41	52	64	76	89	101	113	124	134	142	149	154	158
4	8	495	1	2	4	7	12	18	27	38	53	70	91	114	141	169	200	231	264	295	326	354	381
5	8	1287	1	2	4	7	12	19	29	42	60	82	110	143	183	228	280	337	400	466	536	607	680
6	8	3003	1	2	4	7	12	19	30	44	64	89	122	162	213	272	343	424	518	621	737	860	994
7	8	6435	1	2	4	7	12	19	30	45	66	93	129	174	232	302	388	489	609	746	904	1080	1277
8	8	12870	1	2	4	7	12	19	30	45	67	95	133	181	244	321	418	534	675	839	1033	1254	1509

Reproduced from Table H of Hodges and Lehmann: *Basic Concepts of Probability and Statistics*, published by Holden-Day, San Francisco, by permission of the authors and publisher.

e. Determining Significance

(1) Refer to Table F-3 (Wilcoxon table for unpaired data). Note that it has three columns labeled n_1, n_2, $C_{n_1 n_2}$ and Value of U.

(2) The values of n_1 and n_2 in the data are 6 and 6. Refer to this combination in the first column of the table. The corresponding figure in the $C_{n_1 n_2}$ column is 924.

(3) Run along the U columns to $U = 4$, and then down the $U = 4$ column to the row corresponding to $C_{n_1 n_2} = 924$. The value at these intersecting points is 12. This value represents the number of possible rank totals that would be less than or equal to ($\leq$) 25.

(4) Suppose we consider that if an event occurs less than 5 times out of 100 (5/100 or 50/1000, $p = .05$) it occurred, not as a result of chance, but as a result of the treatment, or in this case, was due to the difference in sex.

The value of 12 in the U column represents a probability of 12 times out of 1000 (1.2/100 or $\frac{21}{4} = 0.012$). This value is considerably lower than the 5/100 that we will accept for considering that the difference in pulse rate is based upon the difference in sex. Therefore, we would reject the hypothesis that

$$PR \, \male = PR \, \female$$

and say that the difference in pulse rates is significant and results from the difference in sex.

REFERENCES

Alder, H. L., and E. B. Roessler. 1968. *Introduction to Probability and Statistics*. 4th ed. W. H. Freeman and Company, San Francisco.

Simpson, G. G., A. Roe, and R. C. Lewontin. 1960. *Quantitative Zoology*. Harcourt Brace, New York.

Snedecor, G. W. 1956. *Statistical Methods*. 5th ed. Iowa State College Press, Ames, Iowa.

Sokal, R. R., and F. J. Rohlf. 1969. *Biometry*. W. H. Freeman and Company, San Francisco.

Steel, R. G. D., and J. H. Torrie. 1960. *Principles and Procedures of Statistics*. McGraw-Hill, New York.

APPENDIX G

Syringe-Type Respirometer

Many of the methods used in the study of respiration depend upon measurement of gas volume or pressure changes. Any variation in the volume or pressure within a closed system in which an organism is respiring represents the net difference between oxygen consumption (which would decrease pressure and volume in the close container) and carbon dioxide production (which would increase pressure and volume). However, if the carbon dioxide produced is absorbed in some way, any changes would be attributed to oxygen consumption.

A simple respirometer that can be used to detect changes in gas pressure and volume is shown in Fig. G-1. This equipment consists of two vessels that can be closed to the outside. Respiring material is placed in one of the containers along with potassium hydroxide (KOH), an agent used to absorb carbon dioxide. Because gas volume is influenced by such physical factors as atmospheric pressure and temperature, the second container (identical except for the living material) is employed as a compensation chamber.

Place the material to be tested in one of the respiration bottles containing a vial of KOH, which is suspended in the bottle by a string held in place by a rubber stopper. To the control bottle add 10 ml of water. Allow both bottles to equilibrate for 10 minutes. After temperature equilibration, level the fluid in the manometer by adjusting the stopper on the temperature control bottle. The system is then closed by attaching a syringe to the needle that penetrates the stopper of the respiration chamber. Record the starting volume of the syringe. The manometer fluid can be restored to its original level by slowly pressing in the syringe plunger until the two columns of fluid are level in the U-tube. At the conclusion of the experiment, weigh the tested materials. The respiratory rate (milliliters of oxygen consumed per gram of unit weight) may then be determined and plotted on a graph.

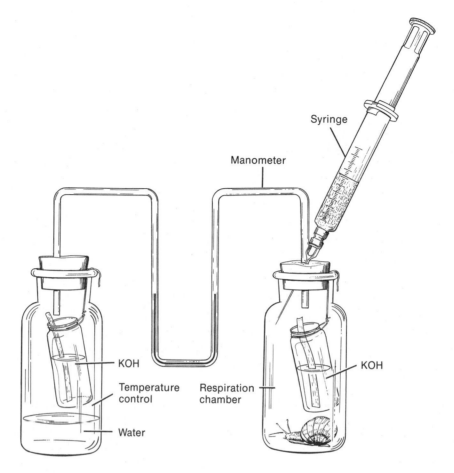

FIGURE G-1
Syringe-type, constant-pressure respirometer.

REFERENCES

Dixon, M. 1951. *Manometric Methods as Applied to the Measurement of Cell Respiration and Other Processes.* Cambridge University Press, New York.

Umbreit, W. W., R. H. Burris, and J. F. Stauffer. 1964. *Manometric Techniques.* Burgess, Minneapolis.

DATE DUE

DEC 2 1 '82			

```
QH                      069860
317   Abramoff
A2    Laboratory outline in biology
      --II
```